# THE MEMORY THIEF

## and the Secrets Behind How We Remember

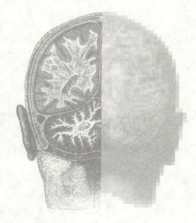

## A Medical Mystery

· · · · · · · · · · · · · · · · · · · · · · · · · ·

# LAUREN AGUIRRE

PEGASUS BOOKS

NEW YORK  LONDON

THE MEMORY THIEF

Pegasus Books, Ltd.
148 West 37th Street, 13th Floor
New York, NY 10018

First Pegasus Books cloth edition June 2021

Interior design by Maria Fernandez

Library of Congress Cataloging-in-Publication Data is available.

ISBN: 978-1-64313-652-3

10 9 8 7 6 5 4 3 2 1

Printed in the United States of America
Distributed by Simon & Schuster
www.pegasusbooks.com

*For my wise mother, Char Seeley,*
*who always remembers what matters.*

# Contents

· · · · · · · · · · · · ·

Preface ix
Prologue xiii

## PART ONE: BROKEN BRAINS 1

1 Case One 3

2 Questions Lead to Answers 14

3 Person, Place, and Time 24

4 Short Shrift 39

## PART TWO: EPISODES 49

5 The Hum of the Restless Beehive 51

6 The Cluster Grows 56

7 No Place in the World 69

8 Maps and Memories 77

9 The Forgetting Disease 87

## PART THREE: OPIOIDS 99

10 Learning to Be Addicted 101

11 Extreme Potency 111

12  World of Hurt                        119

13  Finding Fentanyl                     126

14  Taken as Prescribed                  137

## PART FOUR: INTERVENTIONS                 149

15  Some Part of Me Is Missing           151

16  The Next H.M.                        160

17  Rebirth                              169

18  Time for a Do-Over                   177

19  The Flip Side of the Coin            187

## PART FIVE: OUTSTANDING QUESTIONS         197

20  Finding Lost Memories                199

21  The Case for Tau                     207

22  The Road Ahead                       212


Epilogue                                 223

Acknowledgments                          235

A Note About Sources                     239

Notes                                    241

Glossary                                 267

Credits                                  273

Index                                    275

# Preface

. . . . . . . . . .

This is a story about memory—the brain's almost magical and still somewhat mysterious trick for making sense of the world. You could argue that our memories are as essential to identity as breathing is to life. So when news surfaced in 2017 that a small group of people had lost the ability to form even a single new memory, the story naturally captured public attention. Outlets such as the Associated Press, BuzzFeed, STAT news, the *Boston Globe,* and the *Atlantic* all covered the news about fourteen opioid overdose survivors in Massachusetts who had woken up with amnesia. In the aftermath of the overdose, each of these patients had been left with a strange and precisely defined pattern of damage to the hippocampus, the memory center of the brain. Doctors had never seen anything like it, but after an initial flurry of attention, interest waned. The incident was relegated to the realm of mystery, believed to be limited to a few unfortunate occurrences in one small corner of the globe. At the time, I was on staff at the PBS documentary series NOVA, where we covered the so-called amnestic syndrome in a brief blog post.

I found it difficult to resist the gravitational pull of this story because I know firsthand how terrifying it is to lose one's memories. Years ago, on an otherwise unremarkable early morning while tapping away on my laptop, everything I knew slipped away in a matter of mere seconds. I was still there, in the sense that I could hear the heater blowing and see the deep green walls around me. But I didn't know who I was, where I fit in, or what anything meant. It was a vertiginous plunge into nothingness, so disorienting and unnerving that I lay facedown on the carpet. Thankfully, after a minute or two, the memories returned, and everything made sense once again.

After I had undergone many brain scans, neurological exams, and consultations, doctors determined that this strange episode was an aura, a brief seizure provoked by a collection of errant neurons firing too rapidly. The chief of neurosurgery at a prominent Boston hospital proposed removing an abnormal brain lesion. A neurologist, on the other hand, recommended daily medication to keep the auras at bay. Faced with these two options, I turned to as many other experts as I could for informal advice. One was a neurologist friend named Jed Barash. Barash came to my house, sat down at the computer, and scrolled slowly back and forth through my MRI scan. When he finished looking, he said that he would avoid the surgery if it were up to him. In his pragmatic and cautious view, I should just keep an eye on it. It was good advice. I took the medicine, never had another seizure, and probably never will. The only lasting effect of the whole experience was to leave me fascinated with memory and curious about the mysteries posed by weird brains.

Barash would later be the neurologist who would help identify the first four amnesic patients between 2012 and 2015, leading to a statewide search for more. When the Centers for Disease Control and Prevention (CDC) released the 2017 report describing the syndrome, I began to pester Barash with questions. He and his colleagues suspected that the syndrome was caused by fentanyl, an opioid whose use was surging just as the amnestic syndrome survivors were coming to light. These clinicians feared that there would be other victims across the country in places where fentanyl use was rising. They were also beginning to wonder if this mystery at the intersection of two public health crises—opioid use and memory loss—could turn out to be a tiny piece of the giant puzzle that is Alzheimer's disease.

To cover the investigation into the amnestic syndrome and bring to life some of the pivotal moments through reconstructed scenes, I interviewed doctors, patients, and family members, combed through text exchanges, and gathered several hundred e-mails sent between health experts, the

Massachusetts Department of Health, and other government agencies. (More detailed information can be found in the sources section on p. 241)

The more I looked into the syndrome, the more it seemed that I could illuminate how we know what we think we know about memory and memory disorders through the prism of this obscure and tragic story. I learned that the source for new insights is often patients who generously volunteer to participate in research, even when they know it won't benefit them directly. I wanted to explore how such hard-won insights, combined with fundamental neuroscience, suggest unexplored strategies and new hope for treating memory disorders. And finally, I was interested in what happens when novel ideas bump up against conventional wisdom, forcing medical science to move ever closer to what will always be an incomplete version of the truth.

# Prologue

• • • • • • • • • • •

## *California, March 2018*

Alone in the stillness of his apartment, 25-year-old Owen Rivers hunches over his phone and reads the journal entry he's just written into his notes app.

> *If nothing matters, we contrive meaning for ourselves. So why can't mine just be something that soothes me so that I can live here more peacefully? There are pressures from every direction saying that using drugs is unacceptable. And I understand most of them, but I still think that that's been the best solution for me so far.*

He reviews the paragraph again because just writing it down is not enough—it's never enough. But, temporarily satisfied, relieved that he's captured the thoughts of the day before they slip away, he leans back in his recliner and looks out the window at the view over Los Angeles. Owen is tall and lightly built with dark hair and darker eyes. His apartment is clean, the only decorations a lone cactus on the kitchen counter and a watercolor of a man silhouetted against a green wall, arms stretched overhead, palms pressed against a barred window.

Owen pads barefooted across the polished wood floor, pours a glass of cold water, then sinks back into the recliner to update his memory lists. These are the things he doesn't want to forget, color-coded for "long-term," "soon," and "now." Owen is a successful insurance broker who graduated a few years earlier with highest honors from UCLA. He has no apparent reason to suspect memory problems, but he's been plagued for years with an irrational fear of forgetting. The compulsion to write is embarrassing and unrelenting. Each list has dozens of entries. When he gets to the end of one list, he reviews, reorders, and rechecks before moving on to the next, a process that can take hours. Tonight, the long-term category feels less urgent, but he skims through it anyway, for the sake of completeness.

*Long-term:*
*research memory hoarding on google scholar*
*read "the stranger"*
*visit museum of contemporary art*
*investigate organic farming at urban homestead*
*meditate . . .*

*Soon:*
*read blog post on "the portable Nietzsche"*
*try new fish taco place before mom's visit*
*call Irina 7 pm tues happy 6th bday*
*review leasing options on new car . . .*

*Now:*
*check talking points for 10am mtg with Francesco*
*confirm address*
*finish Chinese takeout for dinner*
*garbage out*
*brush teeth*
*set alarm for 8am*
*check e-mail*

The final to-do is the one he's been waiting for all day. Owen skips past the junk mail hoping to find a message from Dr. Denison at Scripps Mercy Hospital. And there it is:

Hello,

Good news, the scan is not showing a mass or tumor. It seems that the first scan was of poor quality and falsely indicated the presence of a mass.

Regards,
RD

Stunned, Owen reads the e-mail again. Two weeks earlier, he'd gone to the hospital for a follow-up MRI. The first scan had been ordered to investigate the source of two strange symptoms he'd reported to his doctor: disturbing, frequent spells of déjà vu and a distinct, synthetic burning odor that no one else could smell. When his doctor had told him that he'd need to get a second, higher-resolution scan because the first one had detected a possible brain tumor, Owen had been ecstatic.

Cancer would have been the perfect justification to use fentanyl again after eighteen months of sobriety. A sense of obligation to family and friends who saw him through three rehabs has been the only thing holding him back. Ever since the age of eleven, when he discovered a bottle of Vicodin in a friend's medicine cabinet, the only reliable reprieve from anxiety, depression, and his obsession with memory loss has been drugs. All attempts to fix himself through other means—therapy, getting perfect grades—have failed. Owen figures if he has a terminal illness, no one will blame him for using again, and he won't have to feel guilty for letting anyone down. So what Dr. Denison and anyone else would consider good news is, from Owen's point of view, terrible news. The unassailable excuse to use drugs, which has sustained him for the last two weeks, is gone.

He stares out the window at the darkening sky, trying to calm himself. Eventually, Owen stands and begins to pace. He loads a playlist of Noise music—it's not music, really, just distorted, atonal sounds layered with feedback and static—hoping the cacophony will obliterate the urge to use fentanyl. He paces some more, sits back down, goes on craigslist, and finds a place to meet a drug dealer forty-five minutes from his apartment. Owen spends the next hour and a half telling himself, *This is a bad idea* as he drives there and back, but the craving is all he feels, and it's too late to talk himself out of it. The bag and seven syringes are right there on the kitchen counter. He watches the needle enter his vein. The anticipated reward is huge. After he injects the second one, calm descends, like a benediction.

For some period of time—hours? days?—images drift in and out of view. A glimpse of his friend and business partner Francesco's apartment. A handful of vitamins. Light dancing off the rim of a huge glass of water. Checking for his phone in his front pocket. A view of the highway from the passenger seat of an unfamiliar car.

—⁓—

Owen examines a square of white gauze taped to the back of his hand. His mother, Rachel, notices the question in his eyes and says, "Hey, honey, you're awake again."

He turns his attention to her, perched on the edge of a chair pulled close to the hospital bed, her red hair secured in a ponytail, then to his sister, Kylie, standing by the window gazing back at him with eyes as dark as his own, and then to the IV pole with its tangle of tubes.

"What am I doing here?"

His mother and sister exchange an uneasy glance.

"Can you explain it this time, Kylie?" Rachel says.

"Francesco found you. You didn't show up for a meeting you guys were supposed to have, and then when you finally answered the phone you kept asking him the same questions over and over." Owen looks mystified. "You're at Cedars-Sinai because you overdosed," Kylie continues.

"Wait, what? I relapsed?"

Kylie nods. "Yeah, on fentanyl. A week ago. We looked through your phone."

She'd been amazed at how readily Owen, normally so private and secretive, had handed over his phone and passwords, like a child would.

Owen lowers his head, too embarrassed to make eye contact. "I messed up. I'm so sorry. I screwed up so bad." He closes his eyes and appears to drift off. The room is quiet for thirty seconds. Kylie bites her nails and Rachel checks her phone. Then Owen opens his eyes again.

"What am I doing here?"

"Read this," Rachel says, picking up a notepad to buy some time while she waits for the tightness in her throat to loosen. There'd been so many times she'd stayed up all night, listening, checking on him, afraid he'd overdose and die. But this—an adult son who clearly couldn't take care of himself—had not been on Rachel's worry list. She hands him the notepad as she says, "Kylie wrote it all down for you."

Owen reads the paragraph and starts crying. "God, I'm so sorry, I really screwed up. I finally had a good job. I ruined my future. But why am I still here?"

"They did an MRI and a bunch of other tests," Kylie says. "And they said the overdose damaged your hippocampus. You told us you remember that from one of your neuroscience classes."

Owen looks incredulous. "Wait, what? My hippocampus is damaged? So I can't remember anything new?" He registers the distressed look on Kylie's face. "Did I already ask you that?"

Rachel's face crumples, then there's a knock on the door and the doctor walks in.

"I'm Dr. Locke," he says as he shakes Owen's hand. He nods at Rachel and Kylie. "Hello again."

"Hey," says Owen, "you're wearing Doc Martens. I have those shoes too."

"Actually, they're Clarks. But they're very comfortable," says Dr. Locke.

"You look a lot like Woody Harrelson, the actor. Did anyone ever tell you that?"

"I hear that a lot," says Dr. Locke as he walks around the foot of the bed to the other side. As soon as his shoes come back into view, Owen says, again, "You're wearing Doc Martens. I have those shoes too!"

Dr. Locke looks at Owen for a long moment before he says, "So, we've finished all our testing. I wish there were something more definitive to tell you, but unfortunately, we don't have good answers as to why your hippocampus was damaged. It's possible there was some kind of contaminant in the fentanyl."

"Wait," Owen says. "Why am I here?"

"You overdosed," says Dr. Locke. "And we don't know why, but it injured your hippocampus."

Behind Owen's lost expression, his mind is working, waiting for that brief, disconnected moment when you wake up in a strange place to resolve into the relief of knowing how you got there. But the moment persists.

Rachel fiddles with Owen's bedcovers. He can tell that she's having trouble managing her distress, but she'll find a way to pull herself together like she's had to in the past, and be back on Facebook in a few weeks posting motivational hashtags—#fitlife, #believeinyourself, #staycommitted.

Dr. Locke places a hand on Owen's shoulder. "The plan is to discharge you in the care of your family, hopefully tomorrow morning if we can get all the paperwork done in time."

Kylie looks from Owen to Dr. Locke. "What can we do to help him?"

"He should take it easy, drink a lot of water, and follow up with his primary care physician."

A flash of recognition crosses Owen's face. "Hey, did anyone ever tell you you look like Woody Harrelson, the actor?"

"I have heard that," replies Dr. Locke.

Owen inspects the gauze covering the IV port on the back of his hand, closes his eyes, and falls back to sleep.

———

The soft whoosh of a passing car pulls Owen awake. He turns to see Kylie, her hands at 10 and 2 on the driver's wheel, the square set of her shoulders. She should be in St. Louis at her internship. After a few moments of silence, he speaks.

"Why are we here?"

"I'm taking you home to Del Mar, big brother. You'll stay with Mom for a while."

She glances at his watchful face, the narrow planes accentuated by a few days' stubble. Anticipating his question, she says, "Yeah, you overdosed." She takes her hand off the wheel for a moment to take a sip of iced coffee. "It'll be okay, though. We'll figure it out."

Owen looks out at the pale blue sky fading into white over the Pacific. He tries to make sense of what brought him to this point, being carried along down Highway 5 in a car with his sister. Owen feels alone, but he's not. There are dozens of people like him all across the country who've overdosed on fentanyl and woken up with amnesia. He also believes that he's wasted his life, but he hasn't. Out of his tragedy the faintest glimmer of hope will emerge, a way, perhaps, to catch an elusive memory thief.

# PART ONE

· · · · · · · · · · · · · · · · ·

# BROKEN BRAINS

# 1

# Case One

· · · · · · · · · · · ·

## *Massachusetts, October 2012*

When neurologist Jed Barash was growing up in Orange, Connecticut, his father's copy of the *New England Journal of Medicine* appeared in the mailbox every Thursday without fail. The arrival of the signature red-and-white *NEJM* marked the passing of Barash's childhood as regularly as a metronome. Fifty-two weeks a year, 600,000 people around the world receive their copy of the world's most prestigious medical journal. It's considered essential reading for medical professionals, including Barash's father, an anesthesiologist. The journal's standards are exacting, the editorial process excruciating. To be published in its pages is a sign of professional success and a passport to credibility.

In the fall of 2012, the normally staid *NEJM* issue that appears in Barash's mailbox includes an attention-grabbing title: CHOCOLATE CONSUMPTION, COGNITIVE FUNCTION, AND NOBEL LAUREATES. A Swiss-born author presents data revealing "a surprisingly powerful correlation between chocolate intake per capita and the number of Nobel laureates in various countries." Switzerland tops the list. The burden of proof is high for such surprising claims. But this article is no more than a spoof designed to highlight a common mistake: jumping to the conclusion that when two things are associated, one causes the other. Some journalists miss the point and present the findings as proof that eating more chocolate makes you smarter. In this case it would be an inconsequential mistake. But when it comes to making decisions that affect human health, getting it wrong is dangerous. Figuring out the true relationship between two associated phenomena is something Barash will come face-to-face with in the months and years ahead.

—⁓—

On the first Friday of October, Barash leans forward in his chair and stares at the MRI scan on his monitor. He's looking at the brain of a young man admitted to the hospital last night, and the image is so strange and beautiful that he knows something has to be wrong.

"Whoa," he says out loud to his empty office. "This is weird."

Floating brightly against the darker background of the rest of the brain are two C-shaped structures tucked on either side of the central fluid-filled cavity. Together they make up the hippocampus—the place that holds the keys to memory—and the intense glow is a distress signal from many millions of cells. Some mysterious, marauding force has laid waste to just this tiny region, leaving the rest of the brain unharmed. Barash looks out his door to the still-quiet waiting room up on the seventh floor at Lahey Hospital & Medical Center in Burlington, Massachusetts, just outside Boston. Then he looks back at the monitor.

Last night's phone call from nearby Winchester Hospital requesting permission to transfer this patient suddenly makes more sense. The distraught 22-year-old had recently overdosed. He was dragging one leg and repeatedly asking his mother if he was dying. Winchester is a smaller hospital that handles routine emergencies like a broken wrist or an appendicitis, but when patients with complex conditions or unexplained symptoms come in, the staff will often send them over to Lahey, a facility that has hundreds of specialists and more equipment. With the high-quality image in front of him, Barash can see what the Winchester staff could not, and it explains why the patient was acting strangely.

In ten years of medical training, Barash has reviewed thousands of scans—brains shrunken from Alzheimer's disease, brains dotted with tiny broken blood vessels, brains with tumors in different sizes, shapes, and locations. In every case, no matter what the damage looked like, it was pretty clear what was going on. But what Barash sees on the screen in front of him is strange and alien, belonging to no category he can imagine. It looks like someone took a page out of his medical school neuroanatomy textbook and deliberately highlighted the brain's memory center.

He reexamines the MRI, scrolling up from the base of the skull through the familiar soft gray brain structures until the hippocampus comes back into view. It seems certain that this patient will fail the memory tests they'll give him today, and the damage has triggered Barash's interest in strange cases and rare brain diseases. He believes more in chance than in destiny, *but still*, he thinks, it's almost as if his years of study and obsession have guided him directly to this moment, sitting in this office, looking at this startling image.

The care of this patient is now with his colleague, Yuval Zabar, a neurologist Barash admires for his intellectual curiosity and ability to run through every diagnosis that could possibly explain a patient's symptoms. It's Friday, the final day of a week being on call, and Barash is allowed to leave for the weekend at noon to make up for the last seven nights of fragmented sleep. But instead of heading for home, Barash turns off the monitor, pulls on his white doctor's coat, and heads down to 6 West to join the neurology team on their daily rounds. He has to see the patient for himself.

Between the patient, his mother and grandmother, and the doctors, there are more than half a dozen people in the cramped hospital room. Barash chooses an out-of-the-way corner from which to watch the examination. This is his preferred mode, to observe rather than be observed, and it doesn't bother him that he wouldn't stand out in a crowd. Born, raised, and educated in New England, 33-year-old Barash is solidly built and stands just shy of six feet, with a broad forehead, short brown hair, and a face that would look incomplete without his glasses.

James Maxwell Meehan, boyish and handsome enough to play the leading role in a romantic comedy, is sitting up in bed looking puzzled. On the whiteboard on one wall someone has written the following:

"Max, you are at Lahey Hospital in Burlington."

"You overdosed the night before last."

"You are having trouble with your memory."

Max's mother, Laura, explains that Max keeps asking her, his sister, and his friend the same questions, so they've written the answers on the whiteboard so he would stop.

A resident-in-training begins the examination by interviewing Laura and Max to find out what brought him into the hospital. A few nights earlier, Max went to one of his favorite bars in Boston's Back Bay, where he danced and drank for hours. Back at his boyfriend's apartment, he injected what he thought was pure heroin and passed out. It wasn't such an unusual night for Max, but according to his boyfriend, when Max woke up the next morning on the couch he complained that his left leg felt numb, and when he tried to stand up he fell over. A few minutes later, when he decided to stand up again, he fell over. Maybe his leg had fallen asleep, he figured. After another little while, he couldn't say how long, he realized it was morning and he ought to get up. But then he noticed that his leg didn't seem to be working. It occurred to him that maybe he'd slept too long on one side and pinched a nerve. Nothing made sense. Every few minutes, the realization that his leg felt paralyzed seemed like a horrible new discovery. Max started sobbing. It's not unusual for people who have overdosed to be confused and disoriented afterward, but Max found himself in a kind of limbo, a never-ending present.

"It's just very odd," his mother says. "He's fine. He's himself, he still has the same sense of humor. We even showed him that *Ain't Nobody Got Time For That* video on YouTube to distract him, and he laughed like he always does at silly things. But then we showed it to him again, and he laughed like it was the first time."

Zabar, an experienced neurologist in his mid-forties, takes over the examination from the resident. Zabar's thick head of hair is showing the first signs of gray, and he has the harried attitude of a man aware that his phone or pager is likely to buzz again within the next five minutes. Most patients appreciate the care he takes to explain things clearly and the way he doesn't sugarcoat bad news.

"Max, I'm going to ask you to remember a few simple words." Zabar looks him squarely in the eye to make sure Max is paying attention. "The words are *purple*, *velvet*, *honesty*."

Zabar pauses between each word to give Max time to register.

"Can you repeat those words back to me now?" he asks kindly.

"Purple. Velvet . . . honesty," says Max.

"Good."

Zabar gives Max a piece of paper with three figures on it—a triangle inside a circle, a skinny rectangle capped with a semicircle that looks like a mushroom, and the letter *L*. Below each shape is a word—*pride*, *hunger*, *station*.

"I want you to copy these three shapes and three words just as you see them." Zabar needs to make sure Max is paying attention and can follow directions.

Max copies them correctly, but when Zabar gives him a blank sheet of paper and asks him to re-create the shapes and words, he can't remember them.

"Okay. Now, can you tell me the three words I asked you to remember earlier?"

"Purple . . ." Max stops.

"Anything else?"

"That's all I remember."

Barash picks up on Laura's growing anxiety. She's trying to get a read on the room, but the resident, intern, and medical student are all focused on Zabar, who, wearing a poker face, presses on. He hides a pen, a crumpled piece of paper, and a cup around the room while Max watches. Several minutes later, when he asks Max to point them out, Max doesn't even know what he's talking about. Barash finally catches Laura's eye and tries to give her a comforting look. From the damage to the hippocampus that they'd seen on the MRI scan, the official diagnosis confirms what was almost a foregone conclusion: anterograde amnesia. Max still knows who he is and remembers everything that happened to him before he overdosed. But he can't form a single new memory. Every moment of the last few days has disappeared.

Zabar regularly sees patients with memory problems, and they're often elderly. Some have had strokes or brain tumors, but most suffer from dementia. Max's memory loss reminds Zabar of people with advanced Alzheimer's disease, but aside from the fact that he's much too young for that, his brain scan doesn't match the diagnosis. In Alzheimer's, some brain regions, including the hippocampus, tend to be smaller than usual, a sign of slow, insidious destruction. But in his twenty years as a neurologist, Zabar has never seen a case like this one. Aside from the damaged hippocampus, blood flow in the rest of his brain is normal. The only significant health problem Max has is drug use, both legal and illegal: tobacco, alcohol, marijuana, cocaine, ecstasy, LSD, and heroin. The standard urine toxicology screen run on any overdose patient detects opioids, corroborating Max's account that he had injected heroin. After leaving Max's room, Barash and Zabar stop in the hallway to talk over the case for a few minutes. Even though it seems likely that the drug or a contaminant is in some way responsible, neither one of them can fathom why only the hippocampus is injured.

Barash heads back to his office to hang up his white doctor's coat and check his e-mail one last time before heading home. Aside from colorful socks, he dresses traditionally, favoring button-down shirts with the sleeves rolled up, a tie, chinos, and comfortable shoes. He's two months into his first job as a full-fledged attending neurologist. Since he's new and there's no space left in the main neurology pod on 7 West, he's been assigned a tiny windowless office that was recently vacated by the chaplain. It's right off the waiting room, where a loud printer just outside the door ejects copies throughout the day. Visitors often assume he's a secretary and step into his office to ask for directions.

Driving home to Lexington from the hospital that afternoon, Barash is so wrapped up in puzzling over the mystery and reliving the events of the day that he barely registers the familiar suburban landmarks along the way—the mall, the movie complex, the gas station. He imagines how he'll describe the case to his wife, Gillian Galen, how he met a patient who couldn't remember more than the previous few minutes, a guy who can't go home and tell someone the story of his own day.

Dr. Galen, slender, green-eyed, and levelheaded, is a psychologist and an excellent observer of human nature. She understands the mind—and Barash—well enough to understand his fascination. They discuss it in their standing-room-only kitchen as they share a pizza and he drinks his Friday-night ginger bourbon fizz.

"His MRI was just insanely weird," Barash says.

"Weird how?"

"Like something completely scorched his hippocampus."

"Both sides?"

"Both sides, and he was toast. Like, two days after an overdose he should be fine, but he was confused. And he wasn't confused in the classic sense of confusion. He literally was just not . . . he wasn't able to move forward in time. He was stuck. He kept going back to the same thing, telling his mom the same thing, asking Zabar the same question within a few minutes. It's a classic amnestic syndrome." Barash stops to imagine. He's used to seeing people with advanced Alzheimer's whose memories are as bad as Max's, but what would it be like to be so young and just wake up that way one morning?

"That's so sad. Will he get better?" Gillian asks.

"We have no idea. Never seen anyone like him before," Barash says, "but it's not gonna be sunshine and rainbows."

———

One week after the overdose, on Max's twenty-third birthday, he returns to Lahey. As they do in most major teaching hospitals, the neurology team has weekly grand rounds, when doctors present interesting cases. The entire department can attend, including nurses, physician assistants, medical students, interns, and residents, and everyone is encouraged to weigh in. Zabar has decided to present Max's case.

Barash takes a seat among fifty or so other people crowded into the windowless, poorly lit neuroscience conference room. Paintings of distinguished neurologists and neurosurgeons line the wood-paneled walls. Max is seated at the head of the aisle with several rows of chairs on either side for the audience. Zabar stands at a podium and gives a brief overview of Max's case. Then, with Max's permission, he examines him, running through cognitive tests like the ones he'd given him in his hospital room a week earlier. He also tells Max a simple story, which he comes back to now and then, asking Max to repeat it at greater and greater time intervals. Max is cooperative, but his memory remains profoundly impaired. He is, in the words of neurologists, "densely amnestic."

When the exam is over, Max leaves, and Zabar displays his MRI on the screen behind him. The image captures the moment of acute damage, when some uncountable number of neurons would have only recently died, trapping water that would normally have been pumped across the cell membrane. Zabar dives into the case review.

"The thing that is really striking to me," says Zabar, "is that this signal is so bright that it looks like a stroke, but this person is so young. He has no risk factors for stroke. He's not immunosuppressed, and there's no evidence of any other infection like HIV, Epstein-Barr virus, or herpes."

One of the stroke specialists in the audience takes advantage of a break in Zabar's rapid-fire delivery.

"I have never seen a scan like this. I mean, it does look like a stroke. The intensity is there, but I've never seen a stroke that just took out the hippocampi bilaterally and not surrounding . . ."

Barash asks the question hanging in the room. "Yeah, why would it just be the hippocampus on both sides?" No one answers, so he continues in his quiet, rumbly voice. "It's so perfect, it's so localized. Whatever happened, there was clearly an affinity for just that area." Usually grand rounds end with a clear understanding of the diagnosis or, if not, ideas to explore further. But today the discussion goes long and the meeting ends with a conundrum; what happened inside Max's brain?

If anyone at Lahey has the expertise to figure it out, it would be head neuroradiologist Juan Small. He's written four books on radiology and is known for his ability to sort through every possible explanation for an abnormal brain scan. And he's careful about drawing inferences. Small couldn't attend the grand rounds presentation, but word's gotten out about a highly unusual case, and he keeps a list of them to help medical students learn how to make difficult diagnoses. Now he sits in his darkened basement office and calls up Max's imaging data on his large monitor.

Small selects the view that shows Max's brain from the base of the skull through to the crown of his head. Slowly, he scrolls up through the soft gray landscape. The familiar contours of the eye sockets begin to appear. Everything looks as it should until two islands of white emerge, complete, perfect, and intensely bright. *I don't know what this is*, he thinks. He's used to seeing tiny white dots in the hippocampus, each one representing dead neurons. They are the telltale signs of a stroke. But he has never seen the entire territory laid waste, and he reviews some ten thousand scans a year. Small scrolls back and forth a few more times, confirming the scope of the damage. *I don't know what this is*, he thinks again. *And not only that. I don't think anybody knows what this is.* Max's brain scan is in a category all its own. Small adds it to his list, although he won't have an answer when his students ask for the correct diagnosis.

Back in his tiny office after grand rounds, Barash turns his focus to preparing for a full slate of afternoon patients. The complete lack of daylight deprives him of any sense of time. The office is spare, with a desk, two chairs for visitors, a file of scientific papers he'll probably never have time to read, and the computer monitor. Diplomas and three pictures decorate the walls. One is an abstract painting made by his father-in-law of two people on a beach. It's Barash's version of a window with a view. Another is a 1960s-era public health poster with a cheerful bee encouraging people to take the polio vaccine, and the third is an advertisement for a lecture entitled PRION BIOLOGY AND DISEASES: A SAGA OF SKEPTICAL SCIENTISTS, MAD COWS, AND LAUGHING CANNIBALS.

Max's case offers the alluring sense of something unknown and unexpected, just waiting to be discovered, taking Barash back more than a decade earlier to Hamilton College, where he first learned about a rare and baffling disease. The condition, which had become the source of Barash's singular obsession, is called Creutzfeldt-Jakob disease, or CJD. It leaves its victims' brains looking like sponges, pockmarked with tiny holes. Early signs include memory lapses, personality changes, and difficulty walking, followed by full-blown dementia and death, typically within a year. But even though CJD's outward trajectory is swift, the incubation period can be decades. In the mid-1990s, young people in England who'd eaten meat from infected cattle developed a type of CJD known as mad cow disease. Health experts feared that the few who died were just the earliest warning signs of an epidemic still to come.

Barash was a sophomore in college in 1998 when Nobel Prize winner Stanley Prusiner gave the lecture advertised in the poster that now hangs on Barash's wall. It wasn't just the gruesomeness and rarity that captured his interest. CJD appears to be caused by a distorted form of a protein, called a prion, which acts as a sort of template, contorting other normal proteins into its virulent form. Up until this time, the reigning theory of infection held that diseases were caused by one of three things; a virus, a bacterium, or a fungus. Prusiner's discovery showed that a theory that had once seemed inviolable and complete was missing something essential. Everything about CJD appealed to Barash. There was mystery, widespread risk, and a breakout discovery that no one saw coming.

After college, he won a yearlong Fulbright Scholarship to live in Iceland and investigate a mysterious outbreak of a similar disease in sheep called scrapie. In the end, despite weeks spent driving around the cold, rainy, rugged landscape collecting samples from muddy farms, his experiments failed to uncover the source. But the thrill of the hunt had left its mark. From then on, Barash wanted to identify strange patterns that others had overlooked, to trace peculiar diseases to their enigmatic roots.

Medical school solidified Barash's conviction that neurology was the right path for him. Everything else seemed uninteresting in comparison. The heart is essentially a pump, the kidneys are a filter, the skeleton is a scaffold, but how the brain re-creates some facsimile of the world and allows us to imagine a different one is still largely mysterious. And the fact that this organ is as vulnerable as it is powerful only added to its appeal for Barash. He found pleasure in memorizing the brain's labyrinthian, beautiful topography. One of his favorite classes at the University of Connecticut School of Medicine was The Power of Observing at the nearby New Britain Museum of American Art. There he learned the principle that guided his medical practice ever since, of taking the time to truly look at something—a painting, a person, a toxicology screen, a brain scan—without any preconceived ideas as to what one should expect to see. He believes that mind-set—unbiased observation and careful attention to detail—can help lead you to the correct diagnosis.

He knows that every now and then a case comes along that doesn't match any previous pattern, a broken brain that winds up revealing how all the pieces are supposed to fit together. Max's case reminds him of one such patient, a man born in 1926 named Henry Molaison, whose personal tragedy overturned old theories about the mind and laid the foundation for what we know, or think we know, about how memories are formed.

# 2

# Questions Lead
# to Answers

· · · · · · · · · · ·

The heart was once assumed to be the source of intelligence and memory. When Egyptians mummified their dead, they left the heart in place. But they discarded the brain, using a hook to pull it out gently through the nostrils to preserve the contours of the face. It's not hard to imagine why they would have considered the heart to be the more critical organ. You can feel and hear the rhythmic beat of it simply by pressing your ear up against someone's chest. Open up the rib cage, and you can see its pumping action at work. But the brain? Its pink, pulsing, convoluted outer surface crisscrossed with purplish blood vessels was an enigma, and so was the interior.

A cross section of the human brain cut from ear to ear reveals a cauliflower-like slice with a ribbon of gray surrounding the white interior. A groove half an inch wide and several inches deep separates the two halves, with smaller grooves marking the boundaries between four matching roundish lobes on either side. Hanging off the back of the lower lobes is a tightly pleated accordion of tissue. In the center, a clear fluid that cushions the brain against impact, supplies it with sugar, and removes waste products fills the connected cavities. And draped around those cavities is a tangle of curved, interlocking structures, one of which looks vaguely like a sea horse, its tail resting near the base of the brain. This is why a Venetian anatomist in the late 1500s named it the hippocampus—after a Greek mythological sea monster that was part horse, part fishtail.

But naming things was as far as these early explorers could go. The curious anatomist's knife could define the borders between regions hundreds of times in hundreds of cadavers without being able to describe how the pieces worked together to generate the mind and control the body. For nearly another four centuries, until the mid-1950s, scientists had no reason to believe that the hippocampus, which makes up less than one one-hundredth of the volume of the brain, is the birthplace of all our memories. We only know this now because of an unintended experiment on a young man who became the most studied patient in medical history. He was famous yet anonymous, known, until his death, only by the initials H.M.

—⁓—

Henry Molaison was a perfectly healthy boy living in Hartford, Connecticut, on the day he had a bicycle accident and hit his head. He continued to lead a normal life for another year or so until 1936, when the seizures began. They were mild at first, just brief electrical brainstorms that made him stop and stare, absent for a moment before he returned to himself and the world. But on his fifteenth birthday, in the backseat of the car on a family trip, Henry's body abruptly stiffened and then convulsed rapidly for several minutes. This was only the first of many severe seizures in the years to come.

As Henry's epilepsy worsened, his world grew more circumscribed. Although he was polite, smart, and tall, with a handsome face behind thick glasses, he couldn't stand the incessant teasing about his seizures and soon dropped out of high school. At seventeen, he started over as a freshman at another school. Despite the frequent seizures, he managed to graduate, although he was deprived of the honor of walking across the stage with his classmates, forced instead by the superintendent to sit in the audience with his parents in case he had "a spell."

After high school, Henry managed to hold down a series of jobs that capitalized on his skills and intelligence, including one at an electric motor company that required him to calculate voltage and power in electrical circuits. But all the while, despite an increasingly complex cocktail of highly sedating drugs, the electrical circuits in his brain continued to misfire. He often missed work, and when he was able to go, a neighbor had to drive him there and back. Henry hardly ever went out at night or on the weekend, preferring the safety of staying home with his parents and listening to the radio.

By the time Henry was twenty-seven, the Molaisons' family doctor gave up hope that he could do anything more for Henry and transferred his care to William Scoville. Scoville was a highly respected neurosurgeon at Hartford Hospital, who had performed hundreds of what he called "partial lobotomies" to try to cure schizophrenic patients for whom no other treatment had worked. Scoville's technique was to remove the hippocampus and the adjacent amygdala, an almond-shaped structure that processes emotions.

Henry wasn't mentally ill, but Scoville had reason to believe he had a shot at helping him. He'd performed this same operation on a psychotic patient, a woman who'd also suffered from epilepsy. And while she'd remained so debilitated by her mental illness that no one had noticed her memory problems, her seizures had quieted down. Scoville also observed that he could sometimes induce seizures during surgeries by electrically stimulating the front tip of the hippocampus, the end closer to the eyes, and the area most accessible through the circles, one and a half inches wide, he planned to bore above Henry's eyebrows. Perhaps, Scoville thought, the area in and around the hippocampus was the source of Henry's devastating illness.

For Scoville, this would be the first time he performed the procedure on an epileptic patient who was not otherwise impaired, although an operation similar to what he was planning had become almost routine at the Montreal Neurological Institute and Hospital, where surgeon Wilder Penfield had developed this treatment-of-last-resort in the 1930s. Penfield was as cautious as he was brilliant. He never operated without first monitoring brain waves to pinpoint precisely where they became abnormal. That would mark the so-called focus, the origin of the seizures. If, after gently testing the brain with electrodes to see if he could locate the diseased tissue, Penfield failed to find a focus, he would close up the brain and tell the patient there was nothing else he could do. Although no one knew exactly what any particular brain region did, to Penfield's way of thinking, it seemed prudent to assume that you couldn't remove both hippocampi, or both of any structure, and expect the patient to be normal afterward.

On August 24, 1953, the day before his surgery with Scoville, Henry told the office staff at the hospital that he was nervous but hoped that even if the surgery didn't cure his epilepsy, it would someday help others. His scalp would be anesthetized during the operation, but he would remain awake the whole time so that Scoville could continuously monitor Henry's ability to respond to questions.

It's unlikely that Henry or his parents knew just how experimental his surgery would be, but Scoville did. Several earlier attempts to locate the focus for Henry's seizures using an EEG had failed. Brain waves have to travel through thick bone before they arrive at the tip of an electrode that can measure the shape and strength of the signal. If the diseased tissue lies deep inside the brain, the telltale signal may not survive the trip, or it may be so faint as to be indecipherable.

Now, in the operating room, Scoville had one last chance to find his target. He sliced through Henry's scalp, peeling it away from the bone and drilling two holes in Henry's forehead. Scoville placed the electrodes directly on the brain, first on one side, and then the other, waiting and watching for the abnormal spikes that would reveal the source of the seizures. This final attempt was also a failure. A more cautious surgeon, like Penfield, would have abandoned the operation at this point, but Scoville did not. He proceeded methodically, suctioning out the hippocampus on both sides, replacing the circles of bone he'd removed, and sewing Henry back up. Whether Scoville meant to or not, he had performed the ultimate experiment, one that had previously been conducted only on monkeys. Whatever function the two hippocampi served, removing both of them had obliterated it.

By one measure, the operation was successful. Henry's seizures quieted down. But within days, it became apparent that the surgery had created a new problem. Henry's mother was the first to notice that he kept asking for directions to the bathroom near his hospital room, forgetting answers to questions he'd asked a few minutes earlier, and not recognizing caregivers he'd recently met. He remembered who his mother was, but for Henry, just as it would be for Max some sixty years later, the flow of time had stopped. Two and a half weeks after his surgery, Henry was sent home with anterograde amnesia, unable to form new episodic memories—memories for specific experiences or events from one's life. At the time, no one could have predicted Henry's devastating memory loss. Scientists assumed that memory was more or less evenly spread around the brain. Scoville's crude surgery revealed just how crude this theory was.

A year after the operation, in 1954, Scoville and Penfield discussed Henry's devastating outcome, which led Penfield to send his brilliant young PhD student, Brenda Milner, down to Hartford to study the young man. Henry told Milner that every moment of his life was "like waking from a dream. Everything is clear. But what happened just before, that's what I can't tell you." Testing confirmed that his memories lasted at most a minute. He was not the first patient in history to suffer from severe amnesia, but this time, armed with the details of the operation, Milner knew that when Scoville had suctioned out Henry's hippocampi, he'd taken his ability to create new memories along with them. This young man's great misfortune had revealed the essential role of those small C-shaped structures.

Another revelation soon followed. To help define the limits of what Henry could and could not learn, Milner devised an ingenious test. She gave him a pencil and a piece of paper with a five-pointed star on it and asked him to trace the outline as neatly as possible. However, instead of looking at the paper directly, Henry had to trace the star's outline while looking at its reflection in a mirror. So when Henry moved his hand up, the mirror showed it moving down. Although the task was tricky, Henry had no problem following directions as long as he wasn't interrupted or didn't have to remember them for long.

Initially, his attempts were halting and messy. But very, very slowly, after practicing the task fourteen times a day for three days, he learned that up was down and down was up, and he could smoothly and accurately trace the star's outline. In other words, Henry had formed a new memory. He was just as astonished as Milner, if for different reasons, to see how accurate his drawing skills had become. With no memory of having taken the test more than once, he couldn't comprehend how he had performed such a difficult task so well. Milner concluded that some other area of Henry's brain must have created his new, so-called procedural memory.

Several years later, in 1962, a graduate student of Milner's named Suzanne Corkin began working with Henry, as she would until he died. Corkin created a complex tabletop maze hidden behind a black curtain and asked Henry to use his hand to maneuver his way through it from start to finish by touch alone, using a stylus. After four days and dozens of practice trials, Henry learned how to work his way through the maze more quickly. Just as with the star-drawing exercise, he had created a procedural memory. What he couldn't remember was the shape of the maze; he never did find his way to the end. Without his hippocampus, Henry's ability to remember a route, whether it was through a hospital wing or a tabletop maze, had vanished. Milner and Corkin's experiments made clear that it wasn't useful to think of memory as one monolithic category.

Henry hadn't lost all his memories from the time before the operation, a condition known as retrograde amnesia. He talked about the 1929 stock market crash and how his mother hadn't been able to get money out of the bank afterward. Such knowledge about oneself or the world is called semantic memory. Henry could describe the 1937 Hindenburg airship disaster and knew that there had been a Second World War. He remembered that he liked to play the banjo, the names of various streets he had lived on, and that he had attended Windham High School. He even remembered being told that he'd had his first grand mal seizure in the backseat of a car.

His history of himself, however, ended some time before the surgery. And long after his mother developed dementia and moved into a nursing home, Henry believed he lived with her. New facts and new events failed to turn into long-term memories. Even his old memories were oddly colorless and impersonal, lacking details or specific sights, sounds, smells, or emotions. They weren't stories about himself so much as facts about his previous life. Henry himself was missing from the picture.

Aside from amnesia, Henry was just as intelligent as he'd been before, at least as measured by a simple IQ test. In fact, his IQ increased immediately, likely because his seizures were controlled to the point where he no longer had to take high doses of sedating medication. In 1962, nine years after the operation, his score of 120 placed him in the category of "superior intelligence." Henry could still answer questions such as who wrote Hamlet and understand sophisticated vocabulary words like *espionage*.

Over the next four and a half decades, Henry submitted to exhaustive studies. He was good-natured about it, even if mystified by his circumstances. His bewilderment showed up in a phrase he often repeated: "I'm having an argument with myself." Once, over the course of a two-hour interview, he shared this thought twenty-seven times, each time as if it was a new revelation. But despite his amnesia, all the years of being tested and questioned left some inchoate insight into his diminished state and his value to the world. The hope he'd shared with hospital staff before his surgery was realized. An interview at MIT in 1970 with a visiting researcher named William Marslen-Wilson captures Henry's mind-set:

W.M.: How are you feeling? Too many questions?

H.M.: No, I don't think there are too many questions. In a way what I'm thinking right off though is, not too many questions. In a way. Because the questions lead to answers.

W.M.: Questions lead to answers?

H.M.: Yes.

W.M.: That's exactly what we always hope for.

H.M.: I wasn't thinking, just, me giving the answers or asking the questions or anything like that. I was thinking of other people.

W.M.: Yes. How do you mean "other people"?

H.M.: Because possibly it can help you to help others.

W.M.: That's exactly why we are asking you these questions.

H.M.: In a way, I guess you could say I have . . . if it helps others, that's good. Very good.

The answers that research on Henry provided about the nature of memory and the role of the hippocampus come with asterisks. While his vocabulary was above average, his ability to put words together to form new ideas was poor. Although his knowledge of history dropped off at some point before his 1953 surgery, he learned the names of a few people he'd seen repeatedly on television since, like John F. Kennedy and the astronaut John Glenn.

After Henry died in 2008, at the age of eighty-two, his brain was sliced into 2,401 sections to build a precise three-dimensional model. This gave scientists a far more detailed view than the most advanced imaging techniques could have created while he was still alive. The surgical tools available in the 1950s had made the operation unavoidably imprecise. A lesion discovered in the left frontal lobe looked like it might have been created when Scoville used a flat brain spatula to lever it up and out of the way to reach the hippocampus. While Scoville had knowingly removed a tiny piece of tissue attached to the hippocampus called the amygdala, he didn't realize that he'd only removed about half the hippocampus, and that he'd damaged the white matter underneath. The cerebellum—two tightly pleated fist-shaped regions at the base of the brain that coordinate motor skills and balance—had shrunk, likely through years of heavy medication.

Previous research on Henry had assumed that the deficits scientists had uncovered could be chalked up to the loss of the hippocampus, because the rest of his brain was essentially normal. The new, postmortem images showed otherwise and added to debates about the finer details of how different memory systems interact. Nevertheless, Milner and Corkin's fundamental insight into the role of the hippocampus as a memory maker has stood the test of time.

Barash had heard of H.M.'s case in college and studied it in medical school. Very late one night, sitting at home on the couch with the TV on in the background, he pictures Max's scan. The two white quarter-moons of the hippocampus float against the dark background and beckon him with a question—what was it about Max's overdose that had zeroed in on this structure and no other? It occurs to him now that Max's overdose caused an injury so pure and so clean that it's "more H.M. than H.M.," a more neatly circumscribed area of damage than Scoville's suctioning tools had left behind.

The Alzheimer's patients Barash sees week after week are a reminder that such precision is rare. When Dr. Alois Alzheimer met 51-year-old Auguste Deter in 1901 at The Castle for the Insane in Frankfurt, Germany, she told him that she had "lost herself." By then, the illness would have ravaged her entire brain. Auguste could not have been the first to suffer from a disease that today afflicts millions of people around the world. But she was the first to be noticed.

"I show her a key, a pencil, and a book, and she names them correctly," Alzheimer wrote in Auguste's medical notes. "But when I ask, 'What did I show you?' she replies, 'I don't know. I don't know.'" Scrawled on a small scrap of paper discovered among Dr. Alzheimer's notes, isolated words drift down the page in Auguste's handwriting. Alzheimer initially described her as having an "amnestic writing disorder" because she couldn't complete a sentence. She would be case 1, and once she came to attention, Alzheimer and his colleagues began to recognize more patients with the same symptoms.

After Auguste passed away in 1906, Alzheimer examined her brain and sketched the strange plaques and teardrop shapes that had spread throughout it. The misfolded proteins had cut off communication and killed neurons. Nearly one third of the cells in her cortex, the thinking part of Auguste's brain, had died. Because her illness was so far advanced, Dr. Alzheimer and his colleagues didn't know that the disease had begun in the delicate hippocampus. In 1906 Alzheimer presented Auguste's case at a scientific meeting, to little fanfare. Only the title of his lecture appeared in the printed abstracts. The full description was not deemed "appropriate for a short publication." Nevertheless, the toxic accumulations Alzheimer described, now known as amyloid plaques and tau tangles, went on to become the defining signature of the disease.

These two very different patients—Auguste Deter and Henry Molaison—changed the course of neurology and ultimately revealed the essential role and unique vulnerability of the hippocampus. Barash isn't audacious enough to imagine that Max's case could wind up in the pages of a medical textbook, but he does know that any new case report can help move science ever closer to understanding, however incomplete. First, though, from a practical point of view, Barash has to figure out precisely what happened to Max. He suspects that gathering the clues won't be easy.

Fortunately, Barash's training has instilled in him a systematic approach to breaking down mysterious cases. Before accepting the job at Lahey, he had completed a two-year fellowship at Yale in epidemiology. It had been an unusual extra step for a neurologist, but Barash looks at epidemiology as puzzle-solving—which he loves—in the service of public health. At Yale, he'd been an oddball, the neurologist interested in obscure brain disorders. Because of his fascination with CJD, the other fellows had called him the "prion guy."

The disease detective's strategy for solving an outbreak is to break it down by person, place, and time, and this is how Barash analyzes it. The person is a 22-year-old man with severe memory loss. The place is Boston, Massachusetts. And the time is the fall of 2012. Max probably overdosed on heroin, just like hundreds of other people in Massachusetts that year. With only one patient and no way to know for sure what drug Max used, Barash reluctantly concludes that the case is unsolvable, at least for now. He files it away in the back of his mind and switches off the TV before heading upstairs for bed. He's unaware that a handful of other neurologists across the state have seen similar patients with nearly identical brain scans. None of them know what to make of it either.

# Person, Place, and Time

. . . . . . . . . . . . . . . . . . . . . . . . . . . . . . .

## *Massachusetts, 2012–2015*

One month after his overdose, Max quits his job waiting tables at Ole, a Mexican restaurant in Cambridge. He knows how to take an order and how to make guacamole right at the table, but he can't remember who ordered what long enough to bring the correct meals back to the customers. He drops out of UMass Boston, putting his dream of being a writer on hold. Back at Lahey Hospital, several hours of testing—drawing, retelling stories, remembering lists of words or strings of numbers—confirm what's obvious to Max's mother, Laura, and his friends. He still has memory loss.

Over the next two years, he occasionally returns to Lahey for follow-up appointments. Few people recover from amnesia, but very slowly, Max improves. On a warm day in late summer of 2014, he goes up to 7 West with his mother to see Zabar, who shares the results from another MRI scan and the latest round of memory tests. This time Max's memory scores are close to normal, but his attention is still impaired. Zabar takes a look at the scan. The bright signal heralding damage has long since faded, but Max's hippocampus is smaller than it was at the time of the accident. It's a clear sign that neurons have died, and it's probably one of the reasons Max is not back to his old self. He's at sea, still unable to hold down a job, and still abusing drugs.

On the other side of the neurology suite, unaware that Zabar is examining Max, Barash stares at the unpacked box of files from his recent move. His new office is at the far end of the neurology pod near all the other younger neurologists, with a window overlooking the Lahey parking lot. A large replica of a vintage microphone he bought at Home Goods sits on a shelf. In his old office, he would grab the mic when colleagues came by and welcome them onto the imaginary set of *The Tonight Show*. Barash has rehung the picture of the beach, along with the polio vaccine poster and the announcement for the prion disease lecture.

The unrelenting nature of his job so consumes Barash that he has barely had time to think about Max's case since grand rounds nearly two years ago. He's frankly amazed at how quickly the initial thrill of being a full-fledged doctor and not having to do the grunt work of residents and interns wore off. It was, after all, a delusion. The grunt work is still there, just different.

Barash believes the best kind of doctor doesn't have to be unusually smart; it's more important to be compulsive, detail oriented, and a good listener. That's how they catch the things you need to worry about. Sometimes he watches the medical show *House*, whose brilliant and asocial protagonist solves medical mysteries with dazzling flashes of insight, and laughs at how absurdly far removed it is from the truth. The administrators at Lahey appear unrealistic about how many patients a neurologist can see in a day. Still, because Barash refuses to shortchange patients by rushing through appointments, he has no choice but to spend nights and weekends tying up loose ends and e-mailing referring doctors and hospital administrators to make sure he has all the medical records for the following week's appointments.

New patients often come in because they fear they have Alzheimer's disease. Barash has come to recognize that if they make it to the appointment on schedule and on their own, they're unlikely to be suffering from dementia. But many of his patients do show up late, and they tell convoluted stories about their lives, a classic sign of dementia. When a family member accompanies the patient, it helps get the story straight, but by now, Barash has discovered that having too many people in the cramped examining room prolongs the appointment without adding useful details.

Barash's examining room is hot in the summer, freezing in the winter, and the acoustics are poor, which makes it difficult for patients to concentrate and pay attention to cognitive tests. Barash often worries that he's not capturing their deficits accurately. He's learned not to mention Alzheimer's unless someone specifically asks. Partly that's because it can be hard to tell in the first or even second appointment which of the most common forms of dementia is driving his patient's mental decline—Alzheimer's, vascular, Lewy body, or frontotemporal. But just as importantly, the word *Alzheimer's* so terrifies patients and family members that everyone tunes out for the rest of the appointment. More often than not, Barash is breaking bad news to someone, trying to frame the diagnosis in a way they can understand, and wishing there were something more he could do to help.

Barash is allowed to do research if he can fit it in between patients, but lately, he's struggled to find the time or the motivation. His seventh paper on the prion disease CJD, this one on the accuracy of diagnoses in Massachusetts between 2000 and 2008, has just come out. But it will be his last paper on the subject. He knows he's in no position to make an exciting, original contribution to the field of prion disease. The primary reward has been getting to know his coauthor, Al DeMaria, the state epidemiologist at the Massachusetts Department of Public Health (DPH), a quiet, principled man responsible for investigating and tracking diseases.

Barash's mentors—and his father, a world-renowned anesthesiologist— recommend that he abandon his long-standing fascination with prion disease and stop avoiding the obvious. He's a neurologist who treats patients with dementia: he should research dementia. One former supervisor suggests he develop a diagnostic tool to figure out when patients should stop driving, but the project sounds as dull as it is practical. Barash is driven by curiosity, and he's committed to doing what he believes makes sense, despite what anyone else says. He's stubborn but also pragmatic. He knows that the efforts of some of the smartest, highly funded researchers in the world haven't resulted in an Alzheimer's treatment, and neither have the billions of dollars pharmaceutical companies have spent running clinical drug trials. Studying Alzheimer's looks like a blind alley. And besides, while Barash is good at looking for patterns, he's not a basic research scientist.

Now, as he contemplates the unpacked moving box, he begins to question the trajectory of his career, or at the very least, whether he's working at the wrong hospital. The pressure to squeeze more appointments into each day is intensifying, and time with family is taking a backseat. If he is going to spend nights and weekends working, he wants to make a difference. Barash resolves to remain on the lookout for some medical puzzle to present itself among the patients he sees, something that would give him the feeling he has something more to offer them. But it seems like a long shot. There are several internationally renowned hospitals in Boston, the kind more likely to admit unusual cases.

Barash appreciates his good fortune. His wife, Gillian, is an oasis of common sense and equanimity, qualities that serve her well in her practice as a therapist for troubled and often suicidal adolescents at McLean Hospital. The couple has recently moved across Lexington into a house with a large kitchen, a backyard, and enough space for baby Henry and a hoped-for second child. Barash knows he should be satisfied, but, as he tells Gillian on a long drive home from a weekend visiting her mother in Maryland, "It's the same thing every day, week after week. I wanted to see patients, but not so many that I don't have time to stop and think." The year 2014 is winding down, and 2015 promises to be no different. The predictable days stretch out ahead of him, each one the same as the last.

On the other side of the country, in a high-ceilinged, hushed library at UCLA, a student named Owen Rivers is so engrossed in studying the topic of his obsession—memory—that he's temporarily relieved of his desire for drug taking. Owen inhales deeply from a yellow sticky note infused with his favorite cologne. A paper he'd read during his Introduction to Clinical Psychology class showed that people recall information better if they pair learning with a pleasant smell. For reasons he can't explain, he has a deep-seated fear of losing his memory, even though it will be several more years—long after he's done studying for exams—before his premonition comes to pass. Owen checks his phone and sees he has time for one last practice question.

*Which of the following is TRUE of patient H.M. after his surgery?*
  *a. His epilepsy worsened.*
  *b. He could learn new procedures.*
  *c. He could remember the scientists he worked with.*

*The correct answer is b. he could learn new procedures.*

Owen returns the sticky note to the outer pocket of his backpack. Later today, during the exam for his Neurobiology of Learning and Memory class, he'll take out a freshly sprayed sticky note and smell it every few minutes to reactivate the memories he's forming in the library right now. His memory-strengthening trick seems to be working: despite his enormous task load, Owen is earning the same straight-A grades he received in community college, only now in much more demanding courses, and he still has time to read the philosophers Nietzsche and Kant.

But before the exam, Owen is due at the weekly meeting in Professor Wendy Limm's lab, where he's scored a coveted undergraduate research position. Today there'll be a team review of the results for a study on whether daily cocaine use changes how mice learn and what genes are involved. Owen hurries past the row of small operating rooms lining one side of the lab. There, skilled postdoctoral students perform the delicate surgery that makes it possible to capture images showing how cocaine changes connections between neurons.

In the conference room at the end of the lab, Owen takes a seat alongside the other students. Limm turns off the overhead lights and loads a time-lapse video, and Owen watches, enraptured, as a twiggy white dendrite projecting from a single neuron jitters on the screen like a century-old silent film with too few frames. Small nubs that sprout along the length of the dendrite are destined to form new connections with nearby neurons and encode the memory of the cage where the cocaine was administered. When the video ends, a postdoctoral student runs through a series of bar graphs and images that need to be finalized. The statistical analysis bores Owen, so he tunes out until Limm describes her plans for the next project. She wants to do more than prove that cocaine speeds up learning and strengthens the memory of using this drug. She wants to know precisely why that happens and then use that insight to develop new ways to treat addiction.

When the meeting ends, Owen heads to his learning and memory class feeling temporarily hopeful that science might someday help him overcome his daily desire to use drugs. Maybe he'll even go to med school and become a psychiatrist. Owen cruises through the exam, inhaling from the sticky note every few minutes, secure in the knowledge that he is acing the test. But as he walks back to his dorm on the other side of campus, the satisfaction of having done well on the exam fades, and by the time he reaches his dorm room, the old familiar feeling of dread is fully upon him. When Owen sees that his roommate is out, he spreads a thin line of cocaine on his desktop, snorts it, and pulls out his phone.

> *The dread was sort of gone for a minute, but it's back and I don't feel happy because I'm too obsessed with meaning and purpose and the lack thereof. I wanted this so desperately but I actually hate whatever it is that my neurons are learning right now because I hate how I feel and I can't fix myself. Heroin is the only complete solution, the only way I have ever felt at peace in the world. Not a day goes by when I don't crave H.*

Owen moves on to his memory lists and then backs everything up to a hard drive.

—⚬—

It's 2015, and Barash's moving box sits, still unpacked, in the corner of his third office at Lahey. The constant moving feels like a mocking reminder that his career lacks a clear path forward, but unbeknownst to him, the next chapter of his life is about to begin when a 41-year-old construction worker from New Hampshire, named Anthony, walks into his office. Two months earlier, Anthony had decided to stay home alone while his wife and children went out of town for the weekend to visit relatives. When his family returned, he was confused, disoriented, and asked the same questions repeatedly.

In the local emergency department, doctors diagnosed anterograde amnesia. Like Max, Anthony remembered his past, but he couldn't pass the simple *apple*, *table*, *honesty* memory test. He told the medical staff that he hadn't used illegal drugs since the birth of his children, so they didn't request any toxicology screens. They did order a brain scan, but the radiologist didn't notice anything unusual that could account for his memory loss.

In January, still suffering from amnesia, Anthony arrives at Lahey for a second opinion. Barash hasn't been able to collect all the hospital records ahead of time, so he hasn't seen the MRI. Anthony, his wife, and his mother all crowd into the cold examining room. Anthony waits for his wife and mother to sit before he does.

After introducing himself, Barash says, "So, Anthony, I have most of your records here, but can you tell me a little bit more about what brings you to Lahey?"

"I don't . . . I don't honestly understand what's happening to me. I think that's why we came here."

He looks for reassurance from his wife, Johanna, who takes over. She tells Barash that Anthony seemed fine on the Saturday of their weekend away when she called to say hello, but by Sunday morning, something had changed.

"He kept asking me why the kids weren't in school, and I kept telling him it was Sunday, so of course they wouldn't be in school."

"Do you remember that?" Barash asks Anthony.

"No. I don't remember much of anything," Anthony says. "I really don't. Not since that weekend. And I haven't been able to get back to work since—Johanna, how long has it been?"

"Eight weeks," his wife says.

"Has it gotten any better since then?"

"Maybe, but not by much," Johanna says. "I tried to get him to play board games last week, since he doesn't like to watch TV anymore."

"I can't follow the plot," Anthony says.

"And he kept asking me the rules. Every turn," Johanna continues.

After the physical examination, Barash asks Anthony whether he has any history of illegal drug use. Anthony takes a breath while his wife and mother look on expectantly. "I was a heroin addict, but I haven't used in a long time."

"Sixteen years," Johanna jumps in.

Barash moves on. "The notes here say that the MRI was normal, but the hospital in New Hampshire wasn't able to send the records in time for your appointment. I'd like to look at it myself. Were you able to bring a copy?"

Johanna finds the disc in her bag and hands it to him. "Thank you," he says, "and thank you for coming all this way." Barash gets up and shakes everyone's hand. "We'll do our best to figure out what's going on here. Check in with the receptionist on your way out and make an appointment to come back and see me in eight weeks if you can. Let me know if you need anything before then."

Back in his office, Barash puts the disc in his computer and waits for the images to load. He's puzzled by the patient's severe memory loss, although he doesn't expect to find any answers in the scan, which the radiology report described as unremarkable. But Barash doesn't like to believe what he's told. The average radiologist reads about ten thousand scans a year, so they're bound to miss something every now and then. He always reviews scans himself.

The image displayed on the monitor is like a time machine. In an instant, a memory that had lain dormant resurfaces, transporting Barash back to a fall day in 2012, sitting in his old office, looking at Max's scan. Barash has only one monitor, but he doesn't need to see the images side by side to see that the two are almost indistinguishable. The hippocampus on Anthony's scan stands out brightly against the dark, normal structures of the rest of the brain. Barash can't be sure why the radiologist didn't flag this. Maybe he was inexperienced, or tired at the end of a long day, or just not expecting to find anything. Or, perhaps he had seen the perfect symmetry of the pattern but decided it was a false signal because it didn't make any sense. But for Barash, things that don't make sense are what he pays attention to.

Once more he scrolls through the scan bottom-up from the base of the skull. The damage is severe, covering the entire length and width of the hippocampus on both sides of the brain. Leaning back in his chair, watching the cars leaving the parking lot, Barash begins sifting through the possibilities. To rule out a stroke, he'll request another MRI to look for a blood clot. But this pattern doesn't look like any stroke victim he's ever seen. It looks just like Max's. Despite Anthony's denials, the only link between the men appears to be heroin, an opioid banned from medical use in the United States in the 1920s but widely abused ever since. So why would this bizarre injury only show up now? What's changed?

Suddenly an idea occurs to him. Max said he used heroin, but what if he unwittingly used fentanyl? Heroin is often cut with other substances like sugar, starch, or powdered milk. Increasingly, it's being cut with fentanyl, a man-made opioid about fifty times more potent than heroin. Deaths from fentanyl are on the rise. Perhaps Anthony also unknowingly used heroin mixed with fentanyl—or even pure fentanyl. It's an intriguing connection. Hospitals don't test for fentanyl when overdose victims come in, so Barash can only speculate. His next step is to confirm his instinct that the pattern of brain damage is the same. Barash only has a minute until his next appointment, but before the end of the day, he sends the head of neuroradiology, Juan Small, an e-mail asking him to take a look.

Eight floors down in the basement, in a dark room in the radiology department, Small pulls up Anthony's MRI scan on his oversized monitor. He's kept Max's scan bookmarked in the system to make it easy to show medical students, so it takes no more than a minute to find it and toggle back and forth between the two images. Barash was right about this one. Small has no idea what catastrophic event took place inside these two men's brains, two years apart, but it's hard not to wonder if it's more than a fluke. Unfortunately, Barash, Small, and Zabar don't have enough evidence to go on to figure out the connection. And whatever it might be, it seems unlikely that a third case will just happen to turn up in Barash or Zabar's offices, or on Small's radiology reviews. There are dozens of brain specialists at Lahey. No one has the time or the need to learn about each other's patients. Perhaps, Small suggests, they should search back through the hospital records of every patient who visited the emergency department or the neurology clinic in the past few years. Maybe they'll find more cases.

Late one spring night, Barash is lying in bed with the lights out, wondering how he'll find time to do this project, when the solution becomes obvious: P. Monroe Butler, the young resident he's been supervising lately. Like all residents, Butler will have to complete a research project before he can graduate, and this could be it.

Barash has met a lot of smart people, but Butler is more than that. He thinks on a different wavelength. Butler received a master's in philosophy, religion, and ethics before going to medical school, where he invested an extra two years getting his PhD. He's published eight papers, six book chapters, and three abstracts on topics ranging from how Parkinson's disease affects religious beliefs, to Nietzsche's late-life madness, gabapentin for pain therapy, the psychobiology of evil, and the evolution of dreams. Butler's fourth paper, coauthored with an anthropologist, links the spread of a gene associated with novelty-seeking to the migratory patterns of early humans leaving Africa. Barash's instinct is that Butler is the kind of guy who'll find the mystery of the amnestic syndrome appealing.

He knows it's impulsive and maybe even inappropriate, but Barash picks up his phone and sends Butler a text. It's 11:23 P.M. on a Saturday night, and he figures the young resident is out at a bar in downtown Boston with friends, blowing off steam.

> Hey there, Zabar, neurorads (Juan Small) and I have some cases of isolated bilateral hippocampal infarcts, and we're thinking of doing a case series . . . it'd be a good resident project. Since you're interested in cognitive and you're a good writer, it might be a good project for you. Would you be interested?

Butler is lying at home in bed, finishing up a beer in his rental, the former butler's quarters and horse stables at the back of a mansion in Brookline. Although it's the last thing Barash would have imagined, Butler is a single dad, and his young daughter is asleep in bed down the hall, curled up with Thoreau, their Maine coon cat. Butler is used to late-night calls from his superiors and doesn't think it's inappropriate. Three thoughts flit through his mind: *How does he know I'm a good writer? It's nice that he thinks I'm a good writer. Barash said that so that I'll say yes.* He texts back one word: "sure."

The two men will turn out to be a good match. Barash's strong suit is patterns. He looks at the who, what, where, when—the scene of the crime. Butler cares about the *why*: the mechanism that explains the damage, the crime itself. Soon after Barash briefs him on the case, Butler's thoughts turn to a medical mystery from the 1980s involving a bizarre type of brain injury in synthetic heroin users. The solution to the mystery resulted in a significant breakthrough in the study of another intractable neurodegenerative disorder. Barash smiles when he finishes reading a lengthy e-mail Butler sends late one night that describes the parallels. He was right to think that Butler was the perfect person to enlist.

—⁂—

On July 1, 1982, a 42-year-old man from San Jose named George Carillo bought a bad batch of synthetic heroin. The chemist who cooked the drug had made a mistake in the recipe, creating a toxic contaminant called MPTP. What happened next would revolutionize the study of a movement disorder called Parkinson's disease, which usually begins with tremors.

The burning sensation Carillo felt when he injected the drug was the first sign of trouble. Three days later, having been jailed for showing up high on PCP at a court hearing, Carillo could barely move or talk. Initially, the staff at the prison suspected he was malingering. By day seven, Carillo was frozen. Another week later, after a dozen or so doctors had weighed in, a neurologist named Bill Langston examined Carillo at the Santa Clara Valley Medical Center. He, too, was baffled by the man's bizarre symptoms. Finally, three weeks after the overdose, Langston picked up on some slight movement in Carillo's fingers and gave him a pen and a piece of paper. With a trembling hand, Carillo wrote, "I'm not sure what is happening to me. I only know I can't function normally. I can't move right. I know what I want to do. It just won't come out right." Putting those few sentences down on paper took half an hour.

Carillo's mind was working as it should, but the part of his brain that helped control his movements was not. Soon Langston heard reports of other patients south of the Bay Area with identical symptoms, all classic signs of advanced Parkinson's, a slow-moving, inexorable disease that usually strikes patients much older than Carillo and the other victims. It turned out that all six had injected synthetic heroin from the same bad batch.

At the time, scientists already knew that the symptoms of Parkinson's could be traced back to a small bundle of neurons not far from the hippocampus called the substantia nigra, Latin for "black stuff." These neurons are full of a pigment similar to melanin, the molecule that gives skin a dark color. More importantly, they pump out dopamine, a neurotransmitter that, among many other things, guides movement. In people with advanced Parkinson's, the black stuff has been wiped out, along with the ability to control their limbs. But back then, no one knew what was killing the neurons.

As a result, attempts to cure or even treat Parkinson's were at an impasse. And because other animals don't develop Parkinson's, the disease is hard to study. With an animal model—a rodent or a monkey that has artificially been given a human disease—scientists can administer new drugs, sacrifice the animals, examine their brains under the microscope, and see the effect. This research is usually the first step on the road to human clinical trials for new treatments. But despite decades of attempts, there were no animal models for Parkinson's.

When Carillo and the other five drug abusers injected MPTP, they wound up giving scientists two tools in the quest to cure Parkinson's. The first was a long-sought animal model, which researchers could now create merely by injecting MPTP into rodents or monkeys. The other advance was fresh insight into what causes Parkinson's in the first place. MPTP isn't the cause of the disease. But knowing that one molecule could produce Parkinson's symptoms gave scientists their first glimpse of a potential mechanism for the complex cascade of events that leads to the disease. This was invaluable because the holy grail in medicine is to intervene early, before the damage is done, and if possible, to treat the cause, not the symptoms.

Like Butler, Barash is intrigued by the similarities between what took place three decades earlier in California and the circumstances surrounding the two Massachusetts cases, Max and Anthony. In both instances, the patients had taken illegal drugs just before they developed the

full-blown symptoms of a common neurodegenerative disease that
advances slowly and affects much older people. In both instances,
a specific brain region responsible for a particular ability had been
uniquely targeted. In the amnesic patients, it was the hippocampus, and
in the Parkinson's patients, it was the substantia nigra. Just as George
Carillo's locked-in syndrome was akin to the most advanced stages of
Parkinson's, the amnesic patients suffered from the same devastating
memory loss as Alzheimer's patients.

So, could figuring out what happened to the amnesic patients be useful
for tackling Alzheimer's disease? For the past twenty-five years, to test new
medicines, scientists have relied on mice whose genes they've changed to
make them develop an Alzheimer's-like disease. But even though these mice
have been "cured" many times over, none of the drugs that seemed so promising in mice have worked in humans. In other words, these animal models
must be missing something. No animal model will ever be perfect; the human
brain is not a giant mouse brain. But perhaps if the doctors at Lahey could
pin down the substance that targets the hippocampus, it would be the first
step on the long path toward a new animal model for memory disorders.

Barash and Butler realize that speculations like these are easy to come
by and hard to prove. They don't know with any certainty what caused
the amnesia. Because of the limitations of drug testing, they can't tell if it
was heroin or fentanyl. It could even be some kind of toxic contaminant,
although it seems improbable that the two victims would have taken the
same contaminated batch of drugs two years apart. It's clear to the two neurologists that the source of this bizarre brain damage will remain a mystery
unless they can find out if there are more people like Max and Anthony.

In June, after Lahey grants permission to review hospital records, Butler
meets with Barash in his office to talk over the plan of attack. Just a few
years younger than Barash, Butler is trim, and light on his feet. He has a
square jaw with a dimple in the chin and sea-green eyes that crinkle at the
corners like he's on the verge of either a laugh or an idea.

As Barash motions Butler to sit down, he says, "How did you just know
that MPTP Parkinson's stuff, like, off the cuff? You're like the dude from
*House*."

"Well, there's some connection between spirituality and the progression of Parkinson's, and I'm pretty interested in the neuroscience of that stuff," Butler says. "But it's probably going to have to stay a guilty pleasure. For a while. And this is a cool project."

"You're gonna have to roll through a ton of scans. Take a first pass and then turn it over to Juan Small. You have enough time?"

"It's a little bit restricted." Butler glances at *The Tonight Show* mic on the shelf, then back at Barash. "I'm a single dad. But I can fit it in in spurts."

Barash doesn't miss a beat. "We'll get you remote access. Not a problem. You can do the literature search whenever. Maybe we can get this published somewhere influential, maybe *JAMA*. Shoot high and see what happens!"

"Yeah, sure, yeah, that makes sense."

"I can afford to be high on your prospects since I'm not the one doing the work." Barash pauses. "You're a good person, Monroe."

Over the next few months, Butler goes fishing late at night on PubMed, the Google of science research, trying to find out if anyone else has seen cases like Max's and Anthony's. Keywords such as *heroin, opioid, hippocampal,* and *amnesia* turn up a few leads—but nothing with the word *fentanyl.* Butler comes across the case of a 73-year-old man in Korea with carbon monoxide poisoning and damage to both hippocampi, who had severe amnesia. A 33-year-old man who had inhaled heroin rather than injecting it checked into a clinic in Strasbourg, France, in 2013 with amnesia. The damage to his hippocampus was similar, but only on one side. Butler finds three cases of amnesia and hippocampal damage after cocaine use, one in Pittsburgh in 2004, one in Illinois in 2012, and one in Melbourne, Australia, in 2015. There's no obvious opioid connection, and they're a decade and two continents apart. It's hard to know how to put them together.

Toward the end of summer, Butler is logging into Lahey Hospital's password-protected system from home, trying to find scans that match Max's and Anthony's. Butler finds a dozen or so cases of patients with tiny bright white spots of damage in the hippocampus that would have caused amnesia lasting no more than a day. Then, late one night, he finds the records of a man with the striking telltale pattern on his scan. But the patient lingered in a coma for six days and then passed away, taking with him the evidence that he almost certainly had amnesia.

Butler likes to think about the hippocampus the way he thinks about life. There's passion, and there's reason. There's the cold end of the hippocampus, toward the back, that deals with the places, the facts, the lists of words to remember, and there's the hot end, closer to the eyes, that handles more emotional memories. Butler reaches for his third cup of coffee as he considers the possibility that this research project may not amount to much, but he doesn't mind. It's been stimulating to think about the mechanism that might have caused the memory loss, and after all, this research is just a side project, not the focus of his career. He'll soon leave Boston for the next chapter of his life, a fellowship at the University of California San Francisco, the last leg in the long slog of his academic and medical training.

# 4

# Short Shrift

• • • • • • • • • • • • •

## Massachusetts, Fall 2015

It's dark by the time Barash pulls into his driveway on a Tuesday in early October. The pleasing smell of smoke from someone's fireplace hangs in the air, but his mood is low. Butler's search through Lahey Hospital's medical records hasn't panned out, and for Barash, unlike Butler, this is much more than a side research project. Even though he only has time to work on it on nights and weekends, it helps get him through the days while he looks for a new job.

Barash walks through the back door deciding to entertain Gillian with trivia instead of complaining. She works late on Mondays running an evening group for parents of troubled teens, and Barash knows she's always tired the next day.

"You hanging in there?" he asks as he drops his keys in the bowl on the kitchen counter. Henry, now one and a half, is sitting in his booster seat and gives Barash a big smile.

"It was a typical Tuesday," Gillian says. "Still have notes to write up after Henry goes to bed. You?"

"Butler's striking out looking for more cases," he says as he extricates Henry from his booster seat and gives him a hug. "There's just tons of administrative stuff slowing things down. It's brutal, but it's not like giving up two grand slams in one inning."

"What?" Gillian says, taking the bait.

"Dodgers pitcher Chan Ho Park gave up two grand slams in one inning in 1999. He was annihilated. That's a historically bad day. Today was just par for the course."

Shortly before midnight, sitting on the couch with his laptop open and the TV on in the background, Barash suddenly sees a way to move his investigation forward. Al DeMaria, his coauthor on the CJD prion paper, is a respected, well-connected veteran of public health investigations. His position as the state epidemiologist for the Massachusetts DPH gives him the power to investigate disease outbreaks. In an e-mail he sends DeMaria, titled "Unusual Set of Cases," Barash describes Max and Anthony—the sudden, shocking severity of brain damage to the hippocampus, the dense amnesia, the connection with opioids, and his suspicion that more cases are going under the radar. "I think that these cases could represent a more common syndrome," he writes. Barash points out that heroin and fentanyl use is on the rise in Massachusetts and floats the idea that a contaminant or a bad batch of drugs could be the culprit. With no end to the opioid crisis in sight, he fears that there could be many more cases to come, of patients and families whose lives will be upended. And maybe, just maybe, there's a chance that millions more who suffer from memory-related illnesses could someday be helped by whatever he learns as he pursues the investigation.

DeMaria is called "Dr. D" by everyone from the security guards to the scientists at the DPH's William A. Hinton State Laboratory, a squat, concrete, eight-story building. DeMaria is a short, owlish man in his sixties with pants belted above the waist and inquisitive brown eyes magnified by outsized glasses. When asked at age seventeen by his high school counselor what he wanted to study, the words "infectious disease" popped out. DeMaria isn't sure why he said that and still wonders what his life would look like if he'd answered the question some other way. His offhand comment wound up being prescient. After graduating from Harvard Medical School, DeMaria became an infectious disease specialist.

He gravitated toward working with the underserved, starting out at Montefiore Medical Center in the Bronx. There, as a resident, he organized an unsuccessful strike to protest what he considered to be unfair treatment of hospital interns. Many of DeMaria's patients had substance use problems. At the time, he was morbidly obese and hadn't yet recognized that his compulsion to eat was another form of addiction. From the Bronx, he moved to Boston City Hospital, which serves mostly low-income people from the surrounding neighborhoods, then spent four years in private practice in the blue-collar towns of Malden, Melrose, and Somerville. Finally, he took a job at the DPH. Having been employed there for more than twenty-eight years, DeMaria helped write the rules that give the DPH the power to investigate disease outbreaks, epidemics, and mysterious new health syndromes. He's politically savvy and knows who to call to prevent projects from getting stalled in bureaucracy.

DeMaria could choose one of two routes to drive from his home in Melrose to the state lab. One takes him through the well-maintained suburbs of Roslindale and then past the Arnold Arboretum, Harvard's glorious botanical gardens. The other option is to drive past Boston City Hospital near an area rife with drug use known as the "Methadone Mile." But to keep the weight off—DeMaria lost 180 pounds the year he acknowledged his food addiction—he usually takes public transportation, leaving the house at 6:00 A.M., walking two miles to the train station, and arriving at work well before the rest of his staff. The round table in the middle of DeMaria's office up on the sixth floor is piled high with stacks of paper. Mementos and paraphernalia from public health crises he has managed cover every other available surface. A ceramic facsimile of an AIDS awareness stamp sits on a shelf. Safe vaccine storage warnings are stuck to the filing cabinet. A delicate wooden mosquito dangles from the ceiling. On the walls are newspaper clippings, children's paintings, and a red and black picture of Che Guevara, a gift from a visitor.

DeMaria often hears from doctors who think they've spotted something unusual. Most of the time, it turns out to be a false lead or just unimportant. It's been forty years since DeMaria graduated from medical school, so his first thought when Barash describes bilateral hippocampal ischemia is, *What does the hippocampus do again? I remember something about sea horses.* But he takes Barash's concerns seriously because he worked with him closely on the CJD paper. DeMaria knows that Barash's background in neurology and epidemiology makes him the perfect person to pick up on a pattern that is likely to be more than just a coincidence.

And even though there are only two patients, DeMaria's decades working in public health make him hesitant to dismiss Barash's concerns too hastily. He knows it's not necessarily the raw numbers at the outset but the potential implications that matter. These cases remind him of other medical investigations that started small, with just a few people who had unexplained symptoms. In 1999, an observant infectious disease specialist in Queens contacted New York City's Department of Health to report two patients with sudden paralysis and high fevers. The specialist couldn't account for the symptoms, but the men would turn out to be the first known examples of West Nile virus in the United States. Similarly, the AIDS epidemic, which began with a handful of patients with a rare fungal pneumonia, went on to sicken and kill millions. DeMaria's former supervisor initially regarded the phenomenon as a mere flash in the pan.

The next afternoon, Barash hears back from his former colleague. DeMaria agrees that people with drug problems like Max and Anthony may be stigmatized and "get short shrift." He says that if Barash finds more patients, the situation will qualify as an outbreak, giving the DPH enough justification to demand medical records for any suspected cases. In the meantime, DeMaria shares Barash's concerns with a handful of experts at the Centers for Disease Control and Prevention in Atlanta. With the CDC's attention and support, the search for more victims could go nationwide. The experts soon respond with their take: the two cases may not have anything to do with each other. Perhaps heroin is involved, but a stroke, carbon monoxide poisoning, low blood sugar, prolonged seizures, lack of oxygen caused by the overdose, or heatstroke are all reasonable explanations for the amnesia.

Barash is so irked by the implausibility of their final suggestion—amnesic shellfish poisoning—that he tries to put it out of his mind until he has time to ponder the idea while he runs a few errands on the way home. Ostensibly the hypothesis has something going for it, since people with amnesic shellfish poisoning have profound memory loss, and the damage to their hippocampus appears strikingly similar to Barash's two cases. But neither of the victims mentioned eating seafood, nor did they experience the severe vomiting and diarrhea that go along with food poisoning. The timing is wrong too. Outbreaks of amnesic shellfish poisoning are rare and occur in warmer months, when algal blooms take place. These two patients developed amnesia in October and November.

Back home, Barash vents to Gillian. "They're not seeing the bigger picture," he tells her as they put the groceries away. "I know it's way too soon to draw conclusions, but the most obvious connection is opioid use."

"Ugh. I'm sorry. They probably get a lot of weird e-mails. Oh, great, you remembered the avocados."

"Yeah. Stocked up on ginger beer too. . . . It's not like I haven't considered the other options."

"So, are you going to write a long e-mail explaining your line of reasoning?" Gillian asks. "And then send it early tomorrow morning?"

"Pretty much, yeah."

The scientist in him welcomes the pushback, since it forces him to sharpen his thinking about what could have happened—and these experts surely won't be the first or the last to be skeptical. But they haven't given him any new insight, either. Stroke, hypoxia, a virus, a heart attack—any number of things can hurt the hippocampus. But those conditions hurt other areas too. This damage is so specific. Maybe lack of oxygen played a role, but why rule out opioids? Maybe the CDC experts aren't convinced by the connection with opioids because the cases don't fit neatly into anyone's medical expertise. Maybe they're swamped. And maybe they think he's a little bit overzealous.

Barash is in a Catch-22 situation. There aren't enough cases for the CDC or the DPH to justify spending scarce resources to look into it further. But without looking for more cases, they have no way to find out how many other people are suffering from this same type of brain damage. Toward the end of October, DeMaria encourages Barash to keep searching through the hospital records to see what turns up. If he finds a few more cases, he should let DeMaria know.

Within weeks, Butler and radiologist Juan Small unearth the third case. A 33-year-old woman passed through Lahey a year earlier, in 2014. EMTs on the scene found her unconscious with a needle in her forearm.

They administered naloxone, a powerful medication designed to reverse an opioid overdose rapidly. In the emergency department, doctors focused on stabilizing her. Her heart was racing and her high white blood cell count warned of systemwide inflammation. Once she was finally able to respond to commands, the emergency department staff realized her memory was severely impaired, so they ordered an MRI. But although they noticed the strange pattern of damage, they hadn't heard about Max and didn't know what to make of it. Now, one year later, after reviewing her scan, Small and Barash confer by e-mail. They agree that something has targeted her hippocampus with the same precision as the first two cases.

The fourth case, a 52-year-old woman, arrives at Lahey on October 26. She had been unresponsive when she was discovered at home. To help her breathe she'd been intubated before being brought to the emergency department. She has multiple threats to her health, including aspiration pneumonia; it's not uncommon for people to inhale vomit or saliva into their lungs when they overdose.

After doctors remove the breathing tube, her memory loss becomes apparent. They transfer her out of the intensive care unit and onto the neurology floor on 6 West, where Barash sees her. But by now, more than a week has passed since she was admitted and no one has ordered an MRI. So much time has gone by that the stark contrast between the hippocampus and the rest of the brain has likely faded. Hoping it's not too late, Barash requests an MRI anyway, and when he pulls up the image on his computer monitor, he can still detect the weak signal of the telltale pattern. Opioids are detected in her bloodstream. She is case 4, the evidence he needs to go back to DeMaria.

A few minutes before midnight on November 7, Barash sends DeMaria an e-mail. The opening line is characteristically low-key but to the point: "We found 2 additional cases thus far that might change the nature of the syndrome a bit . . ."

This time, DeMaria agrees that the DPH should sound the alarm. But he's mindful that as devastating as this obscure condition is for the four patients, it pales in comparison to the growing toll of opioid overdoses and deaths in Massachusetts. DeMaria oversees a lean budget and can't justify putting his staff on this investigation. He's afraid that it will look like he's worried about the arrangement of the deck chairs while the *Titanic* is going down. But decades at the department have made him adept at getting things done cheaply. He immediately thinks of a young epidemiologist-in-training named Nick Somerville at the Boston CDC office. Somerville has won an elite fellowship as a "disease detective" and was copied on the first round of e-mail exchanges in October with CDC headquarters. He's just finished analyzing autopsy reports of overdose victims at the Chief Medical Examiner's Office, so he has firsthand knowledge of the extent of the crisis. Somerville jumps at the chance to join the team, and since Barash has been working nights and weekends on the investigation, he's also free labor.

With no funding, the bare-bones team—DeMaria, a veteran investigator of public health outbreaks; Barash, a neurologist with an eye for unusual patterns; and Somerville, a young epidemiologist eager to help solve a mysterious outbreak—begin to spell out the case criteria. Their task would seem trivial to an outsider. All they have to do is give the syndrome a name and write one sentence defining what they're looking for. But coming up with an accurate description for something you don't understand is deceptively hard.

A case in point is HIV/AIDS. In the earliest days of the epidemic, no one knew how to refer to the puzzling illness. The term "the 4H disease" was coined because at first, the people who appeared most likely to get it were hemophiliacs, heroin addicts, homosexuals, and Haitians. It was also known as GRID—an acronym for gay-related immune deficiency. In 1982 the CDC used the term AIDS for the first time and published the case definition: "a disease, at least moderately predictive of a defect in cell-mediated immunity, occurring in a person with no known cause for diminished resistance to that disease."

The constellation of symptoms these amnesic patients have requires a name too, and Barash fixates on getting it right the first time. The investigation will be hard enough to pursue without a name that's off the mark. The first three terms are obvious: *acute*, *bilateral*, and *hippocampal*. But the three men go back and forth on the final word, *ischemic*. It's an accurate description of the intense, bright quality of the signal you would see on the MRI of a patient who had had a stroke, but it falls short when it comes to describing the cleanly circumscribed area of damage. Ultimately, for lack of a better option, they decide to go with *ischemic*, hoping the term won't confuse clinicians reading the case criteria.

Next, they need to define person, place, and time. *Time* is 2012 and onward, since that was the year Max suffered his brain injury. It's possible that cases showed up before then, but they don't want to cast the net so wide that they're overwhelmed with false leads, and they suspect this is a new phenomenon anyway. *Place* will be restricted to Massachusetts, because the DPH only has jurisdiction within its own state. Finally, the *person* must be diagnosed with amnesia. This seems self-evident at the moment, but Barash will later come to realize that their decision will inadvertently obscure the true number of people affected.

Finally, they debate how to address the opioid connection. They've seen only four patients, far too few to be sure that the opioids themselves are to blame. They haven't ruled out a heavy metal or some other contaminant. Or what if some rare genetic disease makes some people more vulnerable to loss of oxygen during an overdose? DeMaria, Barash, and Somerville hedge their bets by leaving opioids out of the official case definition, opting instead to include a brief reference to a "possible association."

By December 10, after more than a month of meetings and e-mails to hammer out the details, the public health alert is ready to go. All that's left is for DeMaria to wrangle the e-mail addresses of every emergency department physician, neurologist, and radiologist licensed in the state from the Board of Registration in Medicine. With the holidays upon them, things have slowed down, and DeMaria decides they should send out the alert after New Year's Day.

A few days before Christmas, Barash sees case 3 for a follow-up visit. A year after her overdose, her amnesia is better. But she still has more subtle problems that nearly qualify her for a diagnosis of mild cognitive impairment, which means her thinking abilities aren't completely normal. And her brain scan reveals that her hippocampus is smaller than when she'd come into the emergency department the first time, just like Max.

Barash considers the other patients. Zabar is still following Max. Case 2, the construction worker, had died of a heart attack nine months after his opioid overdose. Case 4 checked herself out of the hospital and was never seen again.

Barash hopes the health alert, when it finally goes out, will turn up some new cases, but only doctors licensed in Massachusetts will see it. Short of CDC support for a nationwide search, he knows that the most effective strategy for casting the net wider would be to publish a case report describing the first four patients, ideally in a prestigious medical journal. That way, any clinician anywhere in the world who winds up with a similar patient and takes the time to search for clues will find the report on PubMed.

So, the day before Christmas, Barash sends an inquiry to *JAMA*. After that, there's nothing else he can do but sit back and wait as DeMaria pushes the alert through the bureaucracy, even though all the while there could be other patients who will leave the hospital without a diagnosis and disappear without a trace. Barash never assumed that solving this mystery would be easy, but the magnitude of the task is coming into focus.

On February 10 at 9:27 A.M. the alert finally goes out to neurologists and imaging specialists. Anyone who sees or remembers seeing a resident of Massachusetts with "acute bilateral hippocampal ischemia on brain imaging" should contact DeMaria. Between all the late-night e-mails and text messages, meetings, permissions, skepticism, and bureaucracy, more than a year has passed since Barash looked at the brain scan of patient 2 to get to this first step.

Minutes later, the first response arrives in DeMaria's inbox.

# PART TWO

· · · · · · · · · · · · · · · · ·

# EPISODES

# 5

# The Hum of the Restless Beehive

. . . . . . . . . . . . . . . . . .

Santiago Ramón y Cajal thinks in pictures. He sits beside a table, half-turned toward the camera, both hands resting on the tool he uses to slice brain samples. A microscope stands off to the side. Cajal knows precisely how many seconds he must hold the camera's gaze to make a crisp photograph. It's 1886, and he is only thirty-four, but the bags beneath Cajal's dark eyes make him look older. He's come a long way from his childhood home in an impoverished village in the foothills of the Spanish Pyrenees. There he was known for his lack of obedience, moving from one school to the next, even briefly being imprisoned when he was eleven for using a homemade cannon to destroy the gate to a neighbor's garden. But creating images—photographs, sculptures, sketches—is Cajal's lifelong passion. Later, in his autobiography, he described his compulsion for drawing as "an irresistible mania for scribbling." His pencil was "a magic wand."

When Cajal took this photograph of himself, the brain was still essentially a black box. It was possible to dissect it, but beyond that, scientists had made little progress since an anatomist gave the hippocampus its name three centuries earlier. When Cajal looked at a brain, he could see the convoluted folds of the soft, gray outer layer and a deep groove running down the middle, from the front to the back of the brain, that divides it into two hemispheres. Shallower grooves subdivide each hemisphere into four lobes; the largest at the front, the parietal across the top, the temporal along the sides, and the occipital at the back. Also visible were the half dozen structures packed together beneath these lobes and around the top of the brainstem. Hanging off the backside of the brain was the accordion-like cerebellum. What neither Cajal nor anyone else could see very clearly, not even through the lens of a microscope, were the individual cells inside his carefully prepared brain slices. There were millions of them, too tightly packed together to differentiate one from the other. With the tools sitting on his laboratory table, Cajal had gone about as far as he could go.

This hurdle would be cleared within a year, thanks to a new technique for staining cells invented by Cajal's contemporary, an Italian named Camillo Golgi. His silver stain has two essential, almost magical qualities. The first is that it impregnates an entire cell, creating a complete and perfect outline of its shape. The second is that it stains only a tiny fraction of cells. Were it not for this curious trick, the whole slice would appear as one dark mass. But with the aid of Golgi's silver nitrate solution, what Cajal saw when he looked through a microscope's eyepiece was a single neuron silhouetted against the pale-yellow background of millions of unstained cells. Scores of dendrites, which resemble both the bare branches of a tree and the roots beneath it, shot out from the neuron's round cell body. A single long, delicate strand, called an axon, extended in the opposite direction. In the bean-sized rat hippocampus alone, twenty or more miles of these axons are packed together.

Cajal soon realized that Golgi's staining technique was time consuming and fickle, so he found a way to speed up the process and make it more reliable. He was so enthralled by the new universe made visible that he built a private laboratory in his home, where he meticulously prepared slides from myriad animals, like dragonflies, trout, cuttlefish, pigeons, frogs, rabbits, cats, dogs, and, of course, people. Cajal routinely worked for fifteen hours at a stretch, as all the while thin slices of dead brain tissue came alive in his imagination. Cajal drew what would ultimately become a collection of twenty-nine hundred pen-and-paper sketches. His revelatory silver-stained images revealed the three-pound lump of fatty tissue between our ears to be an otherworldly place inhabited by multitudes of elaborately shaped and intertwined cells.

As Cajal saw it, every part of the brain was beautiful, like a giant forest in which he and other scientists were lost. Cajal referred to neurons as "the butterflies of the soul, the beating of whose wings may some day (who knows?) clarify the secret of mental life." He compared large neurons in the hippocampus to "plants in a garden—as it were, a series of hyacinths—lined up in hedges which describe graceful curves." The ceaseless flow of the brain's electrical activity seemed to him like "the hum of the restless beehive."

In fact, Cajal and Golgi already knew that electricity is the universal language of the nervous system thanks to a laboratory accident a century before in which a budding anatomist made a frog's leg twitch by discharging electricity onto an exposed nerve. But the central question that both men wanted to answer now was straightforward: How does electricity—and the information it carries with it—travel through the brain? Golgi's theory was that these electrical impulses vibrated within a brainwide spiderweb of neurons that were physically, permanently connected through their axons. The rootlike dendrites protruding from the cell body were just that—roots to provide the neuron with nutrients and nothing more. What Golgi saw using the silver stain didn't change his mind, and most scientists agreed with him.

But Golgi's theory made no sense to Cajal. How could it be possible, he wondered, to learn new things—to literally change your mind—if all the connections between neurons were locked in place? Using the same stain and the same type of microscope, Cajal saw in his mind's eye what others couldn't. There was no continuous spiderweb. Instead, there must be gaps between neurons so that each had the power to act as its own unit. These gaps, which are called synapses, are just one millionth of a millimeter wide, and they wouldn't actually be visible until the invention of the electron microscope fifty years later. At first, many scientists thought Cajal was crazy. How could electricity travel from one neuron to the next if they were physically separated? Cajal had no way to solve that particular puzzle, but he understood that the brain had to be a dynamic system to respond to the ever-changing outside world. And he didn't care if people thought he was crazy.

The second central question Cajal wanted to answer was what direction information moves in as it flows between neurons. If neurons were individual units, there must be an input and an output. The logical place to look, it seemed to him, was where information enters the brain. So Cajal examined neurons in the back of the eye and the olfactory bulb of the brain, where smell is processed, and saw that the neurons were always oriented so that their dendrites were the point of contact with the outside world. Observing that, he inferred that information flows into dendrites, through the cell body, and out the other end along the long, spindly axon. From there, the signal would be passed on to the dendrites of the next neuron. It seemed rational to Cajal that this same organizational plan should be at work everywhere, not just for brain areas that process sensory information.

Three decades after Cajal saw his first silver-stained neurons, he took another self-portrait. His gaze is just as intense, but his hair is silver, and he wears an elegant suit and tie. By this time, he has received the 1906 Nobel Prize in Physiology or Medicine. Camille Golgi shared the prize for inventing the seemingly magical stain, even though he rejected the discoveries that made Cajal the so-called father of neuroscience. Brain researchers around the globe revere Cajal for his genius and devotion to his work. But unlike Darwin or Einstein, Cajal did not become a household name.

Although we now know that the way neurons send and receive information is far more complicated than Cajal could have imagined, his fundamental insights have stood the test of time. He appears to have been wrong about one thing, however. He believed the brain could not create new neurons. Elsewhere in the human body, new cells replace old ones. But once the brain had fully formed, Cajal wrote, "the founds of growth and regeneration of the axons and dendrites dried up irrevocably . . . Everything may die, nothing may be regenerated. It is for the science of the future to change, if possible, this harsh decree." Cajal was right about so much and was so respected that his dogma would not be seriously called into question until the 1990s.

Among Cajal's drawings, one stands out for its relevance to the study of memory. In an image so clear and coherent that it illustrates scientific papers today, he sketched the complex wiring diagram of the rat hippocampus.

Cajal used tiny arrows to show which way the electricity flowed back and forth throughout its folded layers. In the lower right-hand corner, at a point where electrical impulses leave the hippocampus, he drew two neurons whose axons split into two. One end of each axon projects outward, while the other thread loops back into the hippocampus's graceful curves.

Cajal didn't understand the purpose of this design, but he drew what he saw. It would not be until H.M. underwent his radical surgery that anyone even knew what the hippocampus does. And it would fall to still later scientists to realize that this strange backward flow of information, and the inhibitory neurons within the hippocampus, help maintain the finely tuned activity in the brain's memory center. More than a century after Cajal drew his iconic picture, this exquisite balance would be upended in overdose victims across the state of Massachusetts.

# The Cluster Grows

• • • • • • • • • • • • • • • • • • • • • • • • • •

## *February 2016 to January 2017*

Snowflakes drift past neurologist Chun Lim's window at Beth Israel Deaconess Medical Center as he sits in his tiny office at a desk piled high with papers. Lim brushes back a shock of hair that falls across his forehead and checks his e-mail.

> Sent: Wednesday, February 10, 2016 9:27 AM
> Subject: MA DPH—Potential Cluster of an Unusual Neurological Syndrome
>
> Dear Physician: The Massachusetts Department of Public Health (MDPH) has been made aware of an unusual amnestic syndrome in four individuals, possibly associated with opiate/ opioid use.

For the last thirty years, Lim has wanted to piece together a human memory circuit, a sort of map showing how information travels between the hippocampus and other connected brain regions. So he's naturally inclined to pay special attention to people in whom that circuit has broken. One such patient came to his attention a few years earlier, a middle-aged man who overdosed in 2014. In the emergency department, he tested positive for cocaine and opioids, and his hippocampus was severely damaged. The hospital staff suspected that cocaine inflamed the brain's blood vessels, restricting flow into the hippocampus. But Lim wasn't convinced. If restricted blood flow was the cause, why was just the hippocampus affected? The patient was transferred to a rehabilitation center and agreed to enroll in Lim's research project. But then he dropped out, and Lim never saw him again. At 9:47 A.M., a mere twenty minutes after DeMaria sent the public health alert, Lim responds with the first report of a likely case.

> I was involved in such a case of a 49yo patient . . . He was found confused and amnestic. Imaging with bilateral hippocampal injury and tox screen positive for opiates, which he denied taking.

Thirty miles north, at Holy Family Hospital in Methuen, close to the New Hampshire border, neuroradiologist Mara Kunst reads the health alert. She remembers the case of a 46-year-old man with a history of misusing prescribed opioids. Mysteriously, despite an apparent overdose, his toxicology screen had come back negative for any drugs. He had precisely this pattern of brain damage—just the hippocampus. Kunst remembers another case from that same year, 2015. A 19-year-old man with a history of illicit drug use was brought to the emergency department in Lawrence, a former mill town where fentanyl was creeping into the drug supply. He tested positive for cannabinoids but nothing else. A colleague asked Kunst for a second opinion because the patient's MRI was so unusual. Five months later, the patient recovered most of his memory. But he developed a persistent seizure disorder and, like the 46-year-old, was eventually lost to follow-up.

It seems likely to Kunst that this strange, sudden amnesia is under-recognized. She ticks off all the hurdles that have to be cleared before such a bizarre brain scan would land on her desk. The victim would have to survive the overdose. Someone would have to recognize the amnesia, decide it wasn't just post-overdose confusion, and order a scan—not a CT scan, but a more expensive MRI. This would have to happen quickly, before the signal could fade. And then whoever looked at that scan would have to conclude that the results are mysterious enough to ask Kunst for a second opinion. Still, even with all these hurdles cleared, they'd be unlikely to get a complete picture of what substances the victim took. Labs in large, well-resourced hospitals use decades-old toxicology tests to screen for common drugs. They don't have the equipment or the expertise to detect fentanyl, which is often mixed with other drugs.

Kunst e-mails DeMaria a brief description of the two patients.

> One is a 46 y.o. male with bilateral hippocampal ischemia and lesions of the globus pallidus, and the other is 19 y.o. male with some recovery but lost some hippocampus.

At Boston's Brigham and Women's Hospital, neurologist Joshua P. Klein reads the DPH alert as well. Klein is not convinced that the association with opioid use is anything more than an association, and he's been thinking about this for a while. Years earlier, before Barash saw Max in 2012, Klein had come across a similar case. Like Barash, he'd been struck by how unusual the damage had been and had been curious to know what had happened. As a result, he'd scanned the medical records of his and several other Boston-area hospitals and found a few more cases with the same type of brain damage described in the health alert. But not all of Klein's cases have a documented history of using opioids. He hasn't gotten as far as writing an article, but he doesn't believe these drugs are specifically to blame. That seems like a leap of logic, confusing an association with causation. He can't exclude the possibility that opioids caused a direct toxic attack. But it seems much more likely that opioids were merely how the brain was deprived of enough oxygen, which in turn injured the vulnerable hippocampus. Nonetheless, Klein duly reports the cases he's seen that fit the DPH criteria to DeMaria.

At Massachusetts General Hospital in Boston, two floors underground, the director of emergency radiology, Michael Lev, sits in the dark. He focuses on the second sentence in the DPH health alert. "This amnestic syndrome is associated with acute, bilateral hippocampal ischemia on brain imaging." Lev can picture precisely what the e-mail describes. He's seen this pattern before. The cases stand out, as clear in his mind as the day he looked at them, a class unto themselves. The history in the medical records was always the same: Reason for scan: heroin, found down. Lev tried to get some neurologists and neuropathologists involved, but they didn't think it was particularly interesting, and he couldn't pursue it by himself. He continues reading to the end of the e-mail. "If you think you may have seen such a patient, please contact Alfred DeMaria, Jr. M.D." Now, finally, someone else is paying attention. If there's going to be a team of people trying to figure this thing out, Lev wants to be part of it.

DeMaria knows that he needs a plan of attack at the DPH to handle incoming calls and e-mails. "People have technical questions about parts of the brain I haven't been near in a long time," he tells Barash. DeMaria suggests that Barash respond to technical issues and Somerville follow up on data collection. For the next several weeks, DeMaria continues to receive e-mails:

I believe I have seen two such patients in the past year. I will look in their charts and follow up with specific information . . .

I saw a patient fitting this description exactly, in the hospital, last September: 33y.o. Man used heroin at a friend's house and came home a day later . . .

If the amnesia you are describing is more prolonged with larger hippocampal abnormalities, I saw one such individual since 2012.

The connections between TBI/PTSD and dementia are the current rage, but parallel work in substance abuse hasn't begun, to my knowledge. Your findings could be an important break.

But by the end of February, the e-mails drop off. DeMaria isn't surprised that people have stopped paying attention. Everyone is busy, and the syndrome affects stigmatized, marginalized people. If children were affected, it would be different.

—⁓—

On a chilly early spring day, Barash and Somerville meet outside the main entrance to Massachusetts General Hospital, known informally to Bostonians as Mass General. Michael Lev has offered to help with the investigation. The 999-bed hospital admits close to 50,000 patients a year from all over the world. Surgeons perform tens of thousands of operations. And the emergency department, where Lev oversees a daily deluge of scans, records more than 100,000 visits annually. Barash has put in for a vacation day so he can attend a meeting with Lev. Somerville joins the meeting as part of his Epidemic Intelligence Service fellowship. The smell of decades of coffee permeates the air as Barash and Somerville follow signs toward the Blake Building and take the elevator two floors down to the radiology department.

Lev jumps up from his chair, says hello in a loud voice, and ushers them into his office. His face is friendly, his eyes alive with kinetic energy. The plan for the meeting is to figure out how to sort the disorganized data arriving via secure fax and e-mail in DeMaria's office. Lev may be able to give them a jump start. He and a couple of his trainees have noticed that the emergency department is seeing an uptick in patients with substance use disorders (SUDs). They're curious to find out if the number of scans is going up in tandem. A radiology fellow named Efren Flores has created a spreadsheet to organize the data. Displayed on the large monitor hung on the wall across from Lev's desk is a template with scores of column headings—gender, age, imaged body part, type of imaging, SUD-related, non-SUD-related, local complications of IV injections, neurologic symptoms, joint pain and swelling, trauma, admission date, discharge date, cardiac enzymes, liver enzymes, echocardiogram, EKG, EEG, toxicology screens, ultrasounds, blood cell counts, substance use history . . .

"That's a lot of columns," Barash says.

"Exactly," says Somerville. "You guys have captured almost every possible variable I could think of."

"Or that we can realistically gather given the somewhat haphazard nature of the medical records we're getting," Barash adds.

"Sure, this is probably just a starting point for your investigation. You can narrow it down to fit your needs," says Lev.

After a brief discussion, the Mass General radiologists agree to share a template with all the column headings so that Somerville, Barash, and DeMaria can adapt it to their investigation. The discussion has prompted Somerville and Barash to consider adding a few more variables than they'd planned, but they also have to operate within the limits of the authority granted to them by the DPH. This is a public health investigation, not a research project.

Lev invites them upstairs to look at some scans in the reading room, where radiologists triage which images to analyze first so the emergency department can decide whether a patient can be discharged or needs to be admitted.

"Have you tried to get any of these cases published yet?" Lev asks in the elevator on the way up.

"I submitted the first four cases to *JAMA* back in December," says Barash. "And I figured they'd either be very interested or gun it down dead on arrival."

Lev laughs as they head down the hall toward the neurology reading room right off the emergency department. "They gunned it down, right?"

"Right," says Barash.

"I can tell you as someone on the editorial board of a journal," says Lev, "that editors are looking for something that basically has a spin on something they already know. It has to be incremental. If it's a totally new idea like this, people are going to say it's too speculative."

The reading room is dark, with just enough space for three desks, three monitors, and three more screens hung on the walls. Lev and Flores begin pulling up scans with hippocampal damage they've earmarked for today's review.

"How different do these look to you compared to other scans with hippocampal damage?" asks Barash.

"If you threw a hundred of these up on a display monitor and asked me to play Where's Waldo," says Lev, whose enthusiasm has increased to the point where he's on the verge of shouting, "I'm confident I could say, 'here's the one who had the heroin overdose.'"

After the meeting, Barash and Somerville debrief at the corner of Fruit Street and Cambridge Street just outside Mass General. Somerville tells Barash that being on the ground floor, figuring out the source of a previously unknown health threat—like Legionnaire's disease or Ebola—is the reason he joined the Epidemic Intelligence Service. From Barash's perspective, Lev's enthusiasm for coming aboard their team is a marker of both how far they've come and how far they have to go.

—⁂—

Barash says good-bye to the head of human resources, puts his phone facedown on the desk, and smiles. He's in his third office at Lahey. It's large enough that his desk fits in the middle. And the view is better. But despite the bigger room and the better view, little else has changed. If anything, the relentless pressure to see more patients in less time has only escalated. For more than a year, he's been on the lookout for a new job that will provide him enough time to give patients the attention they deserve, and still leave a few spare hours here and there on nights and weekends to pursue the investigation. He looks around the room, notices how convenient it is that he hasn't unpacked the box from his last office move, and calls Gillian.

"The future ain't what it used to be," he says.

"Is that a Yogi Berra quote?"

"You nailed it. I got the medical director job at Soldiers' Home." Barash stands up and looks out the window, savoring the fact that this won't be his view for much longer.

"That is so great!!! What should we do to celebrate?"

"I haven't thought that far ahead."

A month later, on a hot July day, Barash pulls into the parking lot at Soldiers' Home in the densely populated postindustrial town of Chelsea. The large brick complex sits atop the highest hill surrounding Boston Harbor, and the smell of the ocean hits him as he opens the car door. A red-and-white water tower with the words SOLDIERS' HOME stands guard over the main building like a giant six-legged insect. A blue canopy shades the front door, rows of flags on either side fluttering in the warm breeze. The window frames are peeling, and an air-conditioning unit secured to a metal frame hangs from each one.

Originally opened in 1882 to serve poor Civil War veterans free of charge, Soldiers' Home is almost a miniature city, with a post office, barbershop, a dry cleaner, auditorium, recreation rooms, a dining room, and a canteen. Volunteers from the community hold Bingo nights and throw pizza parties. There's a daily mass for Catholics, Jewish services on Friday evenings, and Protestant services on Sunday. Barash loves everything about it.

Inside, Barash gives his name to the attendant in the vestibule by the front door and takes a seat on one of the couches scattered around the lobby to wait for the HR person. Aside from the many flags, the space feels like an old living room. A large gilt-framed painting of Commandant Lawrence Quigley, a World War I veteran who was a director of the facility, dominates the room. A smaller portrait of Quigley's son, the next commandant and a World War II veteran, hangs above the fireplace.

As the medical director, Barash will be responsible for the health and well-being of close to three hundred veterans from World War II to the conflict in Iraq. There's a long-term-care nursing facility with ten people to a ward and a dormitory with individual rooms for residents who don't need ongoing medical attention. Most of the veterans served in Vietnam, so they're older, and the rate of dementia is high. Barash will manage any medical problems that come up, whether it's an upset stomach, chronic pain, depression, the need for occupational therapy, or an outbreak of the flu. For anyone else, Barash's move to the Soldiers' Home looks like a strange career decision. From his perspective, it's the real world, a place where he can get to know his patients, and exactly where he belongs.

Meanwhile, Butler, his earlier trainee, moves to a very different world, a gleaming glass building at the University of California, San Francisco. After a decade of hands-on and instructional training, Butler had a choice to make. Either take a well-paid job as a full-time neurologist or pursue more training as a fellow, setting him on a path to a research career.

Butler has little doubt about what he wants to do with his life. The medical degree is an insurance policy—a useful, well-paid vocation just in case his research career doesn't work out. As a fellow at the Memory and Aging Center, he'll be able to do research and get to see patients with unusual neurodegenerative disorders, some of whom have traveled across the country or even the world to take part in research and clinical trials. And without the business pressures of a hospital to keep office visits short, Butler will have the luxury of spending time with each patient. One five-star online reviewer commends him for not rushing and says he hopes he "keeps his awesome spirit up."

Butler shares an open pod with a half dozen other fellows and grad students seated nearby. Outside, the skeletons of new buildings rise from a vast expanse of concrete. During Butler's medical training, the hierarchy was unmistakable; it was written on the name tags people wore on hospital rounds. Here at the Memory and Aging Center, informally known as the MAC, the academic hierarchy is just as recognizable, even though no one wears name tags. Like any university, the professors with large grants are at the top. But Butler's not complaining. He's surrounded by exceptionally bright and talented people—the grad students, postdocs, and fellows working their way up; the MacArthur genius, Bill Seeley, who supervises him; and the director of the MAC, Bruce Miller, a giant in his field who built his scientific career looking where others weren't—at a rare neurodegenerative disease called frontotemporal dementia. Barash's hero, Stanley Prusiner, who discovered the prion responsible for mad cow disease, has his office in the same building. The research possibilities energize him.

In one of many serendipitous events to come, Butler realizes that the paper on the first four Lahey cases that *JAMA* turned down may find a home with some help from Miller. Many scientists scorn case studies as antiquated tools of a bygone era. But Miller believes they have enormous potential to be the starting point for new insights, so he launched a rapid response journal called *Neurocase* to disseminate them. On July 28, just as Barash and Butler are settling into their new homes three thousand miles apart, "Complete, Bilateral Hippocampal Ischemia: A Case Series," by Small, Butler, Zabar, and Barash, appears online in *Neurocase*. The journal has a small audience, but the publication still marks a step forward. Now, if someone digs deeply enough on PubMed for a link between opioids and amnesia, they'll find something.

Butler's new mentors caution him to straighten out his meandering career path and head in one direction. Research careers are established incrementally, with each study building on the one that came before. Although Butler is fascinated by the neuroscience of religion and passion, he knows that the topic will have to remain a guilty pleasure for now. No one wants to hear about it from some guy in his thirties. Maybe they'll listen to a professor in his seventies who's paid his dues and made a name for himself. Ultimately, Butler wants to make a contribution to human health and disease. The common thread running through his eclectic research is the neurotransmitter dopamine, which plays a role in movement, memory, decision making, emotions, and more. So, from now on, dopamine and neurodegenerative disorders will be his focus. Butler suspects that Barash will soon be on his tail trying to rope him back in to work on the amnestic syndrome, but he resolves to follow his mentors' advice. And he will succeed, for about six months.

In late August, Barash reads a report from the Centers for Disease Control and Prevention—the same organization that dismissed concerns about the amnestic syndrome—detailing a rapid increase in fentanyl overdose deaths in Florida and Ohio. Barash spies an opening and tells DeMaria they should pull together their case series and "strike while the iron is hot." The CDC's flagship *Morbidity and Mortality Weekly Report (MMWR)* reaches nearly two hundred thousand health-care workers—physicians, nurses, public health practitioners, epidemiologists, scientists, researchers, educators, and lab technicians. It would be the ideal place to report on the growing scope of the syndrome and put the word out that new victims should be tested for fentanyl. As it happens, Barash, DeMaria, and Somerville are nearly done sorting through the records.

The three doctors would sit around a table behind closed doors in DeMaria's crowded office whenever they could grab a few hours to meet. Even though they've winnowed Lev's spreadsheet of categories down to the essentials, many of the twenty-five patient case files they've received from Massachusetts doctors are missing key information. In some instances, the emergency department didn't request basic toxicology screens to find out what drugs were on board. Instead of high-quality MRIs, smaller hospitals took less expensive CAT scans, which can't reveal the damage clearly enough to meet the criteria they've used to define the amnestic syndrome. In other instances, too much time passed before the brain was scanned, allowing the bright signal of intense damage to fade. A few cases have to be disqualified because they date back before 2012.

There's one more criterion they must stick to; a clinical diagnosis of amnesia. In some ways, it's unnecessary. Even before meeting Max Meehan in the hospital room, Barash had known that Max would have profound amnesia. That would be true for anyone with severe injury to the hippocampus. Now, as they finalize their analysis, Barash begins to discern the downside of this measure. Most overdose survivors leave the hospital as soon as they're out of danger; if there were any amnesic victims among them, they would never have been tested or counted as part of the cluster.

One last case that can't be included because it's missing an amnesia diagnosis is both odd and unsettling. This middle-aged woman didn't overdose on opioids where people so often do—outside on the street, at home, or at a friend's house. Instead, she went into a coma in the hospital after being given opioids to manage her pain after abdominal surgery. Her MRI met the criteria of acute bilateral ischemia, but she never recovered from the coma, could never be tested for amnesia, and died soon afterward. Barash adds this curious event to the list of items he'd like to follow up on. For now, they're focused on organizing the list of cases that have made the cut in a way that will stand up to the exacting review process of MMWR editors.

In all, they can only include ten patients in the final tally, bringing the total number of victims to fourteen men and women between the ages of nineteen and fifty-two. By the fall of 2016, all but one, Max, have died or disappeared. Barash, DeMaria, and Somerville hammer out the details of how best to organize the paper, what images to use, and whom to credit. All the while, Joshua P. Klein, who reported several cases to the DPH, has also been looking through medical records for more examples of unusual hippocampal damage. But Klein doesn't limit himself to patients with a diagnosis of amnesia. Or to patients with symmetric damage to both hippocampi. Or to people who may have used opioids—because he still believes it's not the point. Klein has collected sixteen cases. Eight overdosed. Of those, six had opioids in their toxicology screen. Some apparently used other drugs such as cocaine or amphetamines—at least according to the toxicology screens. One had suspected Alzheimer's disease. Some had epilepsy. Another passed away six months after abdominal surgery. Two died of cardiac arrest.

When Barash and Somerville realize that Klein contributed a few of the fourteen cases they plan to include in the *MMWR* report, Barash reaches out, offering to credit him on the paper. Once again, Klein makes the case that opioids are not the direct cause of amnesia. "The imaging finding," he tells Barash in a September 27 e-mail, "is not specific to drug intoxication." Nevertheless, the CDC experts are sufficiently convinced by the worrisome health implications of an association between opioid use and amnesia to accept the paper for publication. On January 27, 2017, "Cluster of an Unusual Amnestic Syndrome—Massachusetts, 2012–2016," by Barash, Somerville, and DeMaria, appears in volume 66, number 3 of the *Morbidity and Mortality Weekly Report*. The story is picked up by popular media outlets like the Associated Press, the *Boston Globe*, STAT news, and CNN.

The next day, Barash texts Monroe Butler a link to the paper. Butler is sitting at his desk in the pod with half a dozen other fellows seated nearby, engrossed in their computer screens. A black-and-white picture of a butler with MONROE spelled out in block letters is taped to the bottom of his monitor, a gift from one of the former fellows.

> Butler: Congrats on picking up on this phenomena and pushing the project forward from the beginning.
> Barash: Thanks, we have a shitload more work to do now . . .

Butler and Barash text back and forth about what that work could look like. With all the publicity, maybe cases elsewhere in the country will be discovered, making it easier to pin down the substance that's causing the amnestic syndrome. That in turn could help get them closer to developing a new animal model for memory loss, and then from there to new ways to protect the hippocampus. "That's my long-term vision," Barash texts. "But there's obviously a lot of steps between here and there, not to mention luck."

# No Place in the World

· · · · · · · · · · · · · · · · · · · · · · ·

## *January 2017*

There is no numerical scale for amnesia, just degrees of severity from "mild" to "dense." An Englishman named Clive Wearing suffered one of the densest, most severe forms ever recorded. Before the catastrophic incident that damaged his brain, Wearing was an accomplished musician—a singer, conductor, keyboardist, and musical scholar. In 1981, he chose the music the BBC played on the day Prince Charles and Diana Spencer were married. A year later, he conducted a Renaissance piece called "Music the Gift of God," which was broadcast in five countries. But three years after that, at the age of forty-six, his memories of those days and every other disappeared, stolen by a common cold sore virus that managed to slip past the blood-brain barrier into his brain. For the rest of his life, he would live in a world with no past and no future—just a terrifying perpetual present.

It began in 1985, in the early days of spring, when Wearing took to his bed with what his doctor assumed was a severe case of the flu. His headaches were so debilitating that at first he couldn't sleep for three days, and then he could do little more than sleep. By the time he went to the hospital several weeks after initially falling ill, a spinal tap revealed that he had herpes simplex viral encephalitis. This virus has a particular affinity for the hippocampus. Wearing was near death and slipped into a coma. His brain was swollen and crushed up against his skull, and his body convulsed with a prolonged grand mal seizure. After two weeks, Wearing woke up, apparently out of physical danger.

A CT scan showed widespread damage, but for the first few months, Wearing was strangely euphoric. He pulled pranks on the hospital staff and played hide-and-seek with his family on a visit home from the hospital. He recognized his wife and children, and his brother, who came to visit from New Zealand. He wrote short journal entries that described what had happened during the day.

But then, in mid-summer, Wearing started crying. He sobbed continuously for over a month, and when he finally stopped, it was clear that he had retrograde amnesia—no memories of his previous life—and anterograde amnesia so severe that his world was limited to seven-second fragments. Wearing had descended into a memoryless void. But he hadn't forgotten how to write and he filled the pages of his diary with repetitive, single-line entries describing his place in time. One page, dated January 13, 1990, begins at 7:46 A.M. with the words, "I wake for the first time." The word "awake" appears nineteen times, along with "first thought," "first time," "first consciousness," "alive," and "death." His meticulous attempts to capture his thoughts before they could disappear ends after thirty-five entries, at 10:07 P.M.

And so it went for years. An MRI scan showed more clearly what the lower-quality CT scan had picked up. There was such widespread damage to the hippocampus and the rest of Wearing's temporal lobes and a portion of the left frontal lobe that the fluid-filled cavities nearby had expanded to take the place of lost tissue. If his journal wasn't available, Wearing would write on walls or furniture. He was deeply depressed, terrified by the incomprehensible circumstances he found himself in, and often irritated or enraged by all the questions people asked—questions that were impossible for him to answer. But despite the horror of having forgotten almost every event and every person in his life, he remembered that he was in love with his wife, Deborah.

In another saving grace, Wearing's musical abilities were spared. Documentary footage shot a year or so after he fell ill shows Deborah escorting him to a chapel where members of his former choir were waiting. The organ held the sheet music for the same Renaissance piece he'd once conducted for listeners around the world: "Music the Gift of God." When Deborah asked Wearing if he would like to conduct, he said, "I don't know what might happen, do I?" But Wearing's procedural memory was intact. He put his hands on the keyboard, played the first few notes, counted in the ensemble, and began to conduct. When the music ended, his body shook and he choked, as if being sent back to the abyss.

Some fifteen years after Wearing fell ill, Deborah brought up the possibility of her moving to Bath, and Wearing began to talk authoritatively about the Bath abbey's arcane architectural details. He was also able to unearth a limited repertoire on other narrow topics. He could hold forth on Queen Victoria, the stars and planets, Parliament, or the nature of electricity. Deborah saw these islands of knowledge as his only way to connect with people. It's impossible to know why these memories resurfaced, but it's unlikely that any significant healing had taken place. And twenty years after his illness, Wearing's sense of having just woken up remained. "Can you imagine what it's like to have one night 20 years long? With no dreams?" he told a British documentary producer in 2005. "That's what it's been like. Just like death. No difference between day and night. No thoughts at all . . . It's precisely like death."

Barbara Wilson is a neuropsychologist who began working with Wearing shortly after he fell ill and still sees him occasionally. Her strategy is to focus on what he can do and fix the world around him to make that possible. There's a piano in his small living quarters. He knows where the kitchen is, even if he doesn't know that he knows it. "His implicit memory is not too bad because if I say to him, 'Clive can you show me where the kitchen is?' he'll say, 'Don't know! Just woken up! Like being dead!' If you say, 'Should we go make a cup of coffee,' he goes straight to the kitchen." Now Wilson and his caregivers don't ask him questions he can't answer, and she doesn't test his episodic memory. It upsets him, and the results don't change. Whenever Wilson needs to calm him down, she'll say, "What's the best age for a child to learn music, and what's the best instrument for a child to learn on?" and he answers triumphantly, "Age of five and the piano," as if not realizing she's ever asked it before. And if Deborah asks, "What does love mean?" he answers with a smile: "Zero in tennis, everything in life."

—⁂—

Belgian neuropsychologist Mieke Verfaellie can pinpoint the moment three decades ago when she became forever fascinated with amnesia. One day, while working at a clinic in Florida studying stroke patients, she met a patient whose scuba gear had malfunctioned, depriving him of oxygen. When the man had returned to the surface, he'd been physically unharmed but densely amnesic. Verfaellie was so struck by her encounter with the diver and the specificity of his memory loss that it altered the course of her career.

Today, Verfaellie is director of Boston University's Memory Disorders Research Center and a research scientist at VA Boston Healthcare System. She's a grandmother with a gentle, old-world bearing that puts people at ease. Downstairs, patients wait in the lobby—mostly men, mostly older, some in wheelchairs. Veterans, more than most, suffer from medical problems such as obesity, diabetes, cancer, hearing loss, and post-traumatic stress disorder, or PTSD. They also suffer disproportionately from two conditions that damage episodic memory: traumatic brain injury and dementia, mainly because the veteran population is aging.

The mainstay for helping anyone with episodic memory loss is the approach Barbara Wilson used to help Clive Wearing: when it's impossible to rehabilitate the brain itself, rehabilitate the world around the patient instead. Encourage them to keep their living environment as routine and stable as possible. Teach them how to use memory aids, like whiteboards, sticky notes, lists, alarms, calendars, memory notebooks, or visual cues; even an index card placed on a car dashboard to remind someone where he's going can mean the difference between being stuck at home or going out to visit friends. And connect patients with support groups so they can share what helps them. Although these strategies may not sound like much, they can sometimes make it possible for people with significant memory loss to live independently and hold down a job.

But Verfaellie and other memory experts would like to offer more, and she hopes that her research with amnesia patients will make that possible. Since people with severe amnesia are rare, Verfaellie's staff neuropsychologist and outreach coordinator, Ginette Lafleche, devotes much of her time to finding them. Lafleche maintains relationships with Boston-area hospitals and rehabilitation centers, goes to support group meetings for patients with brain damage caused by stroke or viral encephalitis, and gives talks. At best, she may locate two patients a year. They're usually eager to interact with experts who can see what one test subject described as his invisible wheelchair. Verfaellie says, "You'll have a patient who will say, 'Today is a good day because I feel the fog has finally lifted.' Which is their way of really expressing, 'Today—or this moment—is really disconnected from anything in the past and from anything in the future.'"

"Everyone knows since H.M. that the hippocampus is important for recalling the past," says Eleanor Maguire, an Irish neuroscientist at University College London. "But memory is not about the past. It's about the future." Maguire asked ten normal test subjects and five amnesic patients to imagine lying on a white sandy beach in a beautiful tropical bay. They were told not to think back to past experiences. The healthy, control subjects imagined vivid scenarios that included details like palm trees rustling in the breeze, a "gorgeous, aquamarine" sea, a creaky old fishing boat, "a guy in the front and I wave at him and he waves back. . . ."

The amnesic patients, on the other hand, described scenes that were vague and fragmented.

> Examiner: Imagine you are lying on a white sandy beach in a beautiful tropical bay.
> Patient: As for seeing I can't really, apart from just sky. I can hear the sound of seagulls and of the sea . . . um . . . I can feel the grains of sand between my fingers . . . um . . . I can hear one of those ship's hooters [laughter] . . . um . . . that's about it.
> Examiner: Are you actually seeing this in your mind's eye?
> Patient: No, the only thing I can see is blue.
> Examiner: So if you look around what can you see?

Patient: Really all I can see is the color of the blue sky and the white sand, the rest of it, the sounds and things, obviously I'm just hearing.

Examiner: Can you see anything else?

Patient: No, its like I'm kind of floating. . . .

"They can't imagine what's behind them and they can't imagine what's round the corner," Maguire says. But on the other hand, if you show these patients a picture of someone lying on a white sandy beach, they can describe it. It makes perfect sense to them because they're high-functioning, rational people. "So they're truly stuck in the present, literally, with what's in front of their eyes." Experiments like Maguire's suggest that the hippocampus is essential for building scenes in the mind's eye, whether it's reconstructing a past event or imagining a future one. But, says Verfaellie, "it isn't exclusively about scene construction. It's really about any time that you have to take elements from lots of different areas and put them together in a new way."

Research by many other memory scientists backs up the idea that the hippocampus is essential for imagination and creativity. For example, in another experiment, a normal person who was asked to list unusual things you could do with a cardboard box came up with twenty-six ideas. A person with amnesia and a similar IQ came up with only two. In a drawing test, these same researchers gave test subjects a sheet of paper with an oval in the center and asked them to use the shape as a starting point for a picture. One amnesic patient turned the oval into a bug. A normal participant also used the oval to create a bug—a giant tick-shaped hot air balloon floating over a city captioned "Tickets for the Tick-Mobile."

—m—

In January, when Verfaellie finishes reading "Cluster of an Unusual Amnestic Syndrome—Massachusetts, 2012–2016," in the *MMWR*, her first thought is to ask her outreach coordinator Ginette Lafleche to find out if there are any overdose patients her team could work with.

The *MMWR* paper describes much more cleanly defined damage to the hippocampus than Verfaellie typically sees. Right off the top of her head she imagines the questions she'd like to explore. Perhaps a new source of brain injury can reveal something unknown about the hippocampus. Did the patients have purely anterograde amnesia, or did some have retrograde amnesia and forgot events before the overdose? How rich were their memories? There are hints that some patients recovered their ability to remember, which is highly unusual, but how rigorous was the testing? If Verfaellie can bring in one of the patients from the cluster of fourteen for a functional MRI, she can investigate whether brain connectivity in certain regions changed. Maybe, with a high-quality scan, she can see the anatomy of the hippocampus clearly enough to figure out if every area was equally damaged. And if the damage turns out to be greater in some areas than others, perhaps neuropsychological testing can reveal subtle differences in aspects of episodic memory that correspond to each region.

From her office up on the eleventh floor of the VA Boston Healthcare System, Verfaellie gets on a phone call that Lafleche has set up with Barash. Her office is equipped with solid institutional furniture. Large windows overlook the non-glittery side of Boston.

"Jed, this is a very interesting and unusual presentation," she says. "I would love to be able to study one or more of these cases in detail."

Barash detects a hint of an accent in her lilting voice. "I wish I could help, but I'm not their treating physician, so you'll have to see if anyone who reported a case can help. Unfortunately, I have to warn you that most have been lost to follow-up."

As the conversation continues, Barash mentions that they've never seen a single case in the winter. "It's hard to know where to place that," Verfaellie says. "Why do you think that is?"

"I don't have a good explanation. Maybe there's something different about how people use drugs in the winter. Maybe it's just a coincidence."

Verfaellie checks the acknowledgments on the *MMWR* paper and recognizes the name of Chun Lim, the neurologist at Beth Israel Deaconess Medical Center who reported a few cases. Verfaellie worked with Lim on a paper years earlier about how heart attacks affect the hippocampus. But just as Barash suspected, Lim has lost track of his patients.

At the end of the day, Verfaellie puts on her coat, walks down the hallway to the elevator, and takes it eleven floors down to the now quiet lobby. She braces herself against the cold February wind, turns down Huntington Avenue, and begins walking home. Of course, Verfaellie doesn't need a map, but for anyone with a damaged hippocampus, whether from Alzheimer's, encephalitis, or an overdose, remembering the way home may be impossible. But why is that the case? And how does the hippocampus help you remember how to get where you're going?

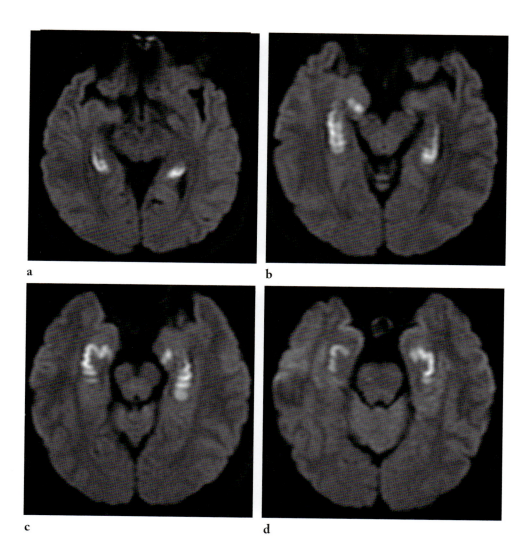

**MRI imaging captured shortly after Owen's overdose.**

Imagine Owen lying on his back, looking up. His hippocampus is white—visible only because it's just been injured. These four sequential slices are taken from the crown of his head, moving down through the hippocampus, starting at the thin tail toward the back of the head (a), all the way through to the fatter end closer to the eyes (d).

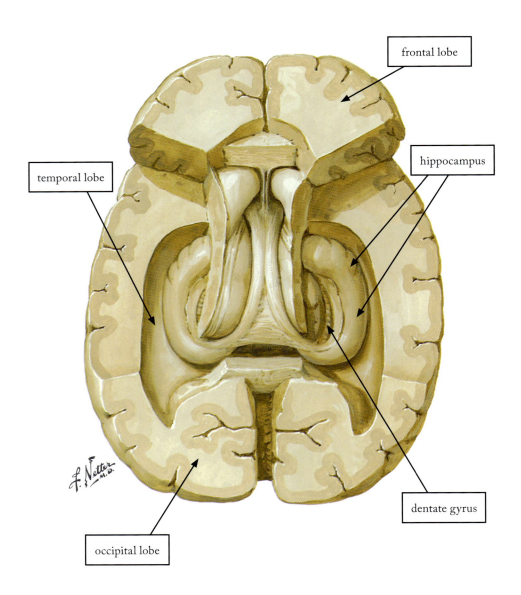

frontal lobe

hippocampus

temporal lobe

dentate gyrus

occipital lobe

**Illustration of the human hippocampus from above.** (This is from the same perspective as Owen's scan on the previous page).

The parietal lobe and the back portion of the frontal lobe are not pictured here. They have been "removed" to reveal the hippocampus, which is located in the lower area of the brain.

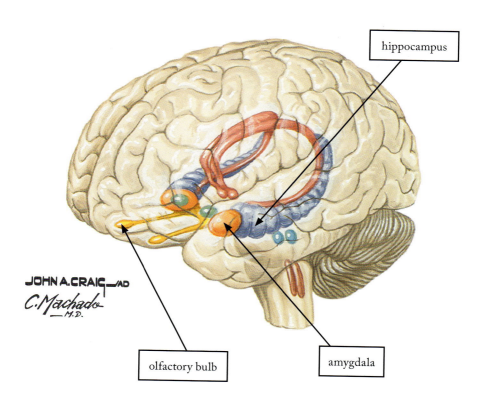

hippocampus

olfactory bulb

amygdala

JOHN A. CRAIG ᴍᴅ

C. Machado
ᴍ.ᴅ.

**Illustration of the human hippocampus and nearby structures.**

The hippocampus is pictured in blue, the olfactory bulb in yellow, and the amygdala in orange.

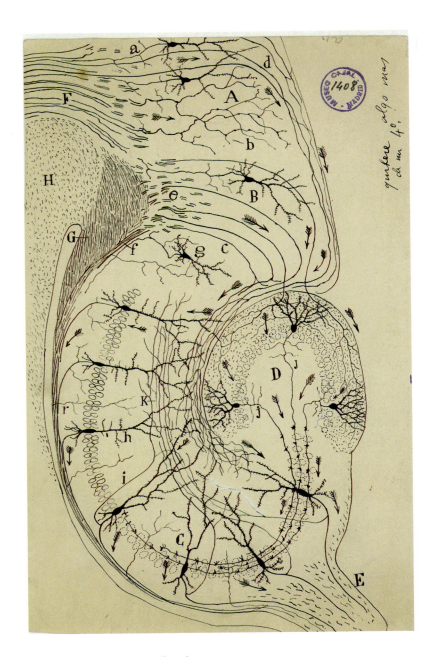

**Cajal's iconic 1911 picture of the rodent hippocampus.**
Imagine a rat lying on its belly with eyes facing left. This is the view from above. In this single slice prepared with a silver stain, Cajal sketched tiny arrows showing how information in the form of electricity loops back and forth through the curvy layers of the hippocampus. (The letter D marks the dentate gyrus. From there, electricity flows down toward the CA3, near the letter C. The letter i marks an axon and h marks a dendrite.)

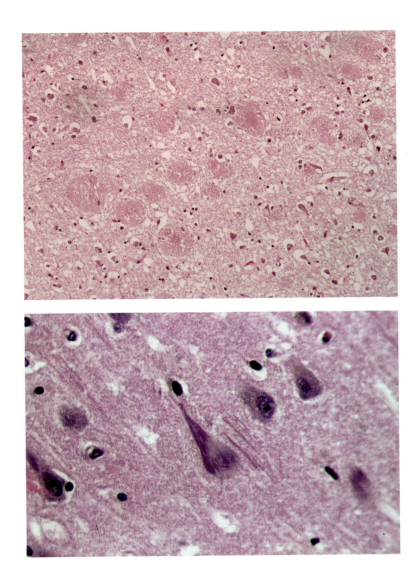

**Plaques and tangles.**

The large round blobs with dark pink centers in the top image are plaques. These sticky clumps of abnormally shaped amyloid beta proteins accumulate between neurons in the brains of people with Alzheimer's disease. The darkly stained teardrop shapes in the lower image are tau tangles in the human hippocampus of a deceased person with Alzheimer's. These tangles have also been found in the brains of chronic heroin users as young as seventeen. Plaques and tangles are the defining hallmarks of Alzheimer's.

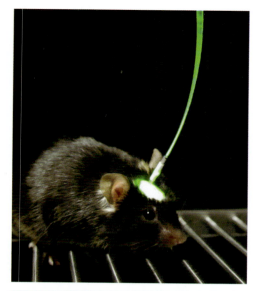

**Light shining down a fiberoptic wire into the hippocampus of a mouse.**

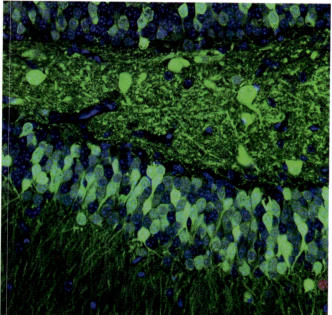

**A fear memory engram.**

The teardrop shaped glowing green cell bodies at the top and bottom of this image belong to an ensemble of neurons in the dentate gyrus that make up a mouse's negative memory of a foot shock. Depending on which end of the hippocampus is stimulated with light—and therefore which neurons in the ensemble are turned on—the memory can be made more or less negative. This insight could one day lead to treatments for PTSD, depression, or anxiety.

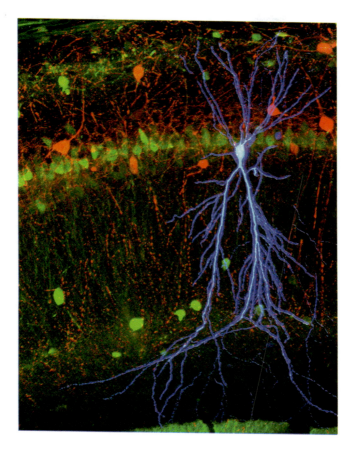

**Inhibitory and excitatory neurons.**

Out of hundreds of excitatory pyramidal neurons in this section of a mouse hippocampus, one has been dyed blue. The inhibitory neurons have been dyed red. These neurons, despite being much smaller and fewer in number, exert an outsized influence by wrapping their axons around the excitatory neurons' cell bodies and tightly controlling if and when they can fire. Without this control, excitation would overwhelm the system and possibly trigger a seizure.

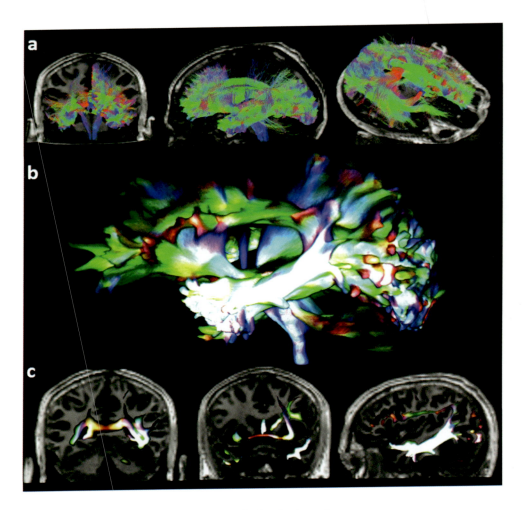

**Connections between the hippocampus and other brain regions.**

This color-coded MRI reveals how neurons in the hippocampus connect to other regions throughout the brain. The brighter and whiter the image, the more densely packed the axons. The large central panel shows a person from the side, looking to the left of the frame. The hippocampus is more or less at the center of the bright white region. Unexpectedly, there is surprisingly little connectivity between the hippocampus and brain areas that process sensory and motor information.

# 8

# Maps and Memories

· · · · · · · · · · · · · · · · · · · · · · · ·

The anesthetized rat lies facedown, its head immobilized in a metal device. On top of the animal's head is a miniature box containing eight tiny, glass-insulated electrodes. It's 1971, and University College London's John O'Keefe is on a mission to catch a memory in the making. Guided by a three-dimensional brain atlas that shows the way to the hippocampus, O'Keefe turns a small knob on the box that gently drives one of the electrodes down through the brain tissue. Compared to a person, a rat hippocampus is easier to reach, being relatively large and closer to the surface. O'Keefe and his graduate student, Jonathan Dostrovsky, move each of the electrodes into the cortex that overlies the kidney bean–shaped hippocampus. Then they attach the box to the skull. A day or two later the rat will be fully recovered and the actual experiment can begin.

In an earlier experiment, when O'Keefe accidentally placed an electrode in the hippocampus instead of the nearby region he was studying, he recorded a cell firing in a way that seemed linked to how the animal moved around its cage—like how fast it was going, or how it moved its head. But the connection, if there was one, was both unclear and surprising. Milner and Corkin's research with H.M. showed that the hippocampus is essential for making memories, not movements. Experiments with rodents suggest the same; if you damage the hippocampus, the animal behaves as if it has forgotten where it is going. But is there something about the actual making of a memory that is inextricably linked to how the rat navigates through space? O'Keefe intends to find out by being inside the hippocampus as memories are being made.

Now they lower the electrode even more slowly, waiting for the clicks of an electrical signal fed through an amplifier to tell them a neuron has just fired. When they hear that noise, it means they have found a target. Even though the rat has a bundle of wires running from the recording equipment to the lightweight box glued with dental cement to its head, it's free to move around its enclosure, a rectangular platform about nine by fourteen inches surrounded on three sides by a white plastic curtain. The fourth side is open, giving the rat a view of the lab.

O'Keefe and Dostrovsky watch while the rat does what rats do. It walks, eats, drinks, grooms, and sleeps. It sniffs things, presses a bar to release food pellets, digs in the sawdust. The scientists expose the rat to different odors, they touch its body, they move the light around, or show the animal a black-and-white striped board. All the while, the electrodes inside the hippocampus record the firing of a handful of neurons, and a miniature amplifier converts the activity into clicking sounds heard over the loudspeaker.

The two men experiment with twenty-two rats. Some neurons don't reveal any discernible pattern of firing. Some neurons don't fire at all, despite the scientists' best attempts to find a stimulus that makes it react. But one day, they locate a neuron they're convinced fires only when the animal is in one part of the testing platform. Every time the rat approaches that part, the clicks pick up speed and then die out when the animal moves away, like popcorn winding down. To make sure the sounds themselves—which the rat hears from the loudspeaker—aren't influencing the neuron's firing, they turn the loudspeaker off and use headphones to eavesdrop. But the pattern remains the same. Of the seventy-six neurons they monitor from twenty-two rats, eight fire in this way, each one reacting only to where the animals are on the platform. O'Keefe has directly matched a specific event in the outside world to the activity of an individual neuron out of the rat brain's some 200 million.

The scientists' conclusion? These cells have one job. Neuron by neuron, as they mark places, the hippocampus builds a map. That's how an animal remembers that there is cheese in one corner and water in another. But they're not "cheese cells" or "water cells." They're place cells. After this preliminary experiment was confirmed many times over, O'Keefe and collaborator Lynn Nadel went further, writing a book called *The Hippocampus as a Cognitive Map*, in which they propose that the fundamental role of the hippocampus is to create mental maps. Space itself is the scaffold for all episodic memories, the dimension to which every other element of an experience is bound.

Over the next decades, through thousands of experiments with rats and mice exploring mazes, running on spherical treadmills, pedaling Flintstone-like cars, standing on platforms, swimming in cloudy pools of water, scampering across electrified floors that gently shock their feet, putting up with irritating little puffs of air, lured by chocolate and cheese and sweet condensed milk, neuroscientists find an entire collection of cells that make up an elaborate mapping system for rodents. In 1984, James Ranck Jr. discovers head-direction cells that act as a compass. In 2005, a husband-and-wife team of Norwegian scientists named May-Britt and Edvard Moser discover cells whose firing patterns create the map's coordinate system. The Mosers name them grid cells. In 2008, they find border cells, which show the rodent how far it is from the edge of an enclosure. In 2015, they identify speed cells. In theory, this collection of cells can provide a rat everything it needs to create a map and encode a memory. The system is flexible too. A place cell is not forever destined to mark one corner of one particular cage. It can remap to new places. Grid cells can remap as well. O'Keefe and the Mosers shared a Nobel Prize in 2014 for the work, but the story is far from over. While animals in cages go about the business of making memories, all these cells influence each other in mysterious ways.

Of course, the idea that our brains attach memories to places isn't new. Before the advent of writing, Greek and Roman orators used places to memorize long texts. According to legend, a grisly accident at a banquet hall in Greece in the 5th century B.C. inspired the technique, known as the "method of loci," or the Memory Palace, when a poet named Simonides recited a victory ode in honor of the host. As any skilled public speaker would have done, he addressed the audience as he spoke, looking around the hall and noticing where each person reclined around the table. When Simonides finished, he was called outside, and moments later, the roof fell in, killing the guests. But Simonides was able to identify bodies crushed beyond recognition because in his mind's eye, each person's image was bound to a specific location. This same strategy is used by modern memory athletes to memorize random numbers thousands of digits long. Similarly, Stonehenge, the Nazca Lines of Peru, the moai of Easter Island, and the Australian aborigines' songlines may be the spaces where ancient people encoded memories vital to their civilizations.

But what about the role of time in memory? Simonides remembered more than a collection of vivid impressions. The sequence of events was fundamental. First, he arrived at the hall. Next, he recited the poem. He was summoned outside. Perhaps on the way out, he brushed his hand along the smooth polished marble of the door frame and caught the scent of roasted pig. Then the roof collapsed, and he went back inside to identify the victims. If he was to go home and tell his wife the story of his day, he would need a beginning, a middle, and an end. A "this" before a "that." Location-specific cells can't be the sole source of information about an episodic memory. Without some cellular mechanism to keep track of time, the poet's story would have been a jumble.

In the late 2000s, Boston University's Howard Eichenbaum began probing the rodent hippocampus to look for cells that track the flow of time. To find them, he had to engineer a simple "first this, then that" routine and test whether rats could learn how to string these events together in the right order. (The fine art of designing such experiments is to make sure the behavior is a window onto the memory making, not something irrelevant like how cold or tired the rodent may be.) Researchers often use either a reward or a punishment to figure out what an animal is "thinking." Eichenbaum and his student Christopher MacDonald used a reward.

In one experimental setup, MacDonald withheld food from a rat long enough to make it hungry. Then he placed the rat in a corridor just wide enough for its body, face up against a wooden panel with a green rubber ball mounted at the level of its nose. After giving the rat a few seconds to register the rubber ball, MacDonald raised the panel and released the animal into another short section of the corridor, a kind of empty holding pen. After ten seconds, the experimenter removed the next panel, allowing the rat to move to the third section of the corridor, where a small terra-cotta clay pot containing sand mixed with cinnamon oil sat atop a platform. A fragment of a Froot Loop was buried inside.

Over many trials, the rat learned that the sequence of events mattered. A green rubber ball followed by a ten-second delay, followed by a pot that smelled like cinnamon always meant there was a Froot Loop reward inside. But if a pot that smelled like basil appeared at the end of the corridor, there was no buried Froot Loop. If the rat didn't bother digging in the basil-scented sand, it was a pretty good bet that it had learned the sequence. After the training, the researchers implanted electrodes in the hippocampus and recorded the neurons' firing patterns to see how the rat remembered. Like O'Keefe's experiment, a pattern emerged.

One neuron fired at the beginning of the waiting period and then shut off. Another fired a short time later, then another, at specific time intervals. The more precise the firing pattern's timing, the more likely the animal was to remember the correct association. Eichenbaum and MacDonald ran the experiment with four rats and found the same pattern. These neurons appeared to track the order of events—first the green rubber ball and then the buried Froot Loop—and bridge the gap between them. Like place cells, these so-called time cells could recalibrate when the time spent in the holding pen changed. Intriguingly, the researchers realized that when the rat was allowed to roam freely, time cells could shapeshift back into place cells. To see how time cells operate on their own when place has nothing to do with the task, MacDonald took the rat out of the maze and put it into a device that keeps it immobilized. The animal wasn't moving or even plotting where to go next. Even so, when MacDonald gave the rat a smell test with a delay period in between, he once again found cells that fired according to the passing of time. So perhaps the dimensions of space, *or* time, or space and time together, serve as a scaffold for episodic memories.

Still, neuroscientists haven't dedicated millions of hours and dollars to teasing out the finer points of how rodents navigate because they care that much about rodents' memory-making machinery. They study them because they're relatively easy to work with. Early symptoms of Alzheimer's include mild memory loss and getting lost, so perhaps illuminating the neural mechanisms of memory in rodents can suggest a way to help people. But how similar are these systems? Neuroscientists can't drill holes in someone's skull, insert electrodes into the hippocampus, allow them to walk around, and then record how neurons fire in response. Or can they?

—⁓—

Fifty years after O'Keefe found a way to listen in to the hippocampus of a freely moving rat, a neuroscientist named Nanthia Suthana figured out how to do that in people. In a warehouse-like room on the fourth floor of UCLA's School of Medicine, she double-checks the wires and equipment attached to one of her test subjects. Cameras mounted on the black walls of the enclosure will track the man's every move. Electrodes surgically implanted in his hippocampus will record brain activity while he memorizes how to find a handful of colored cylindrical halos, scattered randomly across the floor.

Suthana is tired, but she can barely contain her enthusiasm for this experiment. When she was nineteen, H.M.'s story captured her imagination. She wanted to know how our brains can remember a fleeting state of activity in neurons minutes, days, or years later. Or why are some memories forgotten? And is there any way to bring them back? Suthana wants to know how much of what takes place inside the rat brain applies to people, because that's when it will become useful for doctors and patients. Dozens of her grant applications have been rejected over the years because she was, as she says with a smile, "a new nobody." Now her innovative approach to bridging the long-standing gap between human and animal studies is bearing fruit.

Suthana's utilitarian-looking experimental facility at the far end of a hallway is a stark contrast to the rest of the fourth floor. To get to her testing ground, you pass the neurosurgical consultation office, a tasteful waiting area for patients furnished with gray couches, gray rugs, and white tables. Until recently, Suthana's experiments took place in hospitals with the help of epilepsy patients who had electrodes temporarily implanted before surgery. While these patients waited for neurologists to pinpoint the source of their seizures, which often begin in the hippocampus, they used laptops to play games inside simulated cities to test their spatial memory. But the patients were stuck in bed, tethered to brain-monitoring equipment—a poor facsimile for the real world. Even with those limitations, neuroscientists located cells in people that act like place and time cells in other animals. They also found that the way these cells fire corresponds to a brain rhythm called theta, a wave that oscillates about eight times a second. Rodents have theta rhythms too. They're more prevalent when the animal is moving—and the stronger these rhythms are, the better the animals remember. Is the same true for people?

If Suthana wanted to mimic O'Keefe's "freely moving rats" setup, she had to find a way to take her experiments outside their hospital room so her volunteer test subjects could freely move around. Two relatively new technologies that had nothing to do with each other inspired her solution. The first was a treatment for people with drug-resistant epilepsy in which surgeons permanently implant electrodes in the hippocampus to stop seizures. The electrodes are part of a wirelessly controlled device called NeuroPace, which detects abnormal firing patterns, then targets the area with just enough electricity to disrupt the activity before it can spread. It's much like a pacemaker that monitors and responds to abnormal heart rhythms.

Suthana realized she could use the NeuroPace to detect theta rhythms when people walked around exploring a new space; she could then analyze the patterns to see if they matched what happened in rodents. Suthana collaborates with epilepsy surgeons at UCLA to identify patients whose electrodes are implanted right where sensory input from the ears or the nose or the eyes enters the hippocampus.

The other technology Suthana was keeping tabs on was VR gaming, which simulates the experience of being in a three-dimensional world. Like many other people, Suthana is susceptible to motion sickness when she plays VR games, and it was only recently that she decided the technology was good enough to ask her patients to use it. Now, instead of being limited to navigating through a two-dimensional world on a screen while sitting motionless in a hospital bed, patients can walk around. Not only that, Suthana can design whatever world she needs to fit her experiment.

The patient Suthana is prepping today is a middle-aged man named Sam. Before the NeuroPace device was implanted, Sam suffered from intractable seizures that weren't controlled with medication. He wasn't hard to recruit. In fact, Suthana has a long list of patients who want to participate in her experiments. They know that not understanding how the brain works has slowed down the development of effective treatments for epilepsy and other diseases. Even though the results of these particular experiments will never help them directly, Sam and other test subjects are motivated by the idea that someday their contributions can help people suffering from memory loss.

Sam wears a form-fitting black bodysuit that makes him look like he's gearing up for a scuba diving expedition. Sensors attached to his arms, back, hands, and legs communicate with twelve cameras mounted high on the walls. Ten of the cameras sync up the patient's precise location and movements to whatever is happening inside the hippocampus. The other two cameras record video.

A black and navy blue cap with a network of embedded electrodes fits snugly on his head. Even though the signals Suthana picks up from these external electrodes won't be as informative as the recordings she'll receive from the NeuroPace, she considers it free data that she may as well gather. Someday, she hopes, she or other experts will be able to figure out if the measurements of electrical activity captured from outside the brain correlate with what's going on inside. If so, scientists won't be limited to working with the small number of people who have electrodes implanted in their brains.

An iPad tucked into a black pack on Sam's back captures all the data. A VR headset covers his eyes, completing the diving look. But from Sam's point of view, he's standing in a simplified virtual version of this warehouse-like space. Medical staff in white coats, including Sam's neurologist, stand near one wall, outside the region Sam sees. The doctor is on alert, watching Sam's EEG, ready to flip the NeuroPace device back into seizure-stopping mode if he notices unusual firing.

Before today's experiment, Suthana ran a memory test on Sam so she could design a task tailored to his abilities. The test was neither very easy nor very difficult—just the right level so there's room for him to improve. Scattered randomly across the floor are eight yellow halos—glowing translucent cylinders of light standing four feet tall and wide enough for a person to fit inside. Sam's directions are to look around until he spots a halo, walk to the center of it, and stay there while he tries to remember the location. As he moves away, that circle of light will disappear, and another one will pop up. As he navigates, Sam has to learn all eight halos' locations and the order in which they appeared.

Before Sam can reconstruct his path and reveal what he remembers, Suthana asks him to count backward from one hundred by threes. The distraction ensures that Sam can't rely on working memory, which allows people to hold and manipulate information in their heads without having to store it in the hippocampus. After the thirty-second distraction, Sam heads back into the virtual world, and Suthana watches how accurate his memory is.

So far, Suthana has only run this experiment with five patients, but what she found was remarkably similar to what neuroscientists observe in rodents: theta rhythms are more prevalent when test subjects are exploring the environment than when they are standing still. There were some differences between the rodents and Suthana's subjects. Unlike in rats, the theta rhythms came in bursts, rather than continuously, with one notable exception—a blind subject. For him, Suthana designed a virtual world, replacing halos that appeared and disappeared with beeping sounds that grew more frequent and higher in pitch the closer he moved. Like rodents, who use their whiskers to perceive their immediate environment because their vision is so weak, the blind subject had to sample his environment more frequently to update his location. And just like rodents, his theta rhythms were nearly continuous.

But Suthana's real enthusiasm stems from one tantalizing result: when the pattern of theta rhythms during the learning phase matched the pattern during the remembering phase, Sam and the other subjects scored higher on memory tests. Perhaps the nature of this rhythm is what makes memories last longer. Suthana's work bridging the gap between rodent and human studies is now recognized in prominent journals. Grant officers and journal reviewers have gone from dismissing her research to urging her to do more with every project. Using simulated worlds, she and her students can design almost any experiment they want to, as long as it passes what she calls the anti-motion-sickness "Nanthia test." She'd like to take test subjects outside to the streets of Los Angeles, bringing the research that much closer to the real world. But Suthana is as practical as she is eager to expand the work beyond navigation to other aspects of episodic memory and to develop more insights into what makes memories stick. "Now it's a matter of which question to answer first," says Suthana. "I'm hoping others will jump on the bandwagon, because I can't answer all those questions in my lab."

As to the question of whether the hippocampus uses a space-time framework for memories or is an association machine that processes sequences of events in a particular context, Suthana says that "people are coming together to say that it's all kind of the same thing." Her focus is on understanding how theta rhythms help people navigate and remember so scientists can translate experimental results into memory-saving therapies. Perturbing the brain's beautifully organized electrical system safely is a daunting task. Stimulating at the wrong frequency, at the wrong time, or in just slightly the wrong place could easily make memory worse. But deep brain stimulation works for Parkinson's. Someday, instead of pharmaceuticals, we may use electroceuticals to fine-tune brain rhythms and keep memories from slipping away.

# 9

# The Forgetting Disease

*February 2017*

It's 7:05 on a Friday night toward the end of February when Barash decides to text Amar Sahay, his former next-door neighbor. Sahay is a hard-charging, prominent young Alzheimer's researcher at the Center for Regenerative Medicine at Mass General. Barash remembers him as friendly and enthusiastic—Sahay talks so quickly it can seem like his mouth is having trouble keeping up with his brain.

"Can I call? If tonight's bad, how about tomorrow at four? Henry naps at one to three or so, so four is hard, four thirty, five otherwise later in the morning works better. How about Sunday at eleven?"

Networking isn't second nature to Barash, but he understands that the *MMWR* report gives him the calling card he needs to connect with other experts. Sahay, in contrast, is a prolific networker both online and offline. As soon as he saw the *MMWR* report Barash posted on LinkedIn, Sahay sent a congratulatory message.

Barash hopes Sahay can help him figure out how to push the investigation beyond the world of epidemiology and into the lab, where hypotheses can be tested. To go further than the "who, what, and where." If he's right that fentanyl is the culprit, how can he prove it? And why did opioids leave their mark on the hippocampus? Sahay might not have the bandwidth to research these questions, but he's the perfect person to spitball ideas with and make some introductions. His lab is only six years old, but Sahay oversees a team of about ten or so young scientists and has raised many millions of dollars for his research since joining Mass General.

Sahay is open to ideas from any quarter, not just by nature but also because of the dismal state of affairs in Alzheimer's research. More than a century after Alois Alzheimer identified the first patient, scientists still don't know what causes the disease or how to cure it. At best, there is a handful of drugs that may reduce some of the symptoms, at least for a while. And there has been no new drug since 2002. Like any successful fundraiser, Sahay is ready with the elevator pitch for why spending money on basic Alzheimer's research is so essential. Alzheimer's is a disease of aging, and by age eighty-five, the odds of developing it are about one in three. Today, nearly six million people in the United States have Alzheimer's, but as the number of older Americans continues to climb, so too does the number of victims. By 2050, someone will receive a diagnosis every thirty-three seconds, there could be nearly 14 million victims in the United States, and 130 million worldwide. And unless scientists can figure out how to break the impasse against this dreaded disease, it will cost more than $2 trillion in the next decade alone. But there's no way to put a figure on the anguish and loss for patients and their families.

Sahay made a name for himself in 2011, when, as a postdoc at Columbia University, he genetically engineered a mouse to have a better memory. An episode of NPR's *Wait Wait . . . Don't Tell Me!* featured the story. Rattling off the grim statistics is personal for him. Two of Sahay's family members have Alzheimer's. "I'd rather lose my limb than lose my brain," he says. "You lose who you are, you can't identify yourself, you can't recognize your spouse, your children."

Sahay was still in high school when scientists proposed a theory about the cause of Alzheimer's that tied together emerging clues about how the disease begins. The theory, called the amyloid cascade hypothesis, has held sway for thirty years, and clinical trials based on the hypothesis are still in progress. The idea is that the initiating event, the very first thing to go wrong, is the buildup of sticky plaques of abnormally shaped amyloid beta proteins. Then another protein, tau, twists into tangles inside neurons, and the inexorable cascade of degeneration continues—loss of synapses, inefficiently functioning brain cells and circuits, and ultimately dead neurons. By the time most people are diagnosed with symptoms of Alzheimer's, these changes are obvious in brain scans. In this view, amyloid beta plays the leading role, and tau is the supporting actor.

The hypothesis was inspired in part by a clue found in the DNA of people who suffer from a rare, hereditary form of Alzheimer's that strikes a person in his or her thirties to mid-sixties. This condition, called familial Alzheimer's, is unusual in that it is passed directly from one generation to the next, and in the early stages, it damages the parietal or frontal lobes more than the hippocampus. Symptoms such as poor judgment or agitation, rather than memory loss, are often among the first signs. But the amyloid plaques and tau tangles are still there, just as in typical Alzheimer's.

Rare mutations in any one of three genes cause familial Alzheimer's. Because of these mutations, a normal protein that straddles the cell membrane—the amyloid precursor protein (APP)—gets snipped apart in the wrong places before being recycled. This is only the first step in a destructive series of events. The smaller pieces, called amyloid beta, assemble themselves into long, toxic chains, which go on to form the core of sticky, amyloid plaques. The first mutation affecting APP was definitively identified in 1991, and the second and third a few years later. Other clues point to amyloid beta too. People with Down's syndrome have an extra copy of the APP gene, and about half develop Alzheimer's in their fifties or sixties.

These emerging pieces fit together like a puzzle, and scientists coalesced with increasing enthusiasm around the amyloid cascade hypothesis. The search for a cure depends on a theory about the cause, and the amyloid beta cascade idea gave the field a path to follow. As long as scientists could figure out how to remove plaques before they did too much damage, they thought a cure for Alzheimer's might be within reach.

To explore this new hypothesis, scientists would need a lab animal to experiment on—in fact, lots of lab animals. They would need a mouse that "got Alzheimer's" by accumulating amyloid beta plaques. In 1995, scientists in California added a gene into a mouse's DNA that forced it to churn out excessive amounts of the protein that later gets turned into amyloid beta—much more than the mouse could get rid of through the brain's natural house-cleaning process. And, just as the creators predicted, plaques built up between the neurons, and the mice had trouble learning. It appeared these mice had a disease that looked something like Alzheimer's, adding more weight to the amyloid cascade hypothesis.

Before testing experimental drugs or treatments on people, scientists in universities and pharmaceutical companies began to test therapies that cleared amyloid beta from the brain on mutant mouse models. Researchers could fully control the animals' environment—what the mice ate, how much they exercised, and when they went to sleep. The mice in each experiment carried the same genes. In short, researchers could remove the messy complications of human life and isolate the factors they wanted to test.

But did scientists give these engineered mice Alzheimer's—or some other terrible disease? Hints that the story was more complicated and targeting amyloid beta might not work began to surface as early as 2002, when the first major clinical trial in people failed. "That was the tip-off to me way early that amyloid clearance was not going to be the most powerful way of approaching this," says Bruce Miller of the Memory and Aging Center at UCSF. "I don't object to thinking that it might. I do object to the massive billions of dollars narrowly focused on this idea when study after study was coming up negative." New puzzle pieces that didn't fit the amyloid cascade hypothesis were also ignored, such as hints that amyloid beta first begins to build up in the outer layers of the brain before spreading inward to the hippocampus. Or that some people with high levels of the abnormally folded protein could be perfectly normal from all outward appearances—at least for some period of time.

Still, scientists kept at it, testing multiple strategies. They decreased the production of amyloid beta. They tried to stop the process that allows it to accumulate. They tried to speed up the removal process. They mobilized the immune system to attack it. They used antibodies to make vaccines against it. In scores of costly trials, experimental drugs that seemed so promising because they helped mice did not help people. Some drugs worsened cognitive decline and accelerated brain shrinkage. For years, more than half the therapies being tested were designed to target and remove amyloid beta. "Once one of these big lumbering pharmaceutical companies gets started," says Miller, "they really have a hard time being nimble and quick and shifting. They've put so [much] into one story." Sahay describes the focus on amyloid beta as "looking just under the spotlight, which typically happens in all of science."

Nevertheless, amyloid beta is indisputably a core feature of the disease. There are dozens or more genes linked to late-onset Alzheimer's, which afflicts the majority of people who succumb to the disease. But the riskiest gene of all is called *APOE4*, and in addition to its best understood role—carrying fats like cholesterol around the body—it also appears to affect how amyloid beta is recycled. No one dies from Alzheimer's without having accumulated the sticky plaques.

But it can't be the only thing that's wrong. Plaques may be a warning sign of some other problem, not the primary cause of the disease. They may even start out as a way to protect the brain from infection. Michael Yassa, a neuroscientist at the University of California, Irvine, says, "It's almost like having a building that's on fire and having the fire department come—and the only thing they're doing is turning off the alarms as opposed to putting out the fire."

So if going after amyloid beta directly isn't a winning strategy, what is? A list of suspected risk factors for the late-onset variant, which accounts for about 95 percent or more of cases, is long: traumatic brain injury, stress, heart disease, pollution, infections such as herpes simplex virus, inflammation, poor diet, smoking, insufficient sleep, lack of exercise, hearing loss, diminishing sense of smell, diabetes, loneliness, and epilepsy, to name a few. Many of these conditions are intertwined, making it hard to tease out which one could be relevant. Some could be a result of the disease rather than a cause. Some might have nothing to do with how the disease progresses but simply contribute to a toxic stew of insults already simmering away in the aging brain until a tipping point is reached and the Alzheimer's damage is unmasked.

Gender is another mysterious risk factor. Men who get Alzheimer's die faster than women. About two thirds of all cases are women, but they're only at greater risk for developing Alzheimer's at advanced ages. Paradoxically, women appear to have some biological advantages. One is a protective factor on the X chromosome that helps neurons fight the toxic effects of amyloid beta plaques, at least in female mice. Another is better verbal skills, which, while they don't necessarily affect how the disease progresses, make it easier for women to compensate for whatever is going on behind the scenes. Why more women than men have Alzheimer's is actively debated, but the main reason may simply be because women live longer. And if Alzheimer's is anything, it's a disease of aging.

The inability to pinpoint how early in life Alzheimer's disease begins bedevils the field and may be among the most significant barriers to pinning down the cause and finding a cure. Proponents of the amyloid cascade hypothesis point out it's possible that recent trials have failed because the intervention came too late. By the time someone has Alzheimer's and is enrolled in a trial, the disease process has likely been underway for fifteen to twenty years.

Researchers have tried to fix this problem by enrolling people who have what's called aMCI, or amnestic mild cognitive impairment due to Alzheimer's. aMCI can be caused by a variety of diseases, but it is the most common way people with Alzheimer's first show symptoms. Even in those people, whatever it is that causes plaques and tangles to accumulate has been invisibly damaging the brain for a long time. To make matters worse, many people enrolled in the early anti-amyloid clinical trials did not even have Alzheimer's, because it's so hard to diagnose without being able to see what's going on inside the brain. In at least two large early clinical trials, about 30 percent of the participants didn't have plaques and tangles; when their brains were later scanned with an expensive PET imaging technique that had only recently become available, it became clear that they suffered from some other form of dementia. Since then, trials like these require proof of brain amyloid plaques before participants can be enrolled.

The drugs researchers began testing years ago that are failing clinical trials in 2020 are the result of all these barriers. But new tools may help overcome these problems. Diagnostic tests like PET scans, which can detect amyloid beta and tau in living patients, are being incorporated into current trials. And new tests to look for early markers in the blood, cerebrospinal fluid, or urine are coming online and will help researchers make sure they're enrolling the right people early enough to make a difference. One, like a simple blood test for tau, may be widely available in a few years. Researchers can now screen tens of thousands of compounds in a petri dish before they're ever injected into a mouse. More sophisticated genetically engineered mice that come closer to replicating the late-onset type that afflicts most people with Alzheimer's are becoming available.

There's a surprising upside to the fact that so many drugs that cured the simplistic early mouse models did not cure people. Catherine Kaczorowski works at Jackson Labs in Bar Harbor, Maine, which ships millions of mice to some fifty countries around the world every year. She decided to investigate what was wrong with the popular model and discovered that, just by sheer happenstance, there was something about that mouse strain that made it resilient to developing plaques and tangles. That, in turn, made this mouse much easier to cure. What was good luck for the mouse was bad luck for testing Alzheimer's treatments. But Kaczorowski saw a silver lining. "Those are new therapeutic targets that no one's ever considered. People have been completely focused on disease risk. I would argue that if you know how the brain functions superiorly in the face of lots of amyloid, you just want to know what that thing is. And then you want to promote that thing." Now scientists have a new target to go after—resilience genes that may protect people from Alzheimer's even when the deck is stacked against them.

The Alzheimer's field is no longer looking under just one spotlight. While some people still subscribe to the amyloid cascade hypothesis, most agree that the more researchers try different strategies, the better. Kaczorowski is focused on resilience genes, but she argues that the field should "let everybody take as many shots on goal . . . I don't care if we're right or wrong as long as someone ends up being right. Sooner rather than later."

One researcher looking to take another shot on goal is Sahay, who has his sights set on the unavoidable risk factor of old age. He wants to know if there's a way to tweak the processes that lead to many systems failures, including learning and memory. Not everyone who gets old gets

Alzheimer's, but there must be something about aging—especially in the hippocampus—that makes so many of us vulnerable to this particular illness.

Sahay's famous 2011 experiment used a mouse whose memory he'd improved by genetically engineering it to grow new neurons in the hippocampus after the mouse was born. That process, called neurogenesis, declines naturally and precipitously with age, especially in people, although according to the bulk of the evidence, it never stops completely. Until the late 1990s, experts were certain that growing new neurons was impossible. The great Nobel Prize–winning neuroscientist Ramón y Cajal himself said so. Unlike other organs, whose cells are continuously replaced throughout our lives, the brain was different. "Everything may die, nothing may be regenerated," Cajal wrote in 1928. His dogma made sense to most neuroscientists. The structure of the brain, especially the hippocampus, seemed far too elaborate to incorporate new neurons without disturbing the network of interconnected cells that store memories.

The origin story of the discovery of neurogenesis is similar to that of many discoveries that overturn long-standing dogma. In part one, a scientist has a new idea that doesn't fit with accepted wisdom. In part two, they overcome many obstacles to prove their point. And in part three, they receive widespread recognition. Joseph Altman made it through parts one and two. But the attention went to others who confirmed his work decades later.

Altman didn't seek out controversy, but his formative years as a Jewish teenager in Hungary during World War II steeled him to overcome obstacles. When the Nazis invaded, Altman was sent to a forced-labor camp and put to work on a railroad gang. There he began to puzzle over the nature of human behavior. The best way to understand why people do what they do, it seemed to him, was to go back to basics and investigate the brain's inner workings. After the war, as a librarian in Australia and then in the United States, Altman had plenty of time to read books about psychology, human behavior, psychoanalysis, and human brain structure. In New York City, he finally secured the formal training he needed to realize his ambition, and in 1962, he joined the faculty at the Massachusetts Institute of Technology (MIT) as an associate professor.

Altman would seem to have checked all the boxes to be accepted by the research community, but at MIT, his work set him on an unintended collision course with scientific orthodoxy. Altman designed an experiment to compare rats living alone in cell-like enclosures with rats in cages furnished with plenty of food, water, companions, places to play, and other amenities. The brains of the rats who lived in the enriched environment weighed more and contained more new cells, the kind that support neurons and clear out debris. But one result surprised and confused Altman; his experiment also revealed a small number of new neurons. Altman was well aware of Cajal's decree, but he followed the evidence where it led. In a series of experiments in the 1960s, he gathered more evidence for the birth of new neurons in young rats, kittens, and guinea pigs. At first, his discoveries met with little resistance, but over time that changed.

In 1968, Altman was denied tenure at MIT. Undaunted, he moved his lab to Purdue, where he gradually shifted his focus to less controversial topics. But slowly, his funding dried up, the result, he believed, of having challenged scientific orthodoxy. In a brief and sometimes bitter memoir written in 2008, Altman wrote, "Instead of open criticism, there appears to have been a clandestine effort by a group of influential neuroscientists to suppress the evidence we have presented and, later on, to silence us altogether by closing down our laboratory."

But even as Altman was moving on, new studies began to back up the idea that the hippocampus could create new neurons after birth. In the 1980s, scientists discovered that, depending on the season, songbirds produce as many as a thousand new neurons a day. Soon, evidence for neurogenesis in macaque monkeys, tree shrews, and marmoset monkeys emerged. But even then, dogma maintained that while all these animals could grow new neurons, people must be different. And there seemed to be no safe or ethical way to find out.

Then, in the late nineties, a team of Swedish and American scientists hit upon a method. They analyzed the brain tissue of five cancer patients, ranging in age from fifty-seven to seventy-two, who had been administered a diagnostic drug that inserts itself into dividing cells to detect whether cancer cells are proliferating. After the patients died, the drug was found as expected in cancer cells. But it was also found in new neurons in the hippocampus—the first concrete evidence for human neurogenesis.

It would take more than a decade before another international team of scientists found a way to quantify how many new neurons the human brain could create. The technique exploited the aftermath of explosions from Cold War nuclear bomb testing. From 1955 to 1963, these explosions added a heavier version of carbon, called carbon-14, to the atmosphere. During those years, every living plant and animal incorporated more carbon-14 into the DNA of every newborn cell, giving the researchers a way to pinpoint the birthday of any new cell, give or take a year and a half. They hypothesized that wherever neurons were dividing and growing in people, they would detect carbon-14. Their findings? The adult human hippocampus gives birth to seven hundred new neurons a day, a rate comparable to adult mice.

Bolstered by this compelling proof of neurogenesis and other studies since then, a new generation of researchers has taken up the challenge issued by Ramón y Cajal: "It is for the science of the future to change, if possible, this harsh decree." Sahay's goal is to harness this neuroplasticity to fine-tune how neurons communicate with each other to treat Alzheimer's and other illnesses affected by memory, like PTSD, depression, or anxiety.

On a bleak February weekend in 2017, Barash paces the third floor of his house, avoiding the low eaves on either side, sidestepping Henry's growing collection of toys and toddler obstacles while talking to Sahay. Sahay has spent the last ten years developing ever more elaborate animal models, using genetic engineering to tinker with the circuitry that runs through the hippocampus in his lab mice. He describes this in science-speak to Barash as "interrogating the circuits," but it's impossible to interrogate the human hippocampus in the same way. Still, Sahay is intrigued by the precision of the brain damage described in the *MMWR* report—as precise as it would be if a researcher were trying to create an animal model of overnight memory loss in humans.

This thought has been in the back of Barash's mind ever since Butler brought up the connection between the toxic drug contaminant in overdose victims in the 1980s and the first-ever animal model for Parkinson's. But first, Barash needs to find someone who can do laboratory experiments with rodents to see what fentanyl does to their brains. Sahay, unsurprisingly, doesn't have the time. And even then, it would be challenging. Getting clearance from the Drug Enforcement Administration (DEA) to work with fentanyl was already a burdensome process requiring a special license and a separate secure storage box checked by DEA agents. Now Mass General Hospital has decentralized the process, which means that anyone who wants to work with an opioid, including fentanyl, has to request their own license. Sahay suggests Barash track down a collaborator who already has one, perhaps a scientist at the National Institute on Drug Abuse (NIDA). Barash thanks him, hangs up, and decides that as soon as Henry goes down for his nap, he'll e-mail NIDA's director, Nora Volkow.

# PART THREE

· · · · · · · · · · · · · · · · · ·

# OPIOIDS

# 10

# Learning to Be Addicted

. . . . . . . . . . . . . . . . . . . . . . . . . . . . . . .

## *February 2017*

*26 Feb 2017*

*Six months sober but my craving for fentanyl is still there every single day. Remember the benefits of sobriety. Happiness is not the goal. Meditate. Stay away from drug buddies. Take small steps. Can I trick my brain into not wanting to use? Undo maladaptive habits? Use reward/prediction error? This stuff is happening all the time, so why not take advantage? Reality is subjective anyway.*

After finishing his journal entry, Owen spends an hour or two updating and rearranging to-do lists. When he finishes, he feels briefly satisfied. But almost instantly, doubts resurface as to whether the process is complete. The compulsive and embarrassing need to hoard memories dominates his days.

The previous summer marked the lowest point of his life. After graduating with highest honors from UCLA, he returned home to Del Mar, to places and people so inextricably linked with memories of being high that he couldn't resist the urge to use drugs again. Owen moved from ever-higher doses of heroin and then to fentanyl, injecting himself as many as seven times a day. His family's first attempt to get him into rehab went disastrously wrong. The staff gave him medication to block the effects of opioids too early, precipitating a withdrawal so rapid that he had to be hospitalized. Then Owen went back to using, and his family asked him to leave the house. For a while, he lived on the streets of Del Mar, lugging his suitcase around, calling his mother to ask if she could leave food out on the back porch for him.

Owen entered rehab again after his family promised to find treatment for his depression and obsessive-compulsive disorder. But by this point, medicine's repeated failure to help made his dream of going to medical school and helping others look like a waste of time. Medicine is a crock, he figured, and doctors have nothing viable to offer. But Owen doesn't want to let his family down again, and he doesn't want to live on the streets. He takes his pills, gets in bed, and waits for sleep.

—m—

Barash confirms that Henry is fast asleep in his crib before he returns to his weekend e-mail campaign. His strategy for inducing the director of the NIDA to read his e-mail is to compose a message just detailed enough to explain why Dr. Volkow should be interested, but not so long that she decides she doesn't have time to read it. So he lays out the case in a few paragraphs. He floats the idea of studying the effect of fentanyl on the mouse brain ("somewhat analogous to studies of MPTP and parkinsonism"), attaches the *MMWR* paper, and asks if she can point him in the right direction. To his surprise, Volkow e-mails back just a few hours later.

> Dear Dr Barash: Dr Michael Baumann at NIDA is the expert on synthetic drugs and while he himself is not working with fentanyl he can connect u to someone that does best regards, nora d volkow

The next afternoon, Barash hears back from Dr. Volkow's contact.

Hi Jed-

You present a very interesting case series. With regard to the specific involvement of fentanyl, this reasonable hypothesis is difficult to test for a variety of reasons (e.g., lack of forensic confirmation of fentanyl) . . .

Baumann also gives Barash a tip on tapping into a national community of experts who are tracking the opioid crisis.

With regard to NDEWS sharing, you can join the organization and have ready access to all of their information.

All best,
Mike

The National Drug Early Warning System (NDEWS) Network, now known as the CESAResearch Network, is an informal, virtual community of epidemiologists, addiction specialists, first responders, policymakers, and journalists. Its mission is to rapidly disseminate news about what types of drugs are being used across the country. The community shares articles, debates policies, and discusses reports from government agencies like the CDC, the Drug Enforcement Administration in the Department of Justice, and NIH. Because it takes time to collect and collate data, these reports typically paint a picture of drug use as it was a year or two earlier. And there's a lot of data missing, a fact easily obscured by the precise numbers the reports contain.

For example, the CDC reported that there were 67,367 drug overdose deaths in 2018, of which 46,802 involved opioids such as prescription drugs, heroin, and fentanyl. Massachusetts, New Hampshire, Ohio, and West Virginia were particularly hard hit. But these numbers underestimated the toll, because they didn't include all overdose deaths—only those that were reported. Information about which drug caused the death was also murky. If the coroner or medical examiner didn't test for fentanyl, the death certificate might list cocaine as the sole cause of death even if fentanyl was mixed in. In addition, results were reported by individual drug. So a single death that involved three drugs would count as three separate deaths. Network subscribers point out these uncertainties and share timely, anecdotal information about which substances are commonly taken together, the appearance and packaging of synthetic drugs, or hot spots of drug use. Sometimes the information they share sparks debates about the nature of addiction, risk factors, and the potential for relapse.

Fifty years ago, a unique set of circumstances set the stage for a large-scale natural experiment that shed light on that debate. In 1971 Nixon declared heroin the nation's "No 1. Public Health Problem." At the time, the military was concerned about young soldiers in Vietnam who had easy access to heroin that was cheap and potent. One veteran reported that he was offered drugs even as he disembarked from the airplane. Researchers at the Washington University Department of Psychiatry later learned that almost half of all army enlisted men had tried one of two opioids—heroin or opium—and 20 percent had become addicted while there.

This natural experiment's starting point was to figure out who was actively using opioids when they were discharged. So, in September 1971, every soldier scheduled to return home had a urine drug screen. If it was positive, the soldier would be sent for detox, and the military would have identified an active user. A sociologist named Lee Robins, who was hired to head up the research project, selected a random sample of soldiers leaving Vietnam and a sample of those who had tested positive. They had all been in Vietnam for one year, so their exposure to the drug-rich environment was essentially the same.

Six months after the veterans returned home, Robins and her team interviewed them and requested urine samples for drug screening. Almost everyone complied. Robins asked these veterans about their Vietnam experiences, how they thought the military should cope with addiction, and what life was like for them before, during, and after Vietnam. It turned out that it was rare for people to relapse to using heroin again, even among those who had been addicted in Vietnam. The results, published in 1974, were so different from what people expected that the scientific community was skeptical, and the press assumed the study was a Department of Defense whitewash. Some argued that more would relapse after a longer period of time. Therefore, two years later, Robins re-interviewed them and requested urine specimens for analysis. The results were the same.

One of Robins's findings is obvious; the availability of a given drug is key to how much it's used. But most of her analysis ran counter to conventional wisdom, both then and now. For instance, Robins discovered that those who became addicted in Vietnam already had social problems before they enlisted and used other drugs like marijuana, alcohol, and tobacco. This implies that opioids aren't uniquely vulnerable to abuse. Perhaps more surprisingly, the Vietnam veterans study suggests that addiction to opioids is not destined to be a lifelong condition. Robins discovered that only 12 percent of addicted opioid users relapsed after three years. The rest recovered spontaneously without any treatment.

"Certainly our results are different from what we expected in a number of ways," Robins wrote in a follow-up review in 1977. "It is uncomfortable presenting results that differ so much from clinical experience with addicts in treatment." Perhaps Robins's findings didn't make sense for precisely that reason. The data on recovery rates are drawn from a subpopulation of drug users in treatment—in other words, the people most likely to have trouble overcoming addiction. There has yet to be another opportunity to design such a large and rigorous experiment, and no single study can ever capture the messy realities of addiction. But there is evidence that the results may be relevant to the current opioid crisis.

As CESAResearch Network users frequently discuss, polysubstance use is common. Eric Wish, who founded NDEWS and the Network and worked with Robins on the Vietnam study, has analyzed multiple drug use surveys. "You're probably thinking it's different with a prescription opioid issue," he says. "But in every instance, the people who say they have used have also used other drugs." Drug testing backs up the survey results. "What we find without exception when we test people for drugs," says Wish, "is that people testing positive for opioids often have four to six other drugs in their urine, including cocaine. Opioids are the tip of the iceberg, and we should focus on the person, not the drug."

Comprehensive data on recovery rates are lacking, but what little information is available indicates that recovery is more likely than not.

Compared to the twenty-or-so million people in the United States with an active substance use disorder, there are, by one estimate, twenty-three million people in long-term recovery. And only four million of those received treatment, which suggests many people recover on their own. As for opioids, a recent analysis suggests that a little more than 3 percent of people prescribed opioids ever become addicted. However, 3 percent of the two-or-so million people prescribed opioids in a given year translates into an enormous number of people suffering from addiction. And for those who wind up using fentanyl, the results are much more likely to be deadly.

Over the past five to ten years, a slow decrease in the number of opioid prescriptions has probably cut the number of new users who become addicted that way, but these changes haven't ended the opioid crisis or the United States' deadly addiction epidemic. Deaths from opioid overdoses continue to climb, and in 2019 the combined number of overdose deaths from a wide range of illicit drugs was about 72,000—higher than ever before. The devastation is higher still if you include other illnesses caused by drug use, such as circulatory or respiratory diseases. By one recent estimate, close to a quarter of men between the ages of fifteen and sixty-four who died in the United States between 1999 and 2016 passed away directly or indirectly from drug use.

---

Barash follows the NIDA expert's advice and posts a message on NDEWS to ask if anyone has seen or heard about patients with unexplained amnesia. The next day he gets a response from Bertha Madras, the director of the Laboratory of Addiction Neurobiology at Harvard's McLean Hospital.

> Thank you for your very informative and comprehensive overview. . . . Your very significant clinical observations may reflect the tip of a major problem.

As one of six members of the President's Commission on Combating Drug Addiction and the Opioid Crisis, Madras has a bird's-eye view of the epidemic. She keeps a close eye on the network and posts frequently, sometimes multiple times a day. She has spent her professional life examining how addiction changes the brain, especially the young brain. Madras's research focuses on a neurotransmitter called dopamine, a tiny molecule that looms large in the study of addiction.

Stimulants, opioids, cannabinoids, alcohol, or any mind-altering drug can each create different moods and sensations, but from the brain's perspective, people desire them for the same reason. In one way or another, taking these drugs leads to more dopamine in the synapses, the communication gaps between neurons. Dopamine is commonly known as the "feel-good" molecule, but it's probably better thought of as a "take note" molecule, a cue to pay attention because something unusual is happening. Neurons that respond to dopamine are exquisitely sensitive to the difference between what you expect and what you get. The more intense the experience, the greater the error, the more quickly neurons fire, and the more your brain pays attention. It's called reward-prediction error and it speeds up learning and strengthens memories. The phenomenon is at play in gambling, eating, falling in love, or any pleasurable experience. The same is true for aversive experiences, like pain or stress. Drugs create a surge of dopamine that is by some estimates two to ten times greater, and this sometimes leads to compulsive and seemingly intractable behaviors.

But as intractable as the addiction epidemic appears, there are ways to help people. The brain is a complex organ that never stops changing in response to its environment, so, more often than not, compulsive patterns of drug use can abate. Medication-assisted treatment like methadone, naltrexone, or buprenorphine interact with opioid receptors to help ease withdrawal symptoms, keep craving at bay, or both. Harm reduction strategies are built on the principle of helping people "where they're at," like providing free, clean needles for anyone who is actively using. Support from peers, behavioral therapies that help people build skills to avoid drugs, or the surprisingly effective technique of rewarding abstinence with small prizes are all proven interventions.

One nascent effort to treat addiction capitalizes on the notion that it isn't just a disease but a learning and memory disorder. One of dopamine's tricks is to help our brains effortlessly learn to pair pleasure with cues—like the smell of alcohol, the strike of a match, or the presence of a friend who regularly uses drugs. But in theory, anything learned can be updated with new learning. Working in a safe office environment, therapists have tried to help patients unlearn these so-called reward or Pavlovian memories by exposing them to drug cues. But David Epstein, who heads up the Real-world Assessment, Prediction, and Treatment Unit at NIDA, says that relapse rates remain high for these patients, perhaps because their brains learn that those cues don't predict anything that happens out in the real world.

Scientists are now testing ways to turbocharge the unlearning. "Our brains are great at learning around rewards and what predicts them," says Ravi Das, a neuroscientist at University College London. He aims to take advantage of the element of surprise to manipulate drug memories. Das studies alcoholism, which kills more people each year than opioids. Alcohol is widely available, socially acceptable, and the cues that trigger relapse are everywhere—on social media, billboards, television, and at parties.

Das's team recruited ninety people in the United Kingdom who drank heavily almost every day. "The idea," he says, "is to reactivate their memory surrounding alcohol. These are memories that link the taste of beer or the smell of beer and reinforce the desire to drink." On Day 1, Das asks each participant in the treatment group to rate pictures of beer before he places a glass of beer on the table. Next, a series of commands flash onto the computer screen that instruct the participant first to pick up the beer, prepare to drink, and then actually drink. A few days later, Das turns the tables. Just as before, the participant picks up the beer and brings it to her lips. She has every reason to expect that she's about to drink. But this time, instead of the *Drink Now* command, the words *Stop! Do Not Drink* flash onto the screen. Within milliseconds, neurons sensitive to dopamine in the hippocampus are firing rapidly, signaling the difference between what she expected and what's happening now. The theory is that this reward prediction error destabilizes the original memory and forces the participant to pay attention. As a result, her memory for the pleasure of drinking is being updated with new information and reconsolidated into something less powerful.

Das designed the next step to destabilize the association between drinking and reward even further. A doctor infuses the test subject with an anesthetic called ketamine, which interferes with memory formation. The procedure is like opening up and editing a computer file, Das says, and then using ketamine to pull the plug while the memory is reconsolidating. But "it's harder with these reward memories," he adds, "because they're really robust. So you have to get the parameters right during the retrieval bit."

Nine months after receiving the treatment, test subjects were less interested in drinking, anticipated it as less pleasurable, and, according to the diaries they kept, drank half as much as they did before the intervention. In follow-up research with 120 participants, Das changed the protocol. Instead of using ketamine, he upped the ante on the prediction error. Not only were participants surprised by not being able to drink the beer, but they also reviewed disgusting images and consumed multiple small amounts of bitter liquids. Nine months later, participants drank less than they used to, although the intervention was slightly less effective than the ketamine protocol.

"It seems like science fiction," admits Das, but his experiments are some of about ten or so that show promise in treating cocaine, tobacco, and heroin addiction. This experimental memory manipulation has a long way to go before it reaches people who need it. Still, some day it may be another tool to fight an addiction epidemic that has no end in sight.

There are many lenses through which experts view addiction. According to NIDA, addiction is a chronic relapsing disorder, dysfunction, or disease. Genetic risk factors affect how people respond to drugs, and prolonged drug use changes how neurons communicate, weakening some areas and strengthening others. Addiction can also be understood as one solution to an unresolved problem, such as childhood trauma, mental illness, or bleak expectations about the future. And as Das's experiments show, you can treat addiction as a disorder of learning and memory. Epstein believes that multiple theories of addiction are valid. "The questions are," he writes, "when, for whom, and to what extent." James Mahoney, at West Virginia University in Morgantown, an area ravaged by illicit drug use, says, "The way I look at it, we really just need to individualize treatment. People use substances for a variety of reasons. So find out what the root is in each person and target that early in treatment."

———————

Barash takes off his glasses and rubs his eyes. It's Saturday morning, and Gillian is wrangling Henry into his snowsuit. "It kinda sucks that I can't get anyone at NIDA interested," he says.

"Didn't Nora Volkow answer you on a Sunday like two hours after you e-mailed?" says Gillian as she zips up the front.

"Yeah, it's pretty impressive, actually. She turned me over to someone else, but it didn't really go anywhere."

Barash assumes his overture hasn't gone anywhere because people are too busy to take on one more thing—especially an opioid research project suggested by a neurologist with no addiction expertise who is the medical director at a nursing home for veterans. There's another factor that makes Barash's request to test the effect of fentanyl on mice hard to pursue. The Drug Enforcement Agency (DEA) classifies fentanyl as either a Schedule I or a Schedule II drug, depending on which chemical version it is, and sets strict quotas on how much can be produced for medical research. As a result, Volkow and other scientists are stymied in their efforts to research fentanyl. Even the NIH, a government-run research facility, has to wait to get approval for experiments.

"Who was that guy who was so desperate to prove that *H. Pylori* causes stomach ulcers that he drank some?" Gillian asks.

"Barry Marshall. Because no one believed him. Nice trivia." Barash stuffs Henry's feet into snow boots, slings a swim bag over Gillian's shoulder, and holds the side door open for her. "Yeah, I'm obsessed but not that desperate. I wouldn't go that far just to prove my point."

"I'll be back around eleven after Henry's swim lesson," Gillian calls over her shoulder. Barash now has two whole hours to work on Plan B.

# Extreme Potency

· · · · · · · · · · · · · · · · ·

## *February 2017*

Barash reaches into his sweatpants pocket, pulls out his to-do list, and unfolds the well-worn sheet. Gillian teases him about his micrographia because the words are so tiny. Occasionally even he has trouble deciphering the "to-do" items. Some are in caps. Some are in lower case. Some have a line drawn firmly through the middle. Barash unfolds the paper and crosses off *e-mail NIDA*. On the same page, he's written *FENTANYL AND ANESTHESIA*. He calls his father to thank him for making the e-mail introduction to an anesthesiologist named Andrew Kofke at the University of Pennsylvania. Kofke, Barash hopes, may be able to give him some advice.

Paul G. Barash literally wrote the book on anesthesiology. Medical students and practicing anesthesiologists across the country rely on his 2,000-page-long *Clinical Anesthesia* book, now in its eighth edition. The first reference to opioids appears on page 14. "Opioids (historically referred to as narcotics although semantically incorrect—see Chapter 19) remain the analgesic workhorse in anesthesia practice." If you've had an operation, odds are your anesthesiologist has given you fentanyl or a closely related version of fentanyl, which is used in 80 to 90 percent of surgeries.

The original opioid workhorse was morphine, extracted from the seed pods of the poppy flower. It's still widely used today to control pain in the dying. Hydromorphone, oxymorphone, and heroin are versions of this complicated three-dimensional structure made up of four interlocking hexagonal rings. In the late 1950s, a prolific chemist and drug designer named Paul Janssen began the search for a better synthetic opioid, one more fast-acting and potent and with fewer side effects. But morphine's chemical structure was difficult to manipulate, so he chose one ring of a simpler synthetic opioid as his starting point.

Janssen had two missions. First, he had to get the drug into the brain. That meant making a molecule that slipped easily—and therefore quickly—through the blood-brain barrier and into the clear fluid that bathed the cells. Second, Janssen knew there had to be some sort of pain receptor that allowed the drug to activate the cell. A more potent opioid would have to fit more tightly, like a hand in a glove. At the time, no one was aware that neurons even had opioid receptors or that the body makes its own, homegrown opioids. But this didn't change Janssen's task.

For years, the enterprising chemist tinkered, adding and subtracting attachments, twisting its three-dimensional shape, and testing each version as he went. In 1960, he synthesized a new molecule that was dramatically more powerful than anything the world had ever seen. This was fentanyl—one hundred to two hundred times more potent than morphine and capable of taking effect in a minute or two. Instead of being made up of four hexagonal rings directly bound together, fentanyl has only three rings spread out in a line. Its shape is so unlike other opioids that even fifty years later, standard drug screens that detect commonly prescribed opioids and heroin can't recognize fentanyl.

In 1963, fentanyl began to be used as an anesthetic in Europe. In the United States, the FDA held up approval until 1972 due partly to fears that such a powerful drug could be abused. Their concerns turned out to be well founded. In 1980, health care professionals at a methadone clinic discovered that some heroin users seeking treatment had unwittingly taken fentanyl from China instead. Reports also surfaced of anesthesiologists and nurse anesthetists using fentanyl. Nevertheless, fentanyl slowly gained popularity as part of a mix of up to ten drugs given during anesthesia. Its fast-in, fast-out action gives anesthesiologists precise control over timing, allowing them to put patients under only for as long as needed and take them off oxygen sooner.

In the early 1980s, just as the use of fentanyl and related fentanyl analogs was taking off, a young anesthesiologist named Andrew Kofke was finishing his training with a fellowship in critical care medicine at Mass General. Kofke was asked to consult on a patient with such severe and prolonged seizures that they couldn't be stopped with standard anti-epileptic drugs. One last-ditch effort to help such people is to put the patient under general anesthesia. But as Kofke discovered, there was no evidence to suggest which agent was best. This seemed like a problem worth fixing.

The experience spurred him to dig deeper into the effects of some of the many drugs patients receive during anesthesia. He learned that under some conditions, opioids such as fentanyl could cause seizures. So he began to wonder. If opioids can cause seizures, can they also cause damage? Kofke got a lucky break when he moved to Pittsburgh, where his department's head decided his ideas were worth pursuing and funded his research. Even with the money in hand, getting started was a laborious process. Although Kofke used fentanyl every day during his surgeries, he had to submit reams of paperwork to get a new license from the DEA before heading to the hospital pharmacy to pick up the opioids for his animal research. Once over this hurdle, he was off and running, although he soon found this was the least of his problems.

Kofke's first major experiment examined the effect of alfentanil, a fentanyl analog, on rat brains. Because opioids reduce the drive to breathe, which could damage the hippocampus, Kofke kept his animals ventilated throughout the procedure. Kofke discovered that the rats on alfentanil burned through more glucose—a form of sugar that the brain uses to fuel itself—in the hippocampus compared to other regions. Electrodes picked up seizure activity there too. Six of ten rats suffered severe brain damage. The results supported his hunch that fentanyl can have a direct toxic effect on the hippocampus, at least in rodents.

But Kofke got pushback for his research—especially when he suggested that these results could be relevant to patients. A prestigious journal rejected the paper as written. At a professional conference, the editor said to Kofke,

"So you're the one who's going to get us all sued." Briefly, he wondered if this line of research was a good strategy for a young researcher like himself. But Kofke, who had moved to the University of Pennsylvania, refused to change the paper and submitted it to another journal. "We believe that this is the first experimental evidence of histologic brain damage produced by a drug used in the clinical practice of anesthesia," he wrote in the article, published in 1992.

But was his evidence relevant? Rodents and primates might respond differently to the same drug, so Kofke's next step was to examine the effect of fentanyl in three monkeys. He implanted electrodes in the hippocampus of one and detected seizures. He performed PET scans in the other two and confirmed the hippocampus was consuming more glucose—and therefore was hyperactive compared to other brain regions. In 1994 he presented a poster at an annual anesthesiology meeting describing the work. "The data obtained in rodents," he wrote, "may be relevant to humans."

When Kofke's department ran out of funding for him, he had to jockey for research grants. But he was creative and driven and kept finding a way to cobble together enough money to keep going. On rats he tested a range of doses of fentanyl comparable to what humans receive during surgery. He tested multiple types of fentanyl to see if all of them caused damage. They did. Kofke also found drugs that protected rats from fentanyl-induced brain damage. One was naloxone, used to reverse overdoses. Another was phenytoin, an eighty-year-old anti-epileptic. And the third was midazolam, a benzodiazepine that was sometimes given with fentanyl during anesthesia. "If opioid neurotoxicity is clinically relevant," he wrote, "a small change in anesthetic practice might reduce any potential neurologic morbidity."

By 2002, Kofke, now a full professor, had found a way to safely test his hypothesis in people. He wanted to know if the same high level of metabolic, glucose-burning activity in the hippocampus that he'd observed in rats and monkeys also occurred in people. Four healthy volunteers agreed to take a brief, FDA-approved dose of remifentanil, a fast-acting fentanyl analog. Kofke found that all four volunteers burned through more glucose in the hippocampus than other brain areas. At this point, it seemed more than conceivable that in some instances, opioids could be toxic to people, not just rats. "Taken altogether," Kofke wrote, "the available data do seem to consistently support the speculation that, in some very specific situations, such as prolonged sole use of large-dose opioids or with epilepsy, brain injury, or aging, there may be a potential for [opioids] to be neurotoxic in humans."

The burden of proof is higher for an unusual claim, so Kofke ran a larger experiment with fifty rats to closely examine whether the dose mattered. He found that it did. The more remifentanil the rats received, the greater the likelihood that the animals would suffer brain injury—particularly in the area in and around the hippocampus. But this answer led to other questions that had been in the back of his mind all along. Some animals were vulnerable, and others not. Kofke couldn't say why. More importantly, were some *people* more vulnerable than others? And if so, why?

Kofke circled back to people. Scientists had recently fingered the *APOE4* variant as a risk factor for Alzheimer's, a disease that damages the hippocampus. So Kofke wondered if there was a relationship between Alzheimer's and opioids. He partnered with the Duke University Medical Center and won a grant from the Anesthesia Patient Safety Foundation to fund what would become his final study investigating the connection between opioids and brain damage.

Twenty-seven healthy twenty- to thirty-year-olds volunteered to share their *APOE4* gene status and be scanned during a brief, low-dose infusion of remifentanil. Kofke would use the speed of blood flow through various brain structures as a proxy for increased activation. Subjects with the *APOE4* variants showed an unusual pattern—increased blood flow in the hippocampus and less in the surrounding regions. The rest of the volunteers had the opposite pattern. Now Kofke wondered if *APOE4* was the risk factor—or *a* risk factor—for opioid-induced injury. Perhaps with a big enough dose, the potential for brain damage was real.

In addition to presenting his evidence, Kofke described a plausible mechanism for the damage. When fentanyl binds to opioid receptors, it turns off a class of neurons called inhibitory neurons, whose job is to keep excitatory neurons from firing too frequently. Without this control, neurons become hyperactive. At high enough doses, this lack of inhibition could create a storm of seizures and so much unchecked activity that cells die. Not having enough oxygen—either in a patient who wasn't being closely monitored or in someone experiencing an overdose—would only make matters worse. In 2007, Kofke's final paper on the subject invited some future researcher to pick up the baton. "We have provided an essential presupposing element in generating a hypothesis for future studies that opioids have a potential for limbic system neurotoxicity in humans."

Kofke would not be able to follow this road map himself. He wasn't suggesting that anesthesiologists stop using fentanyl. It's safer than earlier opioids and is used day in and day out around the world without any obvious problems. But no drug is completely without risks. What Kofke wanted to find out was whether some people, under some circumstances, are vulnerable to subtle damage. And if so, is there a way to make fentanyl even safer? Returning to where he began his quest, he wrote a paper with recommendations for choosing anesthetics for people with epilepsy, pointing out that remifentanil could lead to seizure activity.

And with that, Kofke was done. He'd been pondering the question of opioids and brain damage for most of his career. He wasn't out of ideas for how to test the hypothesis. He was out of money, frustrated that no one was paying attention, and tired of banging his head against the wall. Three-ring binders full of rejected grants took up an entire shelf in his office. Kofke decided it was time to investigate other, uncontroversial ways to make anesthesia safer. The biggest job of a scientist is convincing other scientists you're right, and he'd failed at that. Someone else would have to stumble across the same question—can opioids damage the hippocampus?—and be determined to find the answer.

—⁓—

Despite the previous week's late February thaw, the temperature this evening is dropping back into the teens. Henry is asleep. Only the tick tick tick of the radiator disturbs the quiet of the house. Gillian, seven weeks pregnant, nauseous, and exhausted, is upstairs in their bedroom, answering texts from a patient. At 8:30 P.M., with a laundry basket full of clothes ready to fold on the floor by the couch, Barash dials Kofke's number. Barash has skimmed through a few of his papers and read the abstracts of the others. Kofke seems like someone who will get what he's talking about immediately. His e-mails come in bursts, short and to the point, haiku-like. This is clearly a person who likes to get things done. That said, Barash is leery of being too excited. He's had enough conversations with people who appear interested, only to drop out of sight.

Barash begins by thanking him for accommodating his schedule, but the anesthesiologist moves quickly through the niceties so he can get straight to the point. "People seem to have this close-minded attitude about opioid-induced brain damage," Kofke says. "But it can happen. I've seen it in rodents." Kofke sketches out two decades' worth of experiments he did to test his hypothesis. Barash listens intently. He soon abandons the pile of folded laundry and starts pacing between the TV room, the living room, and the kitchen. It's hard to get a word in edgewise, but finally, Barash says, "Maybe it's time for a do-over."

"It won't be easy," Kofke says. "I'd put in an animal grant, and they'd say, 'Oh, this couldn't possibly be clinically relevant.' So I'd put in a human grant, and they'd say, 'You need to do animal studies.'" Kofke is picking up steam. He says shellfish poisoning causes the same pattern of damage. He's not suggesting that Barash's patients had shellfish poisoning, but it could be a clue about the mechanism. He says that in 2002, in an attempt to resolve a hostage crisis, the Russians filled a Moscow theater with gas suspected to include a fentanyl derivative, killing a hundred or more people inside. He tried to get a hold of data about the victims' brains to see what happened to them but couldn't make a connection with Russian authorities. Kofke's voice, not quiet to begin with, is getting louder. "I worked on this for almost twenty years and then I had to put it aside. But I knew someday someone like you would come along and put two and two together. If you want to do it, I'll help because I'm still mad as hell."

Barash ends the call and glances at the old wooden pendulum clock hanging on the wall by the kitchen table. It's only 8:50. A lot has changed in twenty minutes.

"How'd your call pan out?" Gillian says as she walks into the kitchen.

"Unreal. This guy has thought it through from every possible angle. At levels I hadn't even considered. I have a ton of reading to do."

"Can he help you?"

"Yeah, he'd help. But there's no money."

Kofke has covered so much ground that Barash hasn't absorbed it all. But the older man's final feisty words stay with him. "Any time you have something a little bit crazy, it's not the discovery that's hard. It's convincing the world that you're right." Barash stoops to pick up a stray Hot Wheels that didn't make it into the car bin and heads back to the couch and the laundry.

# 12

# World of Hurt

· · · · · · · · · · · · · · ·

## *March 2017*

Several weeks later, Barash knocks on Yuval Zabar's door up on the seventh floor at Lahey. Zabar's office is sunny and spacious, with pictures, schedules, family photos, and children's artwork hanging on the walls. Zabar jumps up and shakes Barash's hand. Barash can't help noticing that Lahey has updated the doctors' white coats. The strangely old-fashioned toggle fasteners are gone, replaced by standard buttons. Gillian, who laughed mercilessly when she saw Barash's first coat at Lahey, will be disappointed.

Zabar and Barash take a few minutes to compare notes on interviews they did for an in-depth Buzzfeed News story on Max Meehan and the amnestic syndrome. Barash found the experience somewhat uncomfortable and is relieved the ordeal is over. He knows he's not the most animated public speaker and figures Zabar will be more prominently featured, which is fine by him.

"Let me show you something," Zabar says as he sits back down in front of his computer screen and motions Barash to look over his shoulder. "This paper was published right after our *MMWR* paper came out. They attribute the damage to stimulants, but it looks just like what we're seeing." Zabar clicks on a link and pulls up the MRI images that appear on page two of a paper from a neuropsychologist named Marc W. Haut at West Virginia University School of Medicine.

Barash leans over to inspect the image. "Yeah, it does," he says. "Thanks, Yuval. I'll download it and take a closer look, probably e-mail that lead author tonight."

"Keep me in the loop," says Zabar.

"You got it," says Barash, and heads for home.

—⁓—

Marc Haut lives six hundred miles south of Boston in Morgantown, West Virginia, in Appalachia. The pace is slower, the people are friendlier, and the landscape is beautiful. But the place is also, as Haut calls it, a world of hurt. West Virginia ranks first in the nation for overdose deaths, diabetes, and cigarette smoking and has the third-highest percentage of people living in poverty. Addiction is widespread, and there's not enough access to treatment.

Haut is a calm and encouraging presence in his clinic on Chestnut Ridge Road, but his work is suffused with human tragedy. He wrote a case report about a man who shot himself in the chest after being diagnosed with Alzheimer's. He studied railroad workers whose exposure to solvents damaged communication between the brain's left and right sides. He wrote another case report about a man with a rare and incurable brain disorder who fatally shot himself in the head as a result. If a patient at the clinic causes a disturbance in the waiting room, Haut is there, gently and almost invisibly defusing the situation.

He takes it as a given that everyone he works with is smart. But in his view, being nice is just as essential. His philosophy is that anyone who needs help deserves help, no matter how they got into trouble. When he hears someone say it's a waste of time and money to use naloxone to save someone from an overdose, he points out that no one thinks twice about using a defibrillator to save an obese person eating fried chicken.

This attitude makes Haut the kind of person others turn to first, which is what happened in 2015, not long before Barash came across the third and fourth cases at Lahey. One of Haut's colleagues was worried about a family member, a college-educated professional who'd been abusing drugs since high school and had many other health problems. He was a heavy smoker and had high blood pressure and chronic obstructive pulmonary disease. Every week from Thursday to Sunday, he drank. Twice a week for the previous two years, he'd used MDMA, commonly known as ecstasy. And every six hours, he took a combination of hydrocodone and acetaminophen for chronic pain.

Still, the 55-year-old man was successful enough to own an accounting firm—until a weekend he spent at a hotel using drugs. There, he overdosed, and the woman he was with resuscitated him and left. At the hospital, staff determined he'd also had a heart attack. His drug screen detected cocaine, amphetamines, and opioids. After the patient was discharged from the hospital, he checked into a residential treatment facility for drug abuse. His family soon noticed that he couldn't remember topics they'd discussed five minutes earlier. Ten weeks later, he was still severely amnesic, which is why they brought him to Morgantown to see Haut.

Haut was struck by the unusual MRI taken during the patient's hospital stay. It showed intense damage to the entire hippocampus on both sides. He'd seen a strange MRI just like it a year earlier in a patient who'd overdosed, presumably on bath salts, and Haut had been mystified by that image too. As expected, this new patient's neuropsychological exam showed how poor his episodic memory was. But since the damage extended beyond the hippocampus, to brain structures responsible for controlling movement, Haut wanted to know if his procedural memory was also impaired. If so, he'd have trouble learning new skills. In Haut's view, the standard star-drawing exercise first developed to test patient H.M. could be so frustratingly difficult that it was almost cruel. So he and his colleague, Liv Miller, used an easier experimental test they'd developed that instructs patients to do a sort of connect-the-dots task over and over. If patients get faster with practice, it means their procedural memory is still functional.

Haut was about a decade into his practice as a neuropsychologist when a landmark study gave him more hope about the potential for brains to heal. Back then, the conventional view was that children had enough brain plasticity that they could recover. But for adults, that was it. After a certain age, it was all downhill. University College London's Eleanor Maguire—the same neuroscientist who studies amnesia and imagination—wasn't so sure. Research in other animals revealed that the size of the hippocampus varies by the season and depends on how much they need to rely on their spatial memory. For example, some songbirds must cache their food and then, months later, remember how to find it. Similarly, animals with larger home ranges have larger hippocampi. Perhaps extraordinary demands on the hippocampus are responsible for this remodeling. Maguire had recently discovered that the hippocampus is more active when people are trying to navigate while playing computer games and that modern memory champions who use the ancient Memory Palace technique have even more activity in their hippocampus than ordinary people while doing memory tasks.

Maguire wanted to go a step further. In people, does the mental workout of navigating do more than simply activate the hippocampus? Under extraordinary circumstances, could it also reshape the hippocampus, just as it does in songbirds? This was a question of personal interest, because she is, as she puts it, "so desperately bad at navigating." By the late 1990s, when Maguire began to imagine this experiment, imaging technology had advanced to the point where she would be able to measure the shape and size of different brain regions very precisely. In a eureka-type moment, she realized that the perfect test subjects for a demanding spatial memory experiment drive London's streets every day. They are taxi drivers, and before being granted their licenses, they have to learn the layout of thousands of landmarks and tens of thousands of streets within a six-mile radius of Charing Cross Station. The learning process takes between two and four years, and the final exam is described as acquiring "The Knowledge." Anyone who passes has achieved what is arguably one of the modern world's great memory feats. But, just as is true for modern memory champions, these drivers' IQs are, on average, unexceptional. When Maguire pitched the concept to her advisor, he warned her that the London taxi drivers would never agree. But Maguire thought they would—and she was right.

More importantly, Maguire was right about her hypothesis: every driver had a hippocampus that was different from regular Londoners. Intriguingly, the hippocampus hadn't expanded overall. The slender end toward the back of the head was larger, while the fatter end toward the front had become smaller. The longer they'd spent driving, the more significant the difference. "Our results suggest that the 'mental map' of the city is stored in the posterior hippocampus and is accommodated by an increase in tissue volume," wrote Maguire. She couldn't tell whether the increase was due to neurogenesis or more connections between neurons. But she had shown that experience could measurably remodel the brain, even in an adult, long after it had supposedly stopped changing. Maguire's results ran so counter to dogma that two top-tier journals rejected her first paper. As Maguire recalls, an expert who reviewed both submissions wrote that he would never believe her results under any circumstances.

But the scientific community—and Haut—became believers because with every follow-up study she confirmed and expanded her results. Maguire proved that the London taxi drivers had begun their careers with a hippocampus that looked just like any average person's. In other words, it wasn't the case that people passed the test because they already had a strangely shaped hippocampus that was bigger in the back and made them better equipped from the outset. Instead, the hippocampus had changed in response to training. Maguire also found that in retired taxi drivers in their seventies, the hippocampus reverted to the normal shape, which meant that plasticity remained even in old age. Maguire is now a Fellow of the Royal Society with an entry in *Who's Who*. It lists one of her recreations as "getting lost."

Ever since Maguire's study made headlines, Haut had been collecting papers on brain plasticity and rehabilitation. His hope is always that his patients are capable of more than first meets the eye. One such patient is a woman in her sixties who had a stroke that severely damaged her hippocampus. Her injury was so devastating that at first when her son brought her in for appointments she didn't know why she was there. After she finally understood that she had amnesia, Haut enlisted her procedural memory to help her. He knew the trick would be for her son to teach her

her home's layout and her daily routine very slowly, breaking up the learning into small sections. First, she would repeatedly walk the route from the bedroom to the bathroom until she had it down. Then from the bathroom to the kitchen, from the kitchen to the living room, and so on. Eventually, she could string each piece of her daily route in the right order to get from room to room. And she could connect her daily routines—like taking medicine or making coffee—to each room. She'd take her pills in the bathroom and make coffee in the kitchen. Using Haut's strategy and her son as her teacher, the woman developed enough daily living skills to give her a small measure of independence.

But despite Haut's inherent optimism, he suspected when he met the 55-year-old overdose patient that there was little to offer, and the connect-the-dots test proved him correct. The man couldn't get any faster at the task. Haut figured that the damage was too extensive. But what surprised him was the precise nature of the damage. What drugs had this man taken that could do that? Although there were traces of opioids in the toxicology screen, Haut chalked that up to the hydrocodone prescription. From his point of view, the opioids were inconsequential compared to the cocaine and methamphetamines. Haut decided that the whole case was strange enough to write up, although it would not be until February 2017 that it appeared online and Zabar discovered it. By that time, Haut's patient had passed away.

—⁂—

At home on the couch at 10:00 P.M., recovering from what felt to him like an uncomfortable procedure—talking about his work for the BuzzFeed News story—Barash turns on the New York Knicks versus the Portland Trail Blazers game, mutes the sound, opens his laptop, and composes an e-mail.

"Dear Dr. Haut, With interest, we read your recent paper, 'Amnesia Associated with Bilateral Hippocampal and Bilateral Basal Ganglia Lesions in Anoxia with Stimulant Use.'" Barash goes on to describe the Massachusetts cluster and ends by writing, "I see that your patient tested positive for a prescribed opioid. Is there any chance your patient could have abused other opioids, such as fentanyl? Thanks, Jed Barash."

For the next forty minutes, while the Trail Blazers take a commanding lead over the New York Knicks, the e-mail exchange continues. Haut answers, "There were opioids found on screen but we thought it was from chronic use for pain and not part of the acute picture." Barash explains that fentanyl is often mixed in with other drugs. And it doesn't show up on routine opioid testing. "I understand that the opioid epidemic is a big issue in your neck of the woods as well." Haut promises to keep fentanyl in mind if he sees another case.

A few days later, Barash follows up with a phone call on his way home from work. Haut still thinks stimulants are the real culprit but tells himself, *Marc, don't be defensive. Let's figure this thing out.* The longer he talks with Barash, the more it becomes clear to him that Barash isn't a typical neurologist. He understands memory systems and public health in a way that most neurologists don't. Haut is equally impressed with Barash's training in Boston, a mecca for behavioral neurology. And just as importantly, Barash is nice even as he presses the point that fentanyl doesn't show up on drug screens. Unless Haut's patient knew for sure what he was taking and told the hospital staff, no one would be the wiser.

"I suppose it could have been fentanyl in there, and we just didn't know it." Haut raps his knuckles on the desk in his home office for emphasis. "And we didn't know that we didn't know it."

"What do you think of the odds that you'll see this again?" says Barash.

"I won't be at all surprised to get a call from out there in the wild about another case."

A few days after their talk, Haut e-mails Barash again to tell him he's going to ask for permission to scour hospital records for previous patients who overdosed and got an MRI. Maybe they've missed similar cases. Haut says he'd like Barash's input on how best to go about it and closes by writing, "Thank you again for reaching out, as we have learned some new ways to think already."

# 13

# Finding Fentanyl

. . . . . . . . . . . . . . . . .

## *April 2017*

Butler's phone buzzes with another text from Barash. "Hope you didn't get egged off the stage." It's his way of asking how Butler's recent presentation on the amnestic syndrome went. At the moment, Butler is relaxing in the rental he snagged in a suburb across the bay from San Francisco. The commute by public transportation to the Memory and Aging Center takes an hour, but it's worth it. Butler wants his daughter, now a teenager, in a strong school system. The first year of his fellowship is intense, with high expectations and loads of paperwork. Patients from across the country come to participate in research. Some have unusual disorders such as fatal familial insomnia, Creutzfeldt-Jakob disease, or frontotemporal dementia. Meeting patients like these fuels Butler's desire to understand the mechanisms behind such devastating conditions. Despite the clear-cut academic hierarchy, the center has a collaborative culture of sharing information at weekly meetings for fellows and faculty.

During his presentation, Butler made an unforced error by interchangeably using the words *anoxia* and *hypoxia*. *Anoxia* means the brain isn't getting any oxygen, and *hypoxia* means it's not getting enough. This mistake led the audience to focus more on terminology than the big picture. He was frustrated with himself, but other than that, he tells Barash that the audience "really dug the talk." No one had seen the imaging before. "A few thought it has been happening all along . . . but nobody was looking and collecting the cases." After the weekly meeting, a few people took Butler aside to tell him the syndrome was intriguing. Butler hopes they'll be intrigued enough to let him trawl through medical records in search of overlooked cases, just as he did at Lahey. Epidemiology is not Butler's thing, and the job will be a slog. But finding new cases is a means to an end—if it pans out, it could justify tapping into someone's research grant to tease apart the mechanism.

Even as Butler maps out what he pictures as his foray into the hippocampal forest, the effort to determine if fentanyl is the memory thief has mostly taken a backseat to Barash's family obligations. Gillian's mother, Jane, diagnosed with cancer in the fall, is now in hospice at their home. Barash is on call to field medical questions, pick up supplies she needs, or run out to get food that might appeal to her. After Jane passes away, surrounded by her family, Barash will turn his attention back to the investigation.

The circumstantial evidence—from Kofke's experiments to the appearance of some twenty confirmed or suspected cases in states hardest hit by the fentanyl epidemic—is accumulating, but this thief has left no hard evidence behind. What's more, the number of cases has always seemed too small compared to how many people have overdosed on fentanyl. But the more Barash thinks about it, the more plausible it appears that there's a good explanation for the rarity of reported cases. At least half a dozen hurdles would need to be cleared before the Massachusetts Department of Health heard about a suspected patient.

"First," Barash writes to DeMaria and the radiologist, Michael Lev, "the overdose has to be sublethal in such a way that the patient can actually recover; fentanyl is obviously often lethal." Simply put, many overdose victims die and never receive an MRI. Those who survive often don't want to go to the hospital or have no one to take them. Another hurdle is just the sheer chaos of an emergency department. Busy staff have to prioritize stabilizing and releasing patients unless it's obvious that they need to be admitted or get more tests. "Up or out," the saying goes, or "treat 'em and street 'em." In other words, someone in the emergency department would have to notice the amnesia, realize it goes beyond post-overdose confusion, and investigate further. Another hurdle is convincing the patient to stay—hardly a foregone conclusion, since overdose survivors often sign themselves out even if their doctors recommend against it.

A final hurdle that makes clearing all the others irrelevant is that hospital staff rarely send samples to a national lab to screen for fentanyl. It's expensive, can take a week or more to get results back, and doesn't change how they treat the patient. That means evidence that a patient took fentanyl along with, say, cocaine and MDMA will be missing. None of the amnestic syndrome patients to date were screened for fentanyl. Even if they had been, the very characteristic of the drug that makes it so valuable for anesthesiologists—its fast into-the-brain, fast out-of-the-brain action—makes it hard to detect unless the test is performed quickly. Within seventy-two hours, most of the drug and even its breakdown products have cleared the body. Exactly how long that process takes depends on a person's age, weight, metabolism, overall health, and, of course, how much they used. But time is always of the essence.

Briefly, Barash holds out hope that Boston Medical Center might be the one place in the surrounding area that routinely screens for fentanyl. But as he e-mails back and forth with their staff, he learns that such advanced testing is limited to patients in the Substance Use Disorders clinic, not the emergency department. Time is also of the essence when it comes to capturing relevant brain images. A CAT scan is unlikely to show this type of damage clearly, and on an MRI scan, signs of the immediate aftermath of the injury fade with every passing day. Another potential hurdle is the radiologist's expertise or careful analysis; he or she may ignore the results because the damage is so well defined that it looks like a mistake.

Over at Mass General, radiologist Michael Lev is still on the lookout for cases. The puzzle has bugged him for at least a decade before the DPH public health alert. And for him, the question was never about amnesia; it was about what the MRI looked like. The cases he's seen were victims of cardiac arrest who had a history of opioid use. They had gone into a coma and never regained consciousness so whether or not they had memory loss before they died was unknown. Since victims like this won't meet the

criteria for amnesia that Barash, Somerville, and DeMaria laid out, the
DPH will never hear about them. But a different state agency may.

The Office of the Chief Medical Examiner (OCME) at 720 Albany Street
sits at the heart of a concentrated area of suffering and despair. The Boston
Medical Center's emergency department is across the street. One block to the
southeast lies the Suffolk County House of Correction. People inject drugs
outside and wander obliviously into traffic. Discarded needles litter the
area, sometimes cynically referred to as Methadone Mile. Boston's mayor
has recently allocated a million dollars a year for a shelter for the home-
less and people with substance use disorders. The OCME's building has
windows running the length of it, covered by shades. Out back is a small
parking lot where ambulances bring the bodies.

The OCME's mandate is to investigate "the cause and manner of death
for deaths that occur under violent, suspicious or unexplained circumstances."
It's never been an easy place to work, but the opioid crisis has made it harder.
Chief Medical Examiner Henry Nields was brought on a decade earlier to
fix an office under fire for releasing autopsy reports well after the recommended guideline of ninety days. Some took as long as three years, due, as
Nields explained to state legislators, to understaffed and overworked forensic
examiners, who performed an average of well over four hundred cases a year.
The recommended number is two hundred and fifty.

On Easter Sunday at 7:01 A.M., the beleaguered Nields receives an e-mail
from Barash, with a copy to Nick Somerville. Barash has spotted a convenient excuse to reach out. "Dear Dr. Nields," the chief medical examiner
reads. Two days earlier, the *MMWR* published the results of Nields's and
Somerville's investigation into opioid overdoses. They'd analyzed autopsy
records for six months between 2014 and 2015 and discovered that the
fraction of deaths attributable to fentanyl had increased from less than
half to more than three quarters over this short time. Their report also
confirmed that drug users were often unaware that they'd taken fentanyl,
underscoring hospitals' need to test for it.

Barash hopes that Nields's interest in fentanyl overdoses provides a
natural opening to work together. "We are interested in a potential project
evaluating neuropathology in fentanyl, heroin, and related overdose

cases. . . . Do you think this kind of project might be feasible?" The next afternoon, Nields writes back. "I want to run this by our general counsel before agreeing to anything. If she thinks something could be worked out, perhaps we could schedule a meeting to go over this in greater detail."

In parallel with Barash's efforts with the OCME, DeMaria has been working on a new plan to gather more evidence. The February 2016 public health alert regarding the potential cluster was a onetime request, and he suspects that it's long since been forgotten. Now that the cluster has been confirmed, that tool is no longer available to him. But a second method for gathering information on an emergency basis is to designate an illness or a condition as a "reportable syndrome" for one year. Some conditions, like AIDS, cancer, sexually transmitted diseases, or tuberculosis, are permanently reportable to the DPH. Physicians who don't comply risk being fined. Medical records must be sent by secure fax or e-mail, and printouts are kept locked inside DeMaria's office.

DeMaria is judicious about going through the arduous process of making a disease reportable, and he doesn't want to overwhelm doctors with notifications. Despite his discretion, DeMaria is still famous around the office for making conditions reportable, just as he did with Creutzfeldt-Jakob disease in 2003. In his view, you never know when something unexpected, like mad cow disease, will happen again. The issue with the amnestic syndrome is similar. Without tracking it, there's no way to know how common it is. DeMaria decides to make the syndrome reportable for one year to get a better handle on how many are affected and whether fentanyl is to blame. Now that a case has been identified in West Virginia, it seems increasingly less likely that a contaminant is involved, but they leave open the possibility.

On May 2, at 10:26 A.M., an e-mail entitled "Important message from the MA Department of Public Health" goes out asking physicians to report new cases of the amnestic syndrome and to "consider advanced laboratory testing for the full range of individual synthetic opioids (e.g., fentanyl) and their analogs, as well as for extraneous toxic substances not assessed in these reported cases, as such testing could be a critical component in clarifying an association with substance use." Six minutes later, the first response comes in, from the head of neurology at Brigham and Women's Hospital:

"We have seen some of these." Then, half an hour later: "When I was a hospitalist at Lawrence General Hospital, I saw a patient with a similar presentation. . . . The discharge diagnosis was like 'amnesia,' 'TIA with amnesia,' or something like that." A few hours later: "My brother was an alcoholic and drug abuser for over 35 years. . . . For the last 18 months or so before his death, he exhibited significant memory loss. . . ." Early that evening, from a neurologist at Mass General: "I have such a case—and I believe we salvaged her memory by using high dose antioxidants at the time of presentation. I initiated this therapy for theoretical reasons. . . ." The next morning from a physician assistant in the Mass General emergency department: "I had a patient several months ago. . . . MRI showed bilateral hippocampus necrosis, neurology admitted her."

In answer to a psychiatrist who treats people with substance use disorders and wonders what the presumed mechanism for the damage is, Barash replies, "We have a lot more work to do." In response to an emergency department physician in Manchester, New Hampshire, inquiring about treatment options, Barash writes, ". . . our hope is that we might be able to identify a treatment if we're able to identify the culprit." But it's too late to test any of these patients.

Then, on the afternoon of May 3, just one day after sending out the announcement, DeMaria receives what sounds like the e-mail they've been waiting for, from a doctor at Beth Israel Deaconess Medical Center. "Hi Alfred, My name is Vinod Raman. . . . I have a patient with a history of substance abuse . . . with amnesia as he is unable to recall events leading up to the hospitalization. We are obtaining an MRI, expanded tox screen, and neurology consultation today." For the next few days, e-mails fly at all hours between Barash and half a dozen or more hospital staff in the emergency department, toxicology, pathology, and neurology, as they work out the details of how best to do the testing before it's too late to get meaningful results—or the patient decides to leave.

"Vinod," Barash writes, "not sure if they got the MRI yet, but if not, could a resident try to prioritize with Radiology? My concern is that the signal, if present, fades with each passing day." A few hours later, regarding the toxicology screens, Barash e-mails again: "Might want to move more quickly on that." Some of the Beth Israel doctors appear baffled by the urgency and Barash's perseverance. "He's the neurologist working with the DPH," an emergency department doctor tells the pathologist, "(and maybe a little too invested in the cases)."

But a week and a half later, Barash's nagging pays off. This is the first time anyone has screened for fentanyl in an amnesic patient—and it's positive. Radiology confirms that imaging is a match. The Beth Israel doctors now grasp the importance of looking more closely at overdose patients in the future. Unfortunately, this patient also used benzodiazepines, amphetamines, and cocaine. To be confident that fentanyl is the sole culprit, they need to find an amnesic patient who used fentanyl and nothing else.

On May 11, Barash parks near the old, low-slung brick building at the OCME and walks through the double doors into the lobby. Close to a thousand opioid overdose victims have been autopsied in this building since 2012. If any of them have toxicology screens positive for fentanyl and their brains show the characteristic pattern of damage, the weight of evidence pointing to fentanyl will be much stronger. What's more, a formal pathological review of brain tissue slides might reveal something under the microscope that can't be seen in survivors—what the damage at the cellular level looks like.

Barash joins Chief Medical Examiner Nields and the OCME's legal counsel at a table in a large conference room upstairs. They're trying to figure out if the OCME can reasonably and legally help. Nields has experience going back decades. As a medical examiner in New York during 9/11, he helped identify victims' remains. In 2015, he testified in the Boston Marathon bombing cases, and recently he's been credited with bringing down the backlog of bodies awaiting autopsy. Now his ginger hair has gone mostly gray, and he wears a beard and a mustache. Round wire-frame glasses perch at the end of his nose.

Nields reiterates that he found the *MMWR* paper interesting. Still, medical examiners are restricted in what material they can collect in an autopsy and what they can do with it. If someone is suspected of having died from a fentanyl overdose and the toxicology screen comes back positive, there's little need to investigate further. Just in case new information might surface later, the medical examiner may still decide to remove and store brain tissue samples. But to retrieve those samples a year or more afterward to research the amnestic syndrome would be outside standard operating procedures. It would also mean asking families for permission, which Nields and Barash agree is unwarranted.

As they talk, they hit upon a different solution. Instead of starting with overdose victims, they'll look for cases where a suspected cause of death is blunt force head trauma. Such a scenario will prompt the medical examiner to take brain samples and conduct a formal pathological review that will show the injury in detail. If the medical examiner also has reason to believe the head trauma might have been caused by an overdose victim hitting their head when they passed out, the examiner will likely request a toxicology screen. Nields doesn't know how many cases there are with both overdose and suspected head trauma, but before the meeting wraps up, he agrees to look for cases on file dating back to the beginning of 2012. He tells Barash he'll be back in touch within a few weeks.

That night, Barash waits until Henry is in bed to tell Gillian about his meeting. "Being the chief medical examiner is a terrible job," Barash tells Gillian as they empty the dishwasher. "No one ever notices when something goes right. It can only go wrong." He takes off his glasses, rubs his eyes, puts them back on, and adds, "If this pans out, it'll be a minor miracle."

—m—

"You really need to see this patient," a neurologist at West Virginia Hospital in Morgantown tells Marc Haut. Haut suspects he doesn't have time to examine the thirty-year-old man, who was just transferred from a hospital in Maryland's panhandle. The patient, who overdosed on what he thought was heroin, has amnesia and a strangely damaged hippocampus. *Oh wow, I know what this is,* Haut thinks. He considers his to-do list for another minute, accepts the fact that he can't examine the patient, walks across the hall, and sticks his head into neuropsychologist Liv Miller's office. Haut trained Miller a decade ago and hired her in 2015. "Liv, can you please go see this patient? His name is Christopher."

Miller performs bedside testing and reports back that Christopher has severe amnesia. To make matters worse, according to their connect-the-dots test, his procedural memory has also taken a hit, probably because the injury included a region near the hippocampus that helps coordinate movement.

Haut is busy but not too busy to send Barash an e-mail.

From: Haut, Marc
Sent: Tuesday, May 23, 2017 5:52 PM
To: Barash, Jed
Subject: Re: Complete Bilateral Hippocampal Ischemia
Importance: High

Jed, FYI

We got a new case with bilateral hippocampi and basal ganglia lesions on DWI after OD and found down, tested positive for cocaine at outside hospital 4 days ago; just transferred here, clean urine here trying to find out if they still have the urine at the outside hospital to test for fentanyl etc; . . . . . . he says he usually used heroin but memory bad for event and as you point out who knows what it is cut with. Anything specific we should look for if they have it?

Barash checks his e-mail after dinner and immediately calls Haut. Finding a second patient who took fentanyl adds more weight to their hypothesis. But they'll have to act quickly.

"It's probably too late to do a direct test for fentanyl, but they could look for by-products," Barash tells Haut, skipping the formalities. "Same thing with the urine sample from Maryland. Maybe it's not too late to find fentanyl metabolites."

"I'll keep you posted," says Haut before hanging up.

At midnight, Barash updates DeMaria and Somerville. "Given that cocaine has been involved in relatively few of our cases, I think that it may potentiate the effect of a synthetic opioid like fentanyl, but I don't think cocaine alone could parsimoniously explain the cluster."

The next morning, in the shower, Liv Miller ponders the case. She's seen at least one much older patient who lost episodic memory from a stroke. But Christopher is so young. As long as he doesn't die of an overdose, he still has a whole lifetime ahead of him. But what kind of life? She's often wondered who *she* would be without her memories. On a hunch, Miller decides to check in on Christopher again as soon as she gets to the hospital. She's curious about his procedural memory, even though she's not sure why.

When Miller says good morning to Christopher, he politely introduces himself as if they'd never met. Clearly, his amnesia is as profound as it was the night before. But now, when Miller gives him the connect-the-dots test, he's faster than he was twelve hours earlier. Like patient H.M., Christopher doesn't remember taking the test, but some part of his brain remembers the procedure. Perhaps the damage to the area near the hippocampus that coordinates movement was less severe and is already on the mend. Miller hopes she can leverage his procedural memory to help him regain some independence. She knows Christopher's insurance is unlikely to cover cognitive rehabilitation, but she's also sure that she and Haut can find a way to help without charging him.

A week later, Christopher's advanced fentanyl drug screen comes in:

> From: Haut, Marc
> Sent: Wednesday, May 31, 2017 9:25 AM
> To: Barash, Jed
> Subject: Re: Complete Bilateral Hippocampal Ischemia
>
> Norfentanyl positive here.

> From: Barash, Jed
> Date: Wednesday, May 31, 2017 at 9:47 AM
> To: Marc Haut
> Subject: RE: Complete Bilateral Hippocampal Ischemia
>
> Bingo! Norfentanyl is the metabolite.

Barash has a nagging feeling he'll never see the autopsy results. Henry Nields has announced he'll be stepping down as chief medical examiner. Barash was lucky to get the man's attention in the first place, and he figures Nields was only interested enough to look into it because he'd analyzed fentanyl overdoses. Odds seem slim that the incoming chief medical examiner Mindy Hull will want anyone from her overwhelmed staff to deal with an obscure project that appears relevant to only a few people.

The text exchange continues:

Butler: What exactly are you asking of the pathologists?

Barash: I wanted to get reports on hippocampus and other brain structures in patients who died of overdose. But they are too swamped so they don't do neuropath on those cases unless there is some particular reason to do so. They do sample the brain tissue in those instances but don't turn it into slides.

Butler: Oh I see. Got it.

The waiting continues through July and the first half of August until, finally, a new overdose victim with amnesia appears in the emergency department at Mass General. Barash figures he's badgered Butler enough for the time being. Before getting back in touch he'll wait for the records to arrive in DeMaria's office so he can confirm it's a real case.

The first year of Butler's fellowship, crammed with paperwork for research and clinic patients, has come to an end. From now on he'll only see two or three patients a week, leaving the rest of his time to devote to research. Given his mentors' advice to choose a straight career path, research on dopamine and neurodegenerative disorders appears to be the logical way to go. But even without the texts popping up on his phone from Barash every few months, Butler finds it difficult to stop musing about the curious connection between opioids and hippocampal damage.

While he waits for the Institutional Review Board (IRB) to approve his search through the patient records, people with eerily similar symptoms keep finding him. A teenager comes to the clinic complaining of frequent transient global amnesia spells (TGA), and Butler winds up being the one to see him. Usually, this TGA lasts only a few hours and often happens after head trauma, acute emotional distress, or surgery. This young man, who had several back surgeries and used opioids long term to manage the pain, experienced TGA episodes ten or so times over the past year or two. A woman who took opioids for back pain for several years tells Butler she felt foggy until she stopped using them. For whatever reason—Butler doesn't ask her to—she sends him her 23andMe genetic testing, which shows that she is an *APOE4* carrier. A woman from the Midwest contacts him after reading the *Neurocase* paper about the first four patients at Lahey. Her husband used fentanyl patches and other opioids for ten years, initially prescribed for chronic pain, but ultimately escalating to abuse. As time went on, his memory worsened. Now only in his midforties, he's been diagnosed with early-onset Alzheimer's disease.

As a scientist, Butler doesn't believe in coincidences. But it's also weird. The odds are against his being the only person at the MAC to see or hear about such patients. Maybe what's really going on is that he's primed to pick up on the clues. Or maybe the parallels he's seeing with the amnestic syndrome aren't real, and he's only seeing them because of the lens he's looking through.

Amidst all these tantalizing clues one riddle stands out: if opioids damage the hippocampus, why don't all opioid users develop memory loss? It turns out that this question fits squarely into the research of one of Butler's advisors, neurologist and neuroscientist Bill Seeley. Seeley studies neurodegenerative diseases by looking for mechanisms that could explain why, for any given disease, only certain classes of neurons and brain regions are injured. Researchers hope that answering this question could point the way to more effective treatments. The concept is called selective vulnerability, and the amnestic syndrome seems similar. One brain region, the hippocampus, is selectively vulnerable to damage from opioids. At least in some people.

This time around it's Butler who initiates a text exchange from his home across the bay from San Francisco. It's a Saturday night, and Barash, Gillian, and Henry are on a weekend away in New Hampshire.

8/19/17, 6:25PM

Butler: Still waiting on my IRB approval.

In the meantime, heard about interesting case . . . thinking to invite them to UCSF and to early onset AD program.

Butler goes on to describe the advanced genetic testing and cerebrospinal fluid analysis he would perform if they could bring this man to UCSF or find a case in California. "It's an interesting thought," Barash texts back, "that there could be acute and chronic layers to fentanyl use." Before the conversation wraps up, Butler texts, out of nowhere, "Did I tell you I am going to test Nietzsche's DNA for CADASIL?" "What?" Barash asks. Butler explains he's made an agreement with German archives to analyze the saliva left behind when Nietzsche licked the stamps on a cache of his so-called madness letters. Butler's hypothesis is that Nietzsche, long thought to have died from complications of syphilis, could have succumbed instead to CADASIL, a rare genetic condition that can cause dementia and would account for many of Nietzsche's symptoms. "Butler is like a submarine," Barash tells Gillian. "He just kind of surfaces every now and then, with like four new grants and some new ideas."

Although Butler has multiple projects cued up, the amnestic syndrome occupies most of his attention. A few days later, well after midnight, he sends Barash a thousand-word e-mail that lays out his thinking about the potential for damage in people who use opioids long term. It's full of facts and figures and references to articles that support his hypothesis, but the basic point is this: perhaps there are different levels of brain damage depending on the type of opioid used and for how long. At the severe end are people who use fentanyl and, for some as yet unknown confluence of events, develop profound amnesia or even die. At the other end are people who take less potent prescription opioids for many years. But if he's correct, whether the damage is severe or subtle, it doesn't happen to everyone. Could Kofke's studies with *APOE4* carriers be the explanation? The possibilities are so intriguing that Butler decides to try to enlist some of the younger crowd at the MAC, hoping they'll be just as curious as he is. *I'm glad he roped me in*, Butler thinks. *This foray into the hippocampal forest could keep me going for a while.*

In the first week of September, Butler gets himself invited to a monthly meeting for research coordinators. Some two hundred coordinators serve as liaisons for patients participating in research and tend to be young people in the early stages of their careers. The meeting is their chance to catch up on technology, such as the latest MRI protocols, or to hear from neurology fellows like Butler, who want to share information about their projects. Usually, fifty or sixty people show up. This time the crowd is larger, and there isn't enough space around the large conference room table for everyone. One young woman, Devyn Cotter, grabs a chair in the far corner. Cotter coordinates patients for the Healthy Aging Study and has a master's in neuroscience. Another research coordinator in the audience, Suzy Kwok, coordinates test subjects who have frontotemporal dementia. This type of dementia is much rarer than Alzheimer's. It typically strikes people between their late forties and early sixties and can cause distressing personality changes. Kwok hopes to go to medical school and is considering becoming a neurologist.

Butler begins his presentation with a personal story, an accident he had during his medical internship. On a snowy January evening, after a thirty-hour shift at St. Elizabeth's hospital in Boston, he got into his beat-up rear-wheel-drive. The next thing he knew, he was lying in excruciating pain in a hospital bed, pushing the button on a device that delivered a powerful opioid on demand. His car, which had spun out of control, had been so badly crushed that emergency workers had had to use the jaws of life to pull him out. The damage to his lower leg was so severe that it would require multiple reconstructive surgeries.

"The worst pain I've ever felt in my life was the twenty-four hours after surgery," he tells the research coordinators. "I never wanted to use opioids, but I absolutely needed whatever they would give me." By now, Butler has Suzy Kwok's undivided attention. The opioid crisis has hit a city close to her hometown particularly hard. Through Facebook, she's been hearing about classmates who've become addicted.

"But then a month passes," Butler continues. "I'm in a wheelchair making my recovery, and I call my doctor and ask for some more pain medication. And he was good. He knew what was going on. And this is like in my mental basement. I heard myself say something that sounded so bad. 'No, this is the only thing that works for me.' Classic drug-seeking line. And I just froze, like holy cow. Listen to me."

Six years later, you couldn't tell that Butler survived such a traumatic accident, and he never asked for opioids again. But he got enough of a taste to know how powerful they can be. The toll from the opioid crisis is clear, he tells the coordinators, but now he's beginning to wonder if there's an insidious consequence to widespread opioid use that has gone unnoticed.

"Can these opioids—are they changing the neurocircuitry, leading to memory problems down the line? Maybe it's not just drug-seekers. What about older adults, like some of the patients we see at the MAC, who are taking opioids for years just as prescribed, sometimes high doses? Is that chronic use going to change their cognitive trajectory for the worse? Especially in people already susceptible to neurodegeneration?" Butler checks his phone, sees he's running out of time, and gives a final plug for the project that's taking shape in his mind. He's reverse-engineered an appropriate title, Project Seahorse, an acronym for Susceptibility to Early Atrophy from Hippocampal Opiate Related Substance Effects. "If anybody who has some time is interested, just reach out separately, and we can talk more about it."

*That's a clever name for the project*, Cotter thinks. *And super interesting.* She previously worked in a lab in London with a scientist who used zebrafish as a model to study both Alzheimer's and addiction. Butler's Project Seahorse combines two topics she cares about. After the meeting, she tells him she's on board. Cotter is friendly, petite, and wide-eyed, with a sunny disposition. Kwok follows up with Butler as well to tell him she feels sad for these people, and she's game to help.

Butler doesn't want to waste anyone's time, his own included. They all have day jobs and will have to figure out how to add this research to their official responsibilities. So, he decides to take a quick peek at the data himself to see if the project has any legs. From his cubicle overlooking the parking lot, he can tap into a brain bank of close to two thousand older adults. Some are healthy, some have mild memory impairments, and some have Alzheimer's. The database, run by the Alzheimer's Disease Neuroimaging Initiative, or ADNI, contains cognitive testing, genetic information, brain imaging, and more. Some of the data goes back to ADNI's founding in 2004. Clinicians and researchers from around the country regularly follow and test these participants. Scientists and companies around the world can use the information for free. ADNI is designed to help researchers understand how Alzheimers progresses and to make clinical trials more effective. But in theory, the database also includes the information Butler will need to tease out whether or not chronic opioid use hurts memory.

On his first expedition into the database, Butler looks at the records of a few dozen elderly people who were prescribed opioids chronically. He examines how long they used the drugs, whether or not they had the *APOE4* gene associated with Alzheimer's, and how much each person's hippocampus had shrunk over time. Even in healthy people, the hippocampus begins to shrink very slowly from middle age onward, but the rate of decline increases with age and is significantly faster in people with Alzheimer's disease. Butler's initial calculations suggest that people taking opioids chronically have more volume loss than those who don't. And it appears that people with the major genetic risk factor lose even more volume, an indication that aligns with Kofke's research on fentanyl and *APOE4* carriers. Butler takes just a quick pass—not nearly enough to draw any conclusions, but finds a hint that there could be something there. "I've sunk enough time into it to see if I should sink a ton of time into it," he tells Barash, "and my conclusion is yes, I should do this."

---

The people who work with Cotter on the Healthy Aging Study tend to be an optimistic bunch. They figure that if growing old is the single greatest risk factor for Alzheimer's disease, there must be something valuable to learn about how we age—especially from people who are doing well despite their years on earth. "With aging comes a whole host of pathologies. And that's as true for your brain as it is for your knees," says Joel Kramer, one of the directors of the Healthy Aging Study. "For me, then, the scientific question is, what's this resilience? What's this magic soup?"

On a graph, the average person's cognitive trajectory climbs rapidly from birth until about age thirty, when it levels off and then begins a subtle, gradual decline. Over time, many people think more slowly, misplace things more frequently, have a dwindling attention span, have trouble doing two things at once or finding the right words. If these symptoms are worse than expected for a person's age, it's called mild cognitive impairment, or MCI. If the weakness revolves around memory for recent events or finding your way, it's called amnestic MCI (aMCI), often the earliest stage of the symptoms of Alzheimer's. Only when someone's memory and thinking is so poor that they have trouble living independently do they meet the criteria for a diagnosis of Alzheimer's dementia.

So what is it that makes some people resilient enough to nudge the arc of their cognitive trajectory up above the normal curve? Mounting evidence shows that exercise is one of the most effective ways to change how your brain ages. It reduces inflammation, improves how your body metabolizes sugar, increases blood flow through the brain—all phenomena thought to play a role in how the disease progresses. It's much easier to study this effect in rodents, and it's clear that when rodents run on treadmills more than their peers, they grow more neurons, more connections between neurons, and their hippocampus gets bigger too. Participants in the Healthy Aging Study at the MAC and at research programs around the world are helping scientists dig into these details to find out if the results in mice apply to people.

"What we're trying to understand are the mechanisms and the prescription," says Michael Yassa, director of the Center for the Neurobiology of Learning and Memory at UC Irvine. "What kinds of exercise? How long?" Yassa's research with young adults shows they score higher on memory tests after short, ten-minute bouts of low-intensity exercise. MRI scans taken after these exercises reveal that the hippocampus processes information differently and connects more effectively with other brain regions too. "You don't want to tell an older adult to exercise an hour a day with heavy weights," says Yassa. And maybe you don't have to. "This is very, very light exercise, akin to yoga or tai chi." Yassa's next step is to try this same kind of short, low-intensity exercise routine three times a day with older adults for six months to test the long-term effects.

Ozioma Okonkwo at the University of Wisconsin School of Medicine and Public Health is also looking for a prescription for better brain aging to stave off Alzheimer's. Okonkwo's previous research showed that people who reported exercising more had larger hippocampi, less amyloid beta, better brain metabolism, and better memory than inactive people. But Okonkwo wanted something more rigorous. He decided to single out one measurable aspect of exercise—cardiorespiratory fitness—to find out if it helped late-middle-aged people who were cognitively normal but whose family history put them at risk for developing Alzheimer's. For his pilot study, Okonkwo's team designed a prescription of fifty minutes of aerobic exercise three days a week. The group that followed Okonkwo's prescription had more youthful brain metabolism than people who didn't. In other words, exercise modified a supposedly non-modifiable risk factor for Alzheimer's—aging. In a future trial, he wants to refine the prescription to determine whether there are minimum and maximum amounts of exercise needed to change the likelihood of developing Alzheimer's, or at least altering its trajectory.

Research by Yassa, Okonkwo, Kramer, and many others shows exercise can improve brain health and memory, even in people with known risk factors, and they're working hard to define what type and how much. If they can also figure out how exercise makes people's brains resilient and reduces the risk of Alzheimer's, that could lead to new ideas for treatment—including, perhaps, a treatment that works for people who, for whatever reason, can't exercise.

Joel Kramer's Healthy Aging Group has identified another group of people who fall above the curve of what's cognitively normal—and it's not because they have resilient brains. In fact, occasionally their brains can be filled with amyloid beta, a sign of Alzheimer's disease. Kramer argues that this mismatch between signs of disease and symptoms shouldn't be too surprising. "You can measure arthritis in your knee. You can measure the amyloid and tau in someone's brain," Kramer says. "A separate question is, do you have any symptoms?" A woman from Colombia with a genetic mutation that should have given her Alzheimer's at age fifty only developed symptoms at age seventy-two, despite by then having a brain full of amyloid plaques.

Another patient followed for years at Johns Hopkins University's Brain Resource Center had both plaques and tangles when he died at the age of ninety, but he appeared cognitively normal and was playing chess and doing crosswords until just a few months before his death.

A concept called cognitive reserve helps explain why some older adults outperform their brains. By sifting through the information available in databases that track adults over many years as they age, it seems increasingly less likely to be the case that education, a high IQ, or ongoing cognitive activity prevents brain damage from Alzheimer's, as researchers used to suspect. Instead, it builds up a kind of bank account you can draw on in order to make up for hidden losses. When the bank account is empty, someone who was on a healthy aging curve can decline very rapidly. Women, who tend to have better verbal skills than men and can compensate more effectively, may be older when they finally get the diagnosis—and then fall off the cliff faster. As with physical activity, it's hard to design a study that will result in a prescription for exactly what kind of mental activity is effective. But anything that puts the brain to work—learning a new language or a musical instrument, socializing, reading, crossword puzzles—likely helps.

Researchers estimate that a third or more of cases of Alzheimer's are due to lifestyle factors—such as diet, or physical and cognitive activity—that are largely within our control. If there were a pill that could prevent or delay dementia by that much, it would be the blockbuster drug of the century. Some programs aim to tailor the treatment based on an individual's genetic risk factors, medical history, and current state of health, although this intensive and personalized approach is out of reach for most people. But even in the absence of an individualized prescription, the evidence is clear that healthy lifestyles and mental engagement can make a difference. And the earlier in life these types of interventions are implemented—even in young adults—the better. Perhaps as many as forty-six million people in the United States who have no symptoms are on their way to developing Alzheimer's.

Among the healthy agers Joel Kramer studies are super-agers, a small, elite group of people who have both brain resilience and cognitive reserve. Their memory can be on par with someone thirty or more years younger. Some parts of their brains look like those of young adults, seemingly immune to the normal atrophy that occurs as we age. The study of super-agers and exactly what they do to have such youthful brains is in its infancy. It may be that they work a little harder than the rest of us at physical and mental activity. But for Kramer, one quality stands out. People who age well tend to be optimistic. "They talk about a sense of gratitude, a sense of appreciation for what they have, and not getting frustrated by what they don't have." Of course, the type of person who volunteers to come to the MAC to participate in research may be more likely to have that attitude to begin with. But, Kramer continues, "I really like the idea of optimism as something that's brain-protective."

—⁂—

Barash gets the news on a Thursday in late January—the day of the week when the *NEJM* drops into the mailbox by his front door. But he's at the beach, far from home, on a family vacation with his parents at a professional conference for anesthesiologists. Barash can smell chlorine and sunscreen as he, Gillian, and the two boys head toward the water park. With three-month-old baby Jack strapped into a stroller, Barash takes a minute before chaos ensues to peek at his e-mail, where he sees the notification he's been hoping for:

Subject: New England Journal of Medicine 17-16355.R2

Dear Dr. Barash,
    I am pleased to inform you that your Letter to the Editor entitled, "Acute Amnestic Syndrome Associated with Fentanyl Overdose," has been accepted for publication.

Barash pictures his mother opening the mailbox in front of his child-hood home and pulling out the iconic red and white journal. A few months earlier, doctors at Mass General identified the last of four patients who tested positive for fentanyl. The first was the Beth Israel Deaconess Medical Center patient who took multiple drugs. Another overdosed on fentanyl and was given morphine in the hospital, so they can't rule out the possibility that morphine contributed to the damage. But two overdosed on fentanyl alone. Years earlier, after seeing the first two patients, Max and Anthony, Barash dreamed about publishing a case report in the *NEJM*. This time, with many more cases and expanded drug testing, Barash, DeMaria, and the seven other physicians who helped identify the cases figured they had the goods to aim high. Having their correspondence article published in the *NEJM*, Barash argued in their submission to the journal's editor, would spur the identification of more cases and, in turn, "a better understanding of this emerging condition."

With the shrieks of excited children ringing in the background, Barash shares the news with his parents. "You know what Yogi Berra said," Barash says to his father as he wriggles Henry's arms into a pair of water wings. "You don't have to swing hard to hit a home run. If you got the timing, it'll go."

# PART FOUR

. . . . . . . . . . . . . . . . . . .

# INTERVENTIONS

# Some Part of Me Is Missing

· · · · · · · · · · · · · · · · · · · · · · · · · · · ·

## *California, May 2018*

Owen reaches for his phone and turns off the alarm. Kafka's *Metamorphosis* lies open on the bedside table. The author's severe face, set atop a beetle's body, stares back at him. Was he reading that book last night? It's possible. His phone says it's 9:01 A.M., and the calendar icon on the home screen reads May 1. If that's what it says, it must be true. It's a verifiable external reality. That, and what he writes in his notes. He hasn't trusted his memory for years and certainly doesn't now. This feeling of dread—of not being sure about who he is, what is real, how to feel better, or whether any of it matters—is even more excruciating than it used to be. Maybe reviewing yesterday's notes will help.

*It's like some part of me is missing. I feel scared, hella anxious, depressed, confused. I obviously fucked up but what really happened? Kylie says I overdosed on fentanyl. Makes no sense. I've used so many times. Why this time? Was it a contaminant? I need to figure this out.*

The cloying smell of chai tea hangs in the air. His mother must have gone for her daily run, showered, and left for the day. The muffled sound of a lawnmower buzzes over at the retirement community behind their apartment complex. Otherwise, the bedroom is quiet. Owen inspects the shelves across from his bed, which are lined with his favorite books from college—works by Camus, Kant, and Nietzsche. They represent the beginning of the end of his brief moment of happiness, of loving learning. College was a system he understood. If he worked hard enough, he would get perfect grades. But the philosophers, especially Nietzsche, stirred up a toxic stew of questions about the meaning of life and other unanswerables. Their ideas fed his depression and provided what seemed like the perfect justification to begin using drugs again.

Owen turns his attention back to the phone in his hand.

> *Completed:*
> *e-mailed lauren aguirre (journalist) @ NOVA*
> *e-mailed alfred demaria (mass dept public health)*
> *spoke to dr. barash (neurologist) connected me to dr. butler (neurologist)*
> *vivitrol shot*

That explains the tenderness when he tried to sit up. The pain of the extended-release naltrexone shot is a small price to pay to help avoid relapse. Anyway, he must have done all these things, because his notes say so. Owen crosschecks his notes against the e-mail messages in his sent folder. At 11:01 yesterday morning, he e-mailed a journalist who wrote an article called "Rare Form of Amnesia Linked to Fentanyl Overdoses Is Spreading." At 2:08, he e-mailed someone named Alfred DeMaria, the author of a paper called "Cluster of an Unusual Amnestic Syndrome—Massachusetts, 2012–2016." Owen asked DeMaria about treatment or prognosis and offered to answer any questions. Then he had a phone call with another neurologist named Jed Barash, who told him to call a California neurologist. He doesn't want to get his hopes up too much, but if he could at least get some answers, maybe it would tamp down his anxiety.

> *Tomorrow:*
> *apologize to francesco*
> *join health club*
> *call jake*

*Today:*
*take meds*
*text mom I'm awake*
*breakfast @ 10*
*call dr. butler UCSF*

Long-term lists don't make sense anymore, so he doesn't bother. He feels so limited to the present moment that he can't even picture the future. Owen gets out of bed. A handwritten note in his mother's loopy cursive rests on the black marble kitchen table next to a bottle of ginkgo biloba: *happy birthday! today is the first day of the rest of your life. please don't forget to text me.* Owen sits down, hangs his head, and sighs. He doesn't know how old he is today and has to do the math to figure it out.

———

In early May, Butler answers a call from a Southern California area code, knowing who it must be. Barash texted him to say he'd soon be hearing from a new patient with a strange story. Barash hadn't seen the MRI, but he told Butler that from the way the patient described the scan, it sounded like the real deal. And the guy is smart and articulate. That much is clear as soon as Owen tells Butler he has bilateral hippocampal damage. The more they talk, the more astonishing it seems that a young man who by all outward appearances had such a promising future has come to this. Owen describes his memory hoarding and tells Butler he'll do anything and everything he can to help them figure out what's causing the amnestic syndrome. Butler assures him he'll find a way to get him to the MAC as a research patient and be back in touch with the details. Although Owen will forget the particulars of the phone call as soon as he hangs up, the thrill of feeling like he's finally reached another person who could understand and possibly even help lingers with him long afterward.

Butler is particularly baffled by Owen's description of his obsessive-compulsive disorder, or OCD. The classic symptom taught in medical school is washing one's hands repeatedly. Owen's frequent list making seems more like what early Alzheimer's patients do to compensate when they sense that their memory is beginning to fail. Could it be that Owen's long-term opioid use had already degraded the memory circuits in his hippocampus before the overdose—not enough to prevent him from graduating with a 4.0 GPA, but enough to make him feel he was losing his edge?

Two weeks later, fueled by coffee, Diet Coke, and Barash's texts, Butler goes on a 48-hour data-crunching marathon for Project Seahorse. For the previous month or so, he's been working intensively to find out if the hippocampal volume loss he detected in the few dozen cases he looked at back in September holds up with bigger numbers. It's easy to find a trend in one direction or another if you don't have enough data, and making medical decisions based on small sample sizes can have devastating consequences. For example, from 2005 to 2011, doctors tried to prevent strokes by placing expensive stents inside brain arteries to keep them open. The idea seemed reasonable. Blocked coronary arteries cause heart attacks, and blocked brain arteries can cause strokes. Then a carefully designed study with ten times as many patients found that the practice led to significantly more strokes and more deaths, not fewer.

With this concern in mind, one of Butler's top priorities is to find enough people in the database to feel confident about whatever the results may show. He's been doing most of the work on the project himself because he feels guilty leaning on others when he doesn't have a grant to pay for their time. But Devyn Cotter hasn't needed any pushing from him, and she's created a spreadsheet of about two hundred elderly people who've each had more than one scan. Most of the records are in the ADNI database. Sixty or so patients from the MAC's collection help bulk up the numbers.

Butler creates an opioid group and divides it into people who used them briefly to control pain after surgery and people who took them for many years. His non-opioid group contains people with no pain and people whose pain was treated with drugs like acetaminophen or ibuprofen. To test his hypothesis that opioids make the hippocampus shrink, Butler needs to rule out the possibility that pain itself could be to blame, rather than the drugs used to control it.

Butler has hundreds of MRI scans to process. If he were an expert, tracing the volume of each hippocampus by hand would take him more than a thousand hours. And he's not an expert. Fortunately, in the previous few years, a computer program called FreeSurfer has automated the process. The clear, colorless cerebrospinal fluid that bathes the brain and helps protect it shows up darkest. The white matter is light. And the cortex, or gray matter—the part that does all the information processing—looks gray. FreeSurfer removes the darkest and the lightest part of the image, leaving only gray matter behind. Using healthy brains, radiologists have trained FreeSurfer to zero in on which region of gray matter is likely to be the hippocampus. Then it computes the volume. All of this takes only five minutes. Butler sets up a conveyor belt of sorts, batch feeding five to ten scans at a time to compute the volume. Every hour or two he checks to make sure the program is working before he feeds the next batch.

Butler has a hard time resisting the gambler's feeling of wanting to know if his bet has paid off, so he finally succumbs and takes a peek at how the data are shaping up. Just like the last time he looked, in September, it appears that people who used opioids had smaller hippocampal volumes than those who didn't—especially those who'd taken higher doses for longer times. Surprisingly, though, whether or not the person carried the *APOE4* gene didn't seem to make a difference. Could that mean everyone should be concerned about taking opioids for extended periods? It's a provocative idea, but Butler knows he has a lot more work to do. There aren't that many people with *APOE4* in the database, so perhaps the numbers are too small to tell. He hasn't found enough chronic pain patients who didn't use opioids to rule out the possibility that pain itself shrinks the hippocampus. And he needs to work with one of his cubicle-mates on some sophisticated statistical analysis to make sure he's not seeing things that aren't real. At this stage, his hypothesis remains a hypothesis.

—m—

The day Butler finishes running the available scans through FreeSurfer, DeMaria receives an e-mail written in the early morning hours from a chronic pain patient in the United Kingdom.

Dear Sirs,

Case: Fentanyl Amnesia
Duration: 2 Weeks
Severity: Severe
Source: Internal Medicine Prescription/
Delivery Method: Patch
Composition: Pure*

The patient says he wore a fentanyl patch as prescribed for pain for two weeks and then watched a TV drama, completely unaware that he'd recently binge-watched the whole series. In his e-mail, he describes hours-long conversations with friends that he could not remember:

> It became disturbing as people started to say, "you know what we spoke about last week regarding xyz?" and I would have absolutely no recollection of the conversation whatsoever, not even a déjà vu feeling, this was flat out . . . "what are you talking about, we didn't even speak!"

DeMaria sends a kind e-mail thanking the man for being in touch, but he can't be of much help. As an employee of a public health agency, he's not in a position to respond to clinical questions from patients he's not caring for.

—∞—

The day after Butler sneaks a peek at the Project Seahorse data, he decides to work from home. A lemon tree heavy with fruit stands at the front of the house. Butler's in the kitchen, in jeans and sneakers instead of the stylish clothes he wears at the MAC. An un-plunged French press sits cooling on the narrow counter next to his laptop. Butler perches on a stool, intently reading through a paper by forensic scientists who examined autopsy reports from the brains of longtime intravenous opioid abusers housed at the Edinburgh Brain Banks. When Butler was back at Lahey digging through PubMed, he found a similar study of Norwegian heroin users with brain damage. Butler has a headache; he made the coffee hours ago, but he's been so engrossed that he hasn't gotten around to drinking it. Compared to the brains of people who never used opioids and died suddenly for other reasons like a car accident or a heart attack, the brains of these Scottish opioid users contained noticeably more abnormal tau tangles. The effect was small in the younger group. But in people over thirty, the amount of tau was significant, though it fell short of what would be typical for someone with Alzheimer's. The researchers propose that abusing these drugs for years accelerates the natural aging process of the brain—similar to what happens in Alzheimer's patients.

Butler pulls up an earlier paper from these same Scottish scientists and zooms in on a cross section of the entorhinal cortex, the gateway between the hippocampus and the rest of the brain. Against a pale blue backdrop, a dark, teardrop-shaped tau tangle stands out—the same shape Alois Alzheimer drew more than a century ago when he examined 51-year-old Auguste Deter's brain after she passed away. These opioid users didn't have Alzheimer's, yet their brains contained one of the hallmarks of the disease. There's also evidence of a ramped-up immune system, another signature Alzheimer's feature. Butler dumps the cold French press coffee into the sink, packs some fresh grounds into an espresso maker, and sets it on the stovetop, considering the parallels with Alzheimer's. For Owen, Max, or any other amnestic syndrome survivor, did the initial firestorm of damage during the overdose create large amounts of tau tangles? If so, would it lead to an insidious, unstoppable accumulation of tau tangles? What if they stop using opioids—are their brains young and healthy enough to clear out the toxic debris before it's too late? The possibilities add up to a pile of ifs, whats, and maybes. But it's curious. Very curious.

———

"In 2017, we captured four additional cases, and all four tested positive for fentanyl," says Barash for the fourth or fifth time. It's a Sunday in late May, and he's holed up inside a white van in the parking lot of a trampoline park in Watertown, Massachusetts, with a microphone in his face. Henry is inside with Gillian, baby Jack, and a few friends, bouncing around. A producer from the HBO series *VICE*, who already interviewed him, DeMaria, and Kofke, needs more audio. Barash dislikes the spotlight, but he's willing to do what it takes to get the word out and keep up the momentum. A few days before, he steeled himself, went to the DPH, sat at the table where he, Somerville, and DeMaria had reviewed case reports, and submitted to an on-camera interview. Compared to that experience, the audio interview is a breeze, and Barash is relaxed enough to appreciate the humor in the situation. Satisfied they've captured the voice-over they need to play over the picture of the *NEJM* article, the crew thanks Barash, and he returns to the chaos of the trampoline park. "I took one for the team today," he tells Gillian over the screams of young children.

Barash was right to think that the *NEJM* would bring awareness—and with it, skepticism, especially from anesthesiologists, who use fentanyl safely every day. The chief quality officer for a privately owned national anesthesia practice writes a letter to the editor, questioning the conclusion that fentanyl itself is responsible. His disbelief is understandable. He points out that "millions of surgical patients each year receive fentanyl," yet there are no reports of long-lasting acute amnesia afterward. He asks whether the phenomenon "is due to contaminants rather than to fentanyl itself."

In early June, the *NEJM* publishes Barash and his co-authors' response to the criticism. They point out that overdose victims are often hypoxic for some period of time—which would certainly exacerbate the injury in a hippocampus already revved up and hungry for energy. Surgical patients, on the other hand, are given oxygen. In addition, they hypothesize that midazolam, one of many drugs given during an operation, counteracts fentanyl's potentially toxic effects. Barash and DeMaria are also privy to tantalizing information they can't share that supports their claim. The UK fentanyl pain patch case is based on a single e-mail from a patient, and without medical records, they can't know for sure what happened. They're also aware of similar soon-to-be-published cases that seem to rule out the idea of a contaminant. A middle-aged woman in Ohio developed brain swelling and the hallmark damage to the hippocampus when she was given opioids after surgery. A 63-year-old woman in Canada had used fentanyl patches prescribed for pain before overdosing and developing amnesia. The journal entries she began writing afterward reveal the depth of her loss:

> Feeling even more confused today.
> no sense of time?? Anxious
> Feel like I am in a fog
>
> 3:30PM Have to keep reading my
> journal + white board to
> remember anything -
> memory does not hold

Four months later, the woman was still unable to live without constant supervision.

A decade earlier, in his last paper on the dangers of fentanyl and before the woman's accidental overdose, Kofke had warned that opioids could damage the human hippocampus and suggested a road map for future research. Barash's *NEJM* response to the letter to the editor concludes with a sentence that echoes Kofke's final words:

> New England Journal of Medicine, June 2018
> "Attention to this syndrome may stimulate further investigation
> into the effects of fentanyl on the hippocampus in patients who
> receive this drug for therapeutic purposes."

Barash sends Kofke the *NEJM* link, telling him it's vindication of his decades of research. Kofke posts it on Twitter: "We pub yrs ago on hippocampus injury w opioids. Now neurologist barash discovers amnestic syndrome w opioid od."

# 16

# The Next H.M.

. . . . . . . . . . . . . . . .

## *June 2018*

The alarm goes off on Owen's phone just as he reaches the top of the escalator at the San Francisco International airport. He checks his notes app: PICK UP BAGGAGE. The 2:15 P.M. alarm is the tenth of ten reminders he previously scheduled to go off between landing and leaving the airport. Over his parents' objections, Owen has decided to make the trip alone. It seemed to him that being autonomous was an essential part of his recovery. Glancing up from his phone before reaching the bottom of the escalator, he's relieved to see the baggage carousel. Over the past month, Owen has learned that as long as he stays wholly focused on the current task, he won't forget it. The problem comes when he gets distracted. Bag in hand, the next alarm goes off, reminding him to follow signs to the Bay Area Rapid Transit System. For the next half hour, another series of alarms remind him to stay on the train until he reaches the Embarcadero stop, where his former business partner Francesco will meet him.

The next morning, Owen's elaborate system gets him to the MAC ten minutes early. He rechecks his notes app to confirm the name of the research coordinator who handled all the paperwork and scheduled the testing: Devyn Cotter. He waits on a bench in the black granite-tiled lobby. Facing him are imposing two-story-high walls whitewashed with sunlight. To his back is a soaring, serpentine-shaped exterior wall of windows overlooking a gardened area. His father texts him to make sure he's okay. A feeling of uncertainty creeps up, and he rechecks his notes: Devyn Cotter. When he looks up, a petite young woman with a reassuring smile walks briskly toward him. Cotter ushers him into the waiting area, past a wall decorated with giant, glowing, bulbous blue neurons, and into a small office to sign consent forms.

Owen is nervous about the testing but relieved that he can put his phone away and let someone else tell him what to do for the rest of the day. Following Cotter's instructions, he has held off eating until after the blood draw. Owen averts his eyes until the needle enters his vein, then looks back to watch as the tubes fill up with bright red liquid. Most of the samples will be sent to UCLA to test for genetic risk factors for dementia, including the *APOE4* gene. Bruce Miller has also requested an antibody test at a lab at the University of Pennsylvania to see if a rare autoimmune disease, not fentanyl, could be the source of Owen's brain damage.

Cotter gets Owen a snack and takes him to a small office, where he meets Butler and Kaitlin Casaletto, the neuropsychologist who's agreed to do the cognitive testing. They have decided to get the neuropsychological testing out of the way early, while Owen is still fresh and can perform at his best. Casaletto works with Joel Kramer on cognitive reserve and brain resilience, so she's the perfect person for the job. She knows Owen graduated with highest honors from UCLA, so she's shaped the neuropsychological testing to take into account his intelligence and the fact that he's suffered a significant brain injury. Over the next hour and a half she'll administer more than a dozen tests to probe his memory and capture his overall cognitive strengths and weaknesses.

"Thank you for coming all this way," Butler says. "I can imagine it was pretty overwhelming and anxiety provoking, but you obviously have some extremely effective compensatory strategies."

Owen shrugs apologetically, uneasy with the compliment. "I'm just really grateful that you're able to see me, and everyone went to so much trouble to set up all this testing. It's not entirely altruistic because I'm looking for answers for myself. . . . But if this can help others too, that'd be amazing."

Casaletto, a tall, slim, alert woman, is taking mental notes on Owen's mood and how he interacts with Butler. She knows that anxiety will affect Owen's performance and that Butler is trying to put him at ease before they get started. If she didn't already know Owen's story, it would have taken her a few minutes to notice anything amiss.

The three sit around a circular table, and Casaletto explains that the testing will take about an hour and a half. Owen should let her know if he needs a break at any time. A quick global snapshot of his overall abilities puts him just slightly below average for someone with a college education, but there's a more detailed memory exercise designed to put his hippocampus to the test.

"I'm going to read a list of words, and I want you to try to remember them and repeat them back to me, in any order. Ready?"

"Ready," Owen says.

"*Taxi, cat, surprise, four, run, orange, beautiful, slowly, garden,*" Casaletto says. "Can you repeat those back to me now?"

"*Taxi, surprise, orange, beautiful . . .*" Owen hesitates. "That's all."

Casaletto repeats the test a few more times. By the fourth trial, Owen correctly remembers seven of the nine words. He feels like he might be doing well, although Casaletto's job is to maintain a poker face and never give feedback on the results. So he can't be sure. Casaletto distracts him by having him do other tests, like counting backward from one hundred. Ten minutes later, she asks Owen to repeat the list of words.

He looks lost. "I don't know." He glances at Butler, who's looking on attentively and gives Owen a brief, encouraging smile.

"Give it a shot. See what you come up with," Casaletto says.

"*Uber, money, food,*" he pauses. "That's probably all wrong." Owen sighs, leans back in his chair, and fidgets with the phone in his pocket.

"Don't sweat it, Owen. These tests can be hard for lots of people." Casaletto and Butler maintain neutral expressions, but the fact that Owen can't remember a single word confirms their fears about the crippling impact of the damage to his hippocampus. Later, when Casaletto analyzes all the data, she finds that he has scored below the bottom fifth percentile for episodic memory—like someone with Alzheimer's. But these tests are meant to expose the limits of Owen's memory for new events. As Butler pointed out before they even got started, Owen has some supercharged way of compensating.

"We're going to move on to something different now. I think you might even enjoy it," Butler says.

Casaletto takes out a booklet with blank pages about the size of Owen's phone. She explains that they'll be testing how well he remembers smells. For the past month, Barash has tried to convince Butler to skip this test. To him, it's a minor point that will delay getting this paper published and into the hands of neurologists who might see other cases. But Butler can be just as obstinate as Barash. His hunch is that the test might hint at whether there's more damage to the cold end of the hippocampus toward the back or the hot end toward the front, located right next to the olfactory bulb. Impervious to Barash's texts questioning the "value add," Butler has badgered a neurologist who assesses the loss of smell in patients with dementia to hand over some test booklets. He and Casaletto have refined the test by using Cotter and other students as guinea pigs. Cotter scored so poorly that even Butler briefly questioned the test's value, but she turned out to be an outlier. They've settled on five scents: black pepper, lilac, turpentine, peach, and root beer.

Casaletto rips the first page out of the booklet, scratches it to release the odor, and hands it to Owen. He lifts the paper to his nose and inhales the scent of lilac. The act transports him back to the library at UCLA, the sun shining through the window, his notes on neural plasticity open on the table, lifting a sticky note infused with his favorite cologne to his nose.

"Oh, cool! This is kind of like state-based learning, right? Like pairing a memory with a smell?" he asks.

"Exactly!" says Casaletto. One by one, she rips off the pages, scratches them, and hands them over to Owen. Ten minutes later, after some visual memory tests, Casaletto takes out another booklet and asks Owen to tell her whether each scent is something he smelled earlier or not. Thrown in with the original five smells are some decoys: dill pickle, natural gas, grape, peanut, soap, and lemon. Ten minutes later, Owen correctly remembers all five smells. Butler makes a mental note to tell Cotter that Owen scored better than she did.

The session continues for another hour. Owen is focused, clearly doing his best. He scores in the normal range on a test of so-called autobiographical memory that assesses how well people remember past epochs of their lives. Owen describes how Del Mar was a place where rich kids had nothing to do, so they did drugs. By his own account, he lacked motivation or ambition. He was bullied for being short, because he didn't have a growth spurt until later in high school. He went to funerals for half a dozen friends who'd overdosed.

The testing isn't limited to analyzing Owen's memory. Butler and Casaletto also want to probe the cognitive strengths that make it possible for Owen to research his unexplained amnesia, find experts, arrange all the details to travel by himself to San Francisco, and arrive at the right place ten minutes early. Psychologists call this "executive function," an umbrella term for multitasking, focus, abstract thinking, control, and being able to switch between tasks. The ability relies primarily on the frontal lobes.

One element of executive function is working memory, which is the capacity to hold information in your head and process it. It's sometimes referred to as the brain's "Post-it" note. This type of information doesn't need to be stored, so it doesn't rely on the hippocampus. The common test is called the digit span. Casaletto slowly reads a string of numbers and asks Owen to repeat them back to her. Owen remembers eight in a row—more than the seven digits the average person can remember. The second half of the test is more demanding. Owen's task is to repeat the digits backward, requiring him to not only hold the numbers in mind but manipulate them so he can put them in reverse order.

"I'm going to give you some more numbers, but this time when I stop, I want you to say them backward instead of forward," says Casaletto. "*Eight, four, three, six, five, one, two, nine.*"

Owen listens, completely still, eyes resting on the table, barely blinking. "*Nine, two, one, five, six, three, four . . .*" he says correctly.

Owen remembers seven digits correctly, scoring in the ninety-eighth percentile, near the top of the chart. He fears that people he meets think he's stupid because he asks them the same question repeatedly, but even his combined IQ is high. He is also, according to his testing, severely depressed and anxious.

At noon, Cotter reappears with a cheerful smile and takes Owen for lunch. When they return to the MAC, she ushers him into a small examining room. Butler runs through the standard neurological exam. He tests Owen's reflexes. He asks Owen to walk forward in a straight line, placing one foot directly in front of the other, and then back in the other direction on his heels. He asks him to hold his head still while keeping his eyes fixed on Butler's finger as it moves from left to right. He asks him to close his eyes and touch the tip of his nose with his forefinger. Unsurprisingly to Butler, Owen has no trouble with any of these tasks.

Butler moves on to the so-called mini-cog—three words the patient has to remember after an intervening task such as drawing something simple, like a clock. It's similar to the bedside test that Max took six years earlier at Lahey Hospital and that neurologists around the world use when they want a quick assessment of a patient's ability to think and remember. But Butler doesn't go by the book and use the typical *apple, table, penny,* or even the *apple, table, honesty* trio of words that Barash uses. To add an extra level of effort, he's developed his own version of the mini-cog: *Chattanooga, tennis, Hennessy.* Butler's idea is that Chattanooga places an extra demand on memory because it reminds patients of Tennessee, especially when paired with the rhyming words *tennis* and *Hennessy.* When, a few minutes later, Butler asks Owen to repeat the three words, Owen begins listing states on the East Coast.

After the exam, Butler and Owen have a lengthy conversation about Owen's drug use and what drove his addiction, his worries about memory, and what he knows about the night he overdosed. Owen explains how he moved from a Vicodin pill found in a neighbor's medicine cabinet at age eleven to benzodiazepines, cocaine, heroin, and finally, after college, fentanyl. He describes how, after his first rehab at the end of high school, he went to community college and found a purpose in learning, but at the same time, the unexplainable fear of forgetting that had been slowly building since high school became impossible to ignore. His perfect grades at UCLA seemed to contradict the idea that he had any memory impairment.

Nevertheless, what he describes to Butler as his hypergraphia, his intense desire to write things down, had by then become a debilitating compulsion.

Butler presses him to remember what he can about the night of the overdose. Owen describes the crushing disappointment when he learned that he didn't have a brain tumor—and had, therefore, lost his excuse to go back to using fentanyl. He describes finding an ad for fentanyl on craigslist and driving to meet the dealer. From then on, he remembers only fragmentary images. He says he messed up. He says he's unlucky. He says if only he could go back, all his previous problems are nothing compared to losing his memory. Butler doesn't share his own experience with opioids after his car accident—it would be inappropriate to talk about himself—but he feels a kinship with this articulate, thoughtful young man. The interview is emotionally exhausting for them.

"I feel like I can't really contemplate my future, abstractly, academically, or job-wise." Owen pauses and glances at the framed photo on the wall. It's taken from the perspective of someone sitting in a kayak, the boat's yellow prow in the foreground, a serene lake, and mountains beyond in the background. "I feel limited to right now. And super bored." He laughs, self-deprecatingly. Even without the confirmation from Casaletto's testing, it's clear that Owen is severely depressed and anxious.

"Is this what I'll be for the rest of my life?" he asks.

"The truth is we just don't know," Butler says. "I really appreciate you being so open and honest and taking part in all this testing." Then Butler turns to Cotter and asks if everything is good to go for the brain scan. She nods, and they escort Owen upstairs to the MRI room, Cotter chatting easily along the way. A young neuropsychologist named Renaud La Joie has designed Owen's imaging protocol. If Butler believed in fate, the appearance of La Joie in the same pod as him would seem like a good omen. A Frenchman with a shock of spiky hair, tortoiseshell glasses, and a sweet smile, La Joie developed an MRI scanning technique that measures the size of different regions within the hippocampus. Butler has cajoled La Joie into helping out, although it didn't take much convincing.

Testing his protocol on a patient like Owen is an exciting prospect for La Joie, whose passion is the hippocampus's geography. First, La Joie will take a three-dimensional picture of Owen's whole brain made up of a million and a half or so cubes, or voxels, each measuring one millimeter on all three sides. On the second pass, he'll use this atlas as a reference to zoom in on the hippocampus. For medical purposes, radiologists typically focus on the view from the base of the skull up through to the top of the head. But to get the best possible look at the anatomy of the hippocampus, La Joie chooses the coronal view, from the eyes through to the back of the skull. This time, he'll image the hippocampus in rectangular-shaped voxels, cramming in as much data as possible to optimize the coronal view and capture the iconic seahorse shape. La Joie's custom-made protocol will allow the team to trace the borders of his hippocampus and reveal the fine details of how the overdose damaged his memory center.

In the anteroom just outside the scanner, Owen removes his sneakers, dons a pair of socks with sticky pads on the bottom, and follows Cotter into the scanning room. A large stuffed tiger sits on a box in the far corner. Cotter uses it to show kids where they'll be during the test, or she lets them pet it to put them at ease. The soft whoosh of the MRI scanner's liquid helium pumps blankets the room. A celestial light emanates from within. Owen will have to lie flat and still on his back on a narrow platform with his head inside an enormous, long white tube with a two-foot-wide hole through the center. Cotter hands Owen some earplugs, then he lies down. His head rests in a small white cradle. Cotter stuffs some green foam pads around it to keep it from moving.

"Are you claustrophobic?" she asks as she hands him a small black bulb.

"Sometimes." Owen is extremely claustrophobic and knows he'll need to tamp down his panic throughout the test.

"Squeeze this anytime you need a break or you need to talk to me for any reason. But honestly, just talk to me whenever you want. You don't have to worry about the squeeze ball."

Cotter spreads a warm blanket over Owen. "You probably remember this from your other scans, but hold as still as you can so we can get really clean images. Ready?"

"Ready."

She lowers a white Darth Vader–like mask above Owen's face and snaps it into the cradle beneath. Slits in the mask let him see out. Then she pushes a button, and the platform glides slowly back into the center of the machine. For a few minutes, all Owen hears is the whoosh and raspy, rhythmic, metallic sound of the pumps that maintain the MRI's thirteen-ton magnet at 450 degrees below zero. Behind the smooth white walls of this massive machine, the extreme cold and the densely packed wires that encircle Owen's brain generate a magnetic field sixty thousand times greater than the Earth's. As a result, every proton at the center of every one of his brain's billions of hydrogen atoms is forced to snap to attention and wobble around the same vertical axis, like a spinning top.

Owen is acutely aware that he is encased and immobilized. He notices his jaw, how it feels like it's sinking into his airway. He feels he's not breathing enough, and his heart starts beating faster. And then the scanning begins. He notices the precursor to panic and wills it away, listening to the loud, insistent call-and-response of the giant machine as it proceeds through the sequences La Joie has programmed. Pulses of energy force some of the protons to spin together before they slip back into their original energy states, releasing a signature picked up by the machine. He hears a handful of gentle taps answered by a volley of insistent clangs in a matching rhythm, as three smaller magnets target preprogrammed slices from different angles.

Over the next hour and a half, Owen lies motionless as the machine captures these patterns of electrical currents to reconstruct an image of his entire hippocampus. He hears the peal of a bell tolling, a mournful foghorn, a jackhammer, a leaf blower. It's as loud as a rock concert. The mechanical, rhythmic noise reminds him of one of his favorite industrial music bands, Throbbing Gristle. One sequence makes the platform shake. During short breaks, Cotter asks how he's doing. He's okay, he tells her. He wonders exactly what's happening to all the matter inside his head, how the sounds generate a picture, how the radiologists will know with certainty that what they see is an accurate representation of his brain, a verifiable external reality.

Finally, it's over. The machine disgorges him from its maw. In the anteroom, Owen puts his sneakers back on and collects his phone from Cotter. He thanks her, thanks Butler, and checks his list for instructions on how to get back home to Del Mar.

# 17

# Rebirth

· · · · · · · · ·

*July 2018*

Owen sits quietly at the kitchen table in his mother's apartment, phone in hand, listening. Butler is talking about his MRI scan, his neuropsychological testing, and a prescription for brain health. The word *neurogenesis* comes up. Owen thinks about the notes he wrote in his study packet at UCLA. The idea of neurogenesis seemed so promising and intellectually exciting:

> *so running=more new neurons*
> *what happens to learning?*
> *better learning!*

There'd been a discussion in class about whether it's better to run before or after learning. Maybe, if you run before, you'll create new neurons that will be fresh and ready to be wired up. Or, maybe if you run after you learn, new neurons will wipe out your memories. Owen remembers students in the lab running around like lunatics as if that would make them smarter. He has a sinking feeling. It seemed funny back then. But not now.

—ᨒ—

By current estimates, the human brain contains about eighty-six billion neurons comprised of thousands of specialized types. In the hippocampus, the largest is the pyramidal neuron, with a cell body about the width of a human hair. Some pyramidal neurons have axons long enough to reach brain regions inches away. The smallest neuron, called a granule cell, has a cell body one-tenth the size, about the width of a strand of spiderweb silk. They are the tiny but powerful gatekeepers of the hippocampus, packed into a sideways V-shaped structure called the dentate gyrus. Sights, smells, sounds, tastes, and touch from the rest of the brain converge on this entryway. Sprays of dendrites from granule cells line the outer edge of the V shape and gather this information. Along the inner edge, axons extend in the other direction to send the information on to the dentate gyrus's favorite partner—a larger, interlocking V facing in the opposite direction. This area, called the CA3, is made up of the pyramidal neurons that reminded Cajal of hyacinths lined up in graceful curves. When scientists eavesdrop on the dentate gyrus, it's dead quiet. The cells inside speak softly, but they carry a big stick, preventing the pyramidal cells they talk to in the CA3 from getting overexcited.

Most neurons we're born with have only one fate—to die without being replaced. But the hippocampus gives birth to dentate gyrus cells throughout our lives in a process called neurogenesis. Rivers of these newborn cells flow through the dentate gyrus, ushered along the path to maturity by nearby cells, forming increasingly complex connections. Neuroscientist Amar Sahay at Mass General's Center for Regenerative Medicine hopes to harness the rejuvenating power of these adult-born dentate gyrus cells to stave off memory loss in Alzheimer's. But just as memory has many modes of action, it has many ways to fail. The job of dentate gyrus cells is to prevent memories from blending together. "It's very important for us to keep memories separate," says Sahay. "That's episodic memory: what, when, and where. It is critical to the way we navigate our world." The phenomenon is called pattern separation.

Say you park your car in the same garage every day, but not always in the same spot. How do you remember where you left your car today instead of yesterday? The web of connections between neurons that make up these two memories is unique, but there are many common elements. Your car is always blue, for example, and most of the time you're early enough to park on the first floor. But this morning, the snowstorm made you late, and you had to settle for a spot on the fourth floor. Using just a few cues, like the piles of snow by the garage entrance and the bitter cold, the CA3 fills in the partial picture, reminding you that the car is on the fourth floor today. Neuroscientists call this pattern completion. The delicate balance between pattern separation and pattern completion slowly comes undone with age, so the filling-in-the-blanks part gains the upper hand. As a result, older people are more prone to mix up similar memories—like where they parked their car—than younger people are.

It's logical to imagine that a fresh supply of young, healthy neurons could help the hippocampus make more reliable memories. A decade ago, when Sahay began researching neurogenesis, the evidence he gathered backed up the idea. He chose to test a type of memory that is a staple of rodent studies, the so-called fear memory. The basics are these. Put a mouse into a milk-crate-sized chamber with an electrified floor. When the mouse starts exploring, its feet get gentle shocks. It's not painful, but it is surprising, and it doesn't take long for the mouse to learn to crouch, motionless. How quickly the mouse freezes when it's put back in the cage is a sign of how well it remembers. The memory test is trickier if you create two similar-looking cages, only one of which has an electrified floor. When Sahay blocked neurogenesis in his mice, the mice took longer to learn the difference between the two cages. It appeared that pattern completion had gained the upper hand over pattern separation, making the mice's memories less precise. When he boosted neurogenesis, they were quicker to learn how to separate the safe cage from the foot-shock cage. Theoretically, fresh new neurons in the dentate gyrus were evening out the balance between pattern separation and completion. The experiment seemed to prove that more neurogenesis leads to better memories.

Yet researchers led by Paul Frankland at the Hospital for Sick Children in Toronto came to the opposite conclusion. His team tested baby mice, which naturally have very high levels of neurogenesis, and found that the babies could only remember having received a foot shock for about a day.

But adult mice, who have fewer new neurons, remembered the foot shock for at least a month. If Frankland brought down the amount of neurogenesis in baby mice to a typical adult level, the baby mice didn't forget the foot shock. Frankland then tried giving the adult mice a running wheel so they could exercise whenever they wanted. Neurogenesis increased, and two weeks later, they'd forgotten the foot shock that they'd learned before hopping on the running wheel. Frankland concluded that neurogenesis degrades episodic memories.

How can neurogenesis promote forgetting *and* remembering? Both are probably true. Young neurons are more plastic. They can make connections with other neurons more efficiently, which makes them better at establishing new activation patterns—in other words, learning. That also means that new neurons compete with old ones for connections. Replace enough of the original connections, and the activation pattern that contained an old memory disappears. Until about the age of three, there's so much turnover that all episodic memories are erased. In old age, with fewer new neurons being born, learning doesn't come so easily. The older patterns have the upper hand and are harder to update. Or at least that's the theory. Neither Frankland nor Sahay believes neurogenesis is bad for memory. "Certain levels of forgetting are probably useful for memory function," says Frankland. Some memories may even be detrimental, as with PTSD or depression. On the other hand, you can't forget everything. "The brain is trying to hedge its bets," Frankland says.

The purpose of memory is not so much to reminisce as to make intelligent predictions. It may impress your friends if you remember that South Korea beat Italy to become World Cup champions in soccer in 2002. But it's beside the point if you want to gamble on who'll win next year. In a continually changing world, there's an advantage to forgetting, because what you remember may become irrelevant. Allowing some details to escape one's memory is equally important for developing useful knowledge about the world, so-called semantic memories. For example, a soccer ball is usually black and white. But that's not what makes it a soccer ball. No matter the size or color, the unifying feature is that it's always made up of twelve pentagons and twenty hexagons. The person who remembers every color and pattern of every soccer ball they've ever seen may never abstract its essential nature.

Using this insight, researchers like Sahay aim to tweak neurogenesis to get the balance between new and old neurons just right. Using various combinations of genetic engineering and drugs infused directly into the hippocampus, Sahay can boost newly born neurons' ability to integrate into the dentate gyrus without completely wiping out all the old connections. Then he tracks how well the mice, say, remember the location of a hidden escape platform submerged in a water tank, or which box they need to freeze in to avoid a brief, gentle foot shock. Like the person who successfully finds her parked car, the mice with this altered balance between new and old neurons are better at recognizing the safe place. Sahay found that the intervention helps memory in adult, middle-aged, and even older mice. But as with any proposed treatment for Alzheimer's, earlier is better. "The neurons die at the end. That's the point of no return," he says. If a river of newborn cells is maturing, "they're like, well, where do I belong? How do I connect? Integrate? Our partners are dying up there."

Dr. Rudolph Tanzi, also at Mass General, has been an Alzheimer's researcher since the 1980s, when he discovered the first gene that causes early-onset Alzheimer's disease. The list of tactics he's tested to treat Alzheimer's is long and growing. Recently, Tanzi turned his attention to neurogenesis. Most genetically engineered Alzheimer's mouse models have less neurogenesis than natural mice. If they exercise, neurogenesis increases, and their memory improves. But when Tanzi tried using drugs that ramp up the birth of new cells instead of exercise, it wasn't enough to improve the mice's memory.

Tanzi suspects that new neurons have little chance of survival in a battle zone full of plaques and the inflammation that goes along with it. But, by tinkering with two things at once—artificially boosting neurogenesis and adding a growth factor called BDNF—the mice's memory improved. Tanzi says boosting neurogenesis without BDNF is destined to fail, like trying to grow tomato plants in downtown Boston next to a sidewalk without adding plant fertilizer. "BDNF is like Miracle-Gro," he says. "If you nurture new neurons with BDNF, they survive." But the drugs and gene therapy that

Tanzi's team used in mice can't be used on people. "What we're working on now," he says, "is testing supplements that can achieve the same goal."

—⁓—

Phone glued to his ear, Owen smiles wryly at the bottle of Lion's Mane mushroom extract on the kitchen table that just arrived in the mail. A friend mentioned it was good for memory. Then he turns his attention back to Butler's news. For weeks Owen has been anxiously awaiting his test results, and he's devastated to hear the extent of his brain injury. He's not sure why he's surprised. He can't even say what day of the week it is. And yet, the finality and concreteness of the number Butler just gave him is stunning. Ten percent. His hippocampus shrunk by about ten percent between the MRI scan taken before his overdose in March and his visit to the MAC a few months later.

By some estimates, Owen's hippocampus has shrunk by at least as much as the average sixty-year-old's would in a decade, but Butler is trying to gently break the news. He wants to give him hope and something concrete to do without providing an unrealistic expectation for a full recovery. Butler mentions evidence suggesting that hippocampal health can be improved with supplements like fish oil or turmeric. He recommends vitamins, less sugar, better sleep, meditation, avoiding stress—and exercise to boost neurogenesis. None of this could hurt, and it's all Butler has to offer. There is no proven medicine to heal Owen, just as there is no medicine to heal anyone with Alzheimer's. Still, Owen has a lot of intangibles going for him—the desire to get better, a commitment to stay sober, a supportive family, and time to rest and allow whatever healing might be possible to take place.

After his stressful trip to San Francisco and thousands of research dollars that must have been spent on his testing, the conversation is a letdown for Owen. Butler's prescription for brain health is no different from what anyone else should do. For years Owen tried exercise, meditating, and supplements even as he grew ever more concerned about memory loss. None of it seemed to make any difference. But there it is. Owen is still not sorry he went. Maybe someday it'll help someone else, if not him. He thanks Butler before he ends the call.

Two days later, Butler moves back to Boston to take a job with Biogen, one of the last big pharmaceutical companies still committed to finding drugs to treat Alzheimer's. Most have abandoned the quest. A typical Alzheimer's drug development program can cost more than five billion dollars, enroll hundreds to thousands of participants, and take as long as thirteen years from testing in animals to approval by the FDA. The last drug to make it over the finish line was Memantine in 2003. This drug, along with the handful of others that the FDA has approved, can ease symptoms for a short time in some people. They do nothing to slow down the disease.

In 2018, Pfizer shut down its Alzheimer's division, laying off three hundred researchers and staff. Fortunately, one of those researchers is heading up a new group at Biogen and wants Butler on board. Butler and the other recruits will help the company plan smarter, more sophisticated clinical trials using imaging, genetic testing, and other tools designed to measure what matters for each therapy. As much as Butler has loved the MAC, the move makes sense. He'll be able to take what he's learned and put it to use helping develop treatments that, he hopes, will one day have a significant impact on people suffering from dementia. As a bonus, he'll get to work on a range of dementias, not just Alzheimer's, but frontotemporal, Parkinson's, and even rare genetic dementias like the one he suspects might have felled Nietzsche.

But Butler still has work to do on the amnestic syndrome. First on his list is finishing the analysis of Owen's testing. Biogen has a new three-dimensional VR tool that could give him an extraordinary view of his hippocampus. Bruce Miller, the director of the MAC, supports the work Butler is doing with Owen, and they're aiming to write up a case study and get it published in *JAMA Neurology*. If the editors accept the paper, it will be the first description of this syndrome to appear in a prominent journal for neurologists; after emergency room doctors, these are the specialists most likely to see such patients. After that, Butler wants to get back to Project Seahorse to see if he can gather enough high-quality data. Miller agrees to let Butler continue part-time at the MAC so he can wrap up his work.

Less than a week after Butler moves to Biogen, he runs into Yuval Zabar coming out of the bathroom. Zabar has left Lahey. They speculate about why Owen and all the other amnesic patients had such a dramatic response to fentanyl even though many, or perhaps all of them, had used fentanyl or other opioids before. Zabar has a hunch that this prior use might have subtly weakened neurons in ways that left the victims vulnerable to the final, catastrophic overdose. Without extensive testing in mice, Zabar, Butler, and Barash may never know. But for the patients, the precise mechanism is irrelevant. What matters to them and their families is how the overdose has forever changed their lives.

# 18

## Time for a Do-Over

· · · · · · · · · · · · · · · · · · · · · · · · · ·

### *August to November 2018*

As the months roll by, distressed family members of overdose victims across the country contend with catastrophic memory loss in their loved ones. The medical professionals they turn to are mystified and have no diagnosis or prognosis to offer.

Erica overdosed after years of being addicted to heroin. During her six-day stay in a hospital west of Boston, her mother repeatedly told the staff that Erica couldn't remember anything, but no one ordered tests to investigate further, and Erica's mother feels her concerns were ignored. On Erica's last day in the hospital, her mother raised the memory loss one last time, and the doctor asked why she hadn't done so sooner. Erica was discharged without further tests. A few years later, Erica reads about the Massachusetts cluster through local news reports, tracks Barash down, and speaks to him on the phone. She's working at Starbucks and living at home with her mother. Barash can't help her, but at least she understands what happened and that she's not the only one.

No one knew how long Joshua was in the parking garage before he was found and given several doses of Narcan. When the police told his mother there were needles in his car, she suspected fentanyl, although the toxicology screening came back negative for all drugs. Doctors released

Joshua and said his memory would return after the brain swelling went down. Barash speaks to Joshua's mother on the phone, verifies that the records match up with the amnestic syndrome, and adds the information to his informal tally of possible cases. The MRI was taken eight days after the overdose—too late to detect the damage.

Doctors at an acute neurology unit at a hospital in North Carolina told Tiffany's mother they weren't sure what to do for her. All they knew was that she overdosed on fentanyl, woke up with amnesia, and had brain lesions in the hippocampus. Barash explains to Tiffany's mother that the MRI, taken eight days after the overdose, was too late to pick up on the classic pattern of damage. Barash and his colleagues are struggling with how to define the amnestic syndrome to help doctors provide an official diagnosis. But developing the criteria has been difficult, because so many of the cases, like Tiffany's, are missing information. He adds her to the list of possible cases.

It wasn't the first time Kyle overdosed. On earlier occasions, he'd seemed to struggle for a while, and then he'd gotten better. This time was different. His memory was so unreliable that his mother had to stop working to take care of him. After Butler speaks with the mother and looks at the records, he adds Kyle to his list.

Ben was getting ready to start graduate school. Then he overdosed on heroin, possibly laced with something, and developed amnesia. A month later, his MRI seemed normal, but his memory was not. What's the prognosis? asked the father, who is also a doctor. Were there any recommendations for cognitive rehabilitation? Barash wished he knew.

David was having seizures, so he was taken to the emergency department, where a drug screen detected cocaine and fentanyl. David kept asking his brother and wife the same questions. After six days he was discharged back home. David's brother was petrified about what the future held. Could Barash give him any guidance as to where to turn or what steps to take?

Family members e-mail Andrew Kofke or call Marc Haut. More often, they e-mail Al DeMaria or leave voice mails on his home phone, and he passes the messages on to Barash. Mothers contact Barash on Facebook. Barash calls them back at night after dinner or on the weekend. The heartbroken mother or father or brother tells him their doctors don't know what it is. They say they knew their children could die, but they never imagined that this was how it would end. People find him in such a happenstance way, and these are only victims with friends or family who are able to support them. It's hard not to imagine that the cases he hears about are a small fraction of the real number.

—⁂—

Henry is playing outside a few doors down at the neighbor's house. Baby Jack is eating spinach. Barash walks into the kitchen in his Yogi Berra Museum T-shirt.

"Jack, you look like you murdered Shrek!" he says to Jack.

Gillian smiles, then looks at the wall clock.

"What was that call you were on?" she asks as she wipes Jack's cheeks. "The bits and pieces I heard on your end sounded pretty heartbreaking."

"It was. Another mom. We talked for about an hour. She said she knows how awful it sounds, but in some ways, it's worse than a fatal overdose. Then she thanked me for caring about a heroin addict."

"How many cases is that now," Gillian asks, as she maneuvers the spinach bowl out of Jack's hands.

"Depends who you include. There's so much missing information. There are some confirmed, probables, and possibles. We're up to eleven states."

Barash looks thoughtfully at Jack as he squirms and Gillian tries to wipe the last traces of spinach off his face.

"Is there enough time before his nap for me to get in a run with him in the jog stroller?" he asks.

"Yeah, that'd be great, I can get a few things done."

—⁂—

It's time, DeMaria decides. He's been at the Department of Public Health for forty years. Technically, most full-time state employees work 37.5 hours a week, but DeMaria is on the road so often and so busy that it's hard to keep on top of the basics of life outside of work. For weeks he dealt with a broken light switch in his kitchen by turning it on and off at the circuit breaker in the basement. There was no time to call someone in to fix it. He's decided to scale back to a paid half-time consulting position. Sort of half-time. DeMaria is always thinking about work, and that's unlikely to change. But going into the office only a few times a week is a big change for him. Now that he's not walking the two-mile trip to the train station every day, he'll have to whittle down his allotted daily calories. The upside is that after he's done everything the bureau director needs him to do, he can spend his own free time keeping track of the amnestic syndrome.

DeMaria is struck by the fact that he, Barash, and the small circle of other health professionals aware of the syndrome will so often hear from family members rather than doctors. Even after having sent out two official e-mails—a health alert and a reportable disease notification—few medical professionals seem to be in the know. And so, like Barash, DeMaria is looking for any way to bring more awareness to Massachusetts and beyond. One avenue is the Food and Drug Administration (FDA). Barash and DeMaria have been tracking illicit use of fentanyl, but the pharmaceutical version of the drug is FDA-approved and it falls under the agency's oversight. In fact, soon after the *MMWR* report came out in 2017, a pharmacist and FDA safety evaluator named Samantha Cotter contacted DeMaria to gather information. Since then, DeMaria, Barash, Cotter, and a neurologist at the FDA have been in touch on and off.

When the FDA evaluates the safety of medications or medical devices, the process can easily take more than ten years. But it's not uncommon for safety concerns to crop up long after approval, when tens of thousands or even millions of people have been exposed. The FDA has two systems to help monitor any problems that may surface. Since 1993, the agency has run a program called MedWatch, which allows anyone—a patient or a health-care provider—to submit information about any medical product they believe might have harmed themselves or someone else. Filing a report is as simple as filling out a short form. To cast the net as wide as possible, the FDA encourages people to report events even if they only suspect a problem. It's a free-for-all system with no expectation that anyone has proven cause and effect. But the thinking is that the more data, the better.

Every week a team of pharmacists and doctors reviews the MedWatch reports and puts them into a public database called the FDA Adverse Event Reporting System (FAERS). If the FDA suddenly detects a rising number of adverse reports, they can switch into active surveillance mode and alert providers to be on the lookout for more. This newer, active surveillance system, called Sentinel, includes data from FAERS and anonymous patient records pulled from a growing number of participating hospitals and health-care systems. DeMaria and Barash decide to recommend that doctors report cases to MedWatch any time pharmaceutical-grade fentanyl is used, either as prescribed or illicitly.

Perhaps unsurprisingly, the FDA staff hasn't found any reports about possible harm from fentanyl in MedWatch. Because fentanyl is usually given with other drugs during anesthesia, if a patient has a problem during or after an operation, anyone entering the adverse event into the database would be hard-pressed to specify fentanyl specifically. Nevertheless, Cotter asks Barash to send her any information about adverse reactions—including the Canadian fentanyl patch user—and to enter the information into Med-Watch. At this point, there's not enough information to warrant switching to active surveillance mode. But, Cotter tells them, they'll continue to keep an eye on MedWatch and the medical literature.

—⁓—

It's another beautiful morning in Del Mar. After reviewing his lists and backing them up to a hard drive, Owen puts on a T-shirt, shorts, and running shoes, and gets into the car to head for the hills. He's been going there for years, long before the accident, and the route is familiar. In the car, he turns on the Clash. Owen has discovered the music of the eighties. It appeals to him more than the atonal noise he used to listen to. He parks at the highest point and runs along a path that winds between dusty rocks and sweet-scented chaparral. Afterward, Owen sits on a bench and looks out through the haze to the houses on the hills below and the Pacific beyond.

He should go mountain biking with friends, but lately, he's been avoiding them, despite how supportive they've been. It's embarrassing that he can't remember what they've told him about their lives since his accident. They could be devastated by a breakup with a longtime girlfriend, and he wouldn't have a clue even though they told him the day before. He might even ask about the girlfriend as if they were still dating. It's much easier to spend time with people he's not as close to so he can keep the conversation superficial.

Last week he got a job at a local coffee shop. But when the line backed up and the pressure was on, he became anxious and couldn't remember anyone's orders. He had to quit after two days. Still, he was pleasantly surprised that he learned how to make a few drinks. Maybe he should find work that's simpler and more structured, like being a cashier or bagging groceries at Costco. He'll have to get over his ego about taking a job so different from what he'd imagined for himself, although the idea of going to medical school now is laughable.

A rare red-tailed hawk soars overhead, scanning for prey. Owen wishes there were someone he could talk to who understands what he's going through. He feels so alone. It would help if there were a support group for people with the amnestic syndrome. He's read the *MMWR* paper and knows there are more patients out there, but the only name he has is Max Meehan, from Massachusetts. Owen realizes that he had it wrong about Nietzsche in college—that he was using a superficial explanation of the philosopher's work as a justification for taking drugs. The idea that life has no intrinsic meaning doesn't make it meaningless. A better way to look at it, he thinks now, is that he should construct the meaning for himself. Owen gets back in the car, updates his list, turns on some classical music, and heads back to his mother's apartment. First on his list is getting in touch with Max.

—⁓—

On Halloween, Barash sits on the porch steps with Henry on his lap, enjoying the evening and musing about how hard he'll have to pester

Butler to resurrect Project Seahorse. The more time goes by, the less likely it appears that Butler will have the bandwidth to turn his attention back to the project. Flashing lights from Henry's Batman costume cast a pool of blue around them. Excited voices carry in the unseasonably warm air as trick-or-treaters caravan down the street away from Barash's end of the cul-de-sac. It's almost exactly six years since Barash looked at Max's brain scan.

"Can I have a candy?"

Barash kisses the top of Henry's head. "How many have you had?"

"Four. Cuz I'm four!"

"Dude, that's enough. You'll be hyped up on sugar and chocolate."

"Will I be too excited to sleep?"

Out of nowhere, the word *naltrexone* floats unbidden to the surface. But before Barash can wonder why, Henry interrupts his train of thought. "Dad?"

"Yeah. Too excited. But we can sit here for another ten minutes. Then it's bathtime and one story."

*Naltrexone*, Barash thinks again. Naltrexone is a so-called opioid antagonist that works by binding to opioid receptors without turning them on. It's a cheap, safe, nonaddictive pill that's been on the market for more than thirty years. Kofke's two decades of research showed that opioids damage the hippocampus by overexciting it, sometimes to the point of causing seizures. So could an antagonist like naltrexone do the opposite and protect the hippocampus by turning down its excitability? Naltrexone probably wouldn't help the amnesic patients, or someone with advanced Alzheimer's. But what if it's taken chronically early on in the disease, before there's too much damage? Could it buy someone a few more years? The idea seems so sensible. He figures he can't be the first person to have thought of it.

"Ten minutes are up, Doodle. Time for Batman to get ready for bed." Barash shifts Henry off his lap, stands up, and blows out the candle in the jack-o'-lantern.

As soon as Henry and Jack are asleep, Barash settles into the couch, TV on in the background, computer on his lap. PubMed beckons. Barash types *dementia naltrexone* into the search field. "Effect of naltrexone on senile dementia of the Alzheimer's type," a small clinical trial published in

1985, is the first result. Barash is off and running. In this trial, Alzheimer's patients given naltrexone daily for one week did no better on cognitive testing than those who didn't. The authors concluded, "In the dosage used, naltrexone appears not to be useful in Alzheimer-type dementia." For three more hours, Barash follows the leads, uncovering decades-old trials, some disappointing, others encouraging. The one that puts patients on higher doses for the longest period—two weeks—seems most promising. "Manipulation of the opioid system may yet prove fruitful in Alzheimer's disease . . ." wrote a guy named D. S. Knopman in 1986. The name sounds familiar. Barash looks him up. Professor Knopman is a neurologist at the Mayo Clinic in Rochester, Minnesota. He's also the deputy editor of the journal *Neurology*, sits on an FDA advisory panel, and has hundreds of papers to his name.

Gillian walks through the study with a bundle of sheets in her arms. "Henry threw up. I'm going to throw these in the washing machine and go finish as many patient notes as I can before I pass out."

She waits a moment for him to register and respond, but he's hunched forward, seemingly mesmerized by his laptop.

"If you're staying up late, can you put these sheets in the dryer?"

"Sorry, yes. This is pretty interesting."

"What's interesting?"

"I have an idea for an Alzheimer's treatment that could work, at least in theory."

"Oh, wow. I thought you said hardly anything ever works."

"That's true. But you can't not even try. I can at least see if it's worth a shot."

"So . . . what's your next move?"

"Make an Excel spreadsheet with whatever I can find on PubMed. Probably send a few e-mails."

"Sounds about right," says Gillian, with gentle sarcasm.

Barash smiles, takes off his glasses, and rubs his eyes. After she goes upstairs, he makes a spreadsheet. A few hours later, he composes an e-mail to Kofke and Butler laying out his ideas, along with summaries of the relevant trials he's found so far.

"It might be interesting to pursue a modern trial of naltrexone on amnestic MCI and/or AD. . . . larger n, placebo controlled, longer duration, with appropriate biomarkers, etc. . . . It would be a lot of work, but I think it would be worth a go," Barash writes. The washing machine beeps, signaling again that the cycle is complete. "This line of research seemed to fade away after that. I might e-mail Dr. Knopman to see when/why the trail went cold." At 11:46 P.M., he hits Send, moves the sheets into the dryer, and heads upstairs to bed.

Over the next week, Barash pursues this idea single-mindedly. The irony is not lost on him that he's taking the first steps down the road to Alzheimer's research, a direction his father suggested years ago but which he had ignored. He gets in touch with Knopman, who is surprised to hear that anyone is looking at research from the very beginning of his career. Knopman tells Barash that back then, clinical trials were poorly designed. Even so, it would only be worth looking at naltrexone again if there was a solid hypothesis about why it would work.

Barash e-mails Butler and Kofke again, adding DeMaria to the chain. DeMaria comments that there was a lot of interest in the 1980s in the idea of repurposing naltrexone to treat a variety of conditions. DeMaria was part of a study that looked into whether naloxone—a drug similar to naltrexone only much faster-acting—could be used off-label to treat septic shock. It didn't work, DeMaria tells Barash, who replies, "Well, then here's your chance at a do-over. :)" Barash finds the website for Alkermes, the Boston-area company that makes naltrexone, and sends an e-mail to the senior vice president for clinical development, laying out his hypothesis on naltrexone and asking if she'd be interested in participating in a clinical trial.

―⁂―

Matt Damon, in an MIT sweatshirt, looks down on the Cambridge Brewery from a mural painted onto the brick wall. He shares the wall with other celebrities—Dustin Pedroia, Mindy Kaling, Elizabeth Warren, and a very young Bill Murray. There are far trendier and upscale options in the booming Kendall Square area, but Barash has chosen this low-key spot, known mostly for its beer and burgers, to meet up with Butler and get his take on the naltrexone idea.

"It feels kind of dead-end-y," says Butler. He pauses while the waiter clears the remains of his fish and chips and leaves a fresh glass of beer in front of him. "You'd need a better handle on the mechanistic evidence to put it all together. A bit of a teaser plate of information."

"I feel like we have enough circumstantial data," says Barash. "And at the right dose it's safe. And it's a dollar a pill."

"I don't know. It seems like tweaking. Maybe someone will be a touch sharper for a few weeks or a few months."

"Okay, but with any of these amyloid beta drugs patients are going to have to go in for monthly infusions. And they're expensive. Do you really think Biogen's new drug is going to make a major difference?"

"It may turn out to be more effective in the early stages of Alzheimer's," Butler says.

Barash smiles and finishes his last fry. "You make me feel like I'm at a comedy club where people come to try out their new material."

An hour or so later, the two men head out into the cold. Barash gets in his car, checks his e-mail, and doesn't see what he wants to see; a response from the Alkermes contact. He'll wait a week and bug her again if she hasn't e-mailed. In his experience, if you don't hear back within a day or two, you either don't hear at all, or it's bad news. But the next day, Alkermes schedules a phone call between Barash and two of their senior staff. He knows the odds are against naltrexone working, given all the other failed clinical trials. Despite Butler's lack of enthusiasm, it still seems to Barash like a reasonable enough idea to be worth looking into—and pounding the pavement to see if anyone else agrees.

# 19

# The Flip Side
# of the Coin

. . . . . . . . . . . . . .

A few more of Barash's late-night couch-surfing trips yield a new trove of papers from a researcher named Michela Gallagher, who began to explore the connection between opioids and memory long ago, as a PhD student at the University of Vermont, publishing her first paper on the topic in 1978. Back then, opioids were merely her window into how memory worked. Finding a cure for Alzheimer's was not part of her thinking.

In early experiments, Gallagher gave rodents naloxone and discovered that this anti-opioid improved the animals' ability to remember the location of foot shocks. But opioids affect how the body manages pain, which could also affect the animals' behavior. What would happen if she took pain out of the equation? Instead of a fear memory, she trained the rodents where to find food pellets in a maze. Gallagher manipulated the cues to ensure that the rodents couldn't rely on procedural memory to help them find their way. Instead, they had to use the visual and auditory clues that she placed around the maze. When she gave them naloxone or naltrexone, they were better at finding the food pellets. In other words, these opioid antagonists improved spatial navigation—a key component of episodic memory. Gallagher soon became interested in the flip side of this coin; if opioid antagonists enhanced memory, could opioids themselves harm memory? In another experiment, she found that old rats naturally have more of their own opioids floating around in the hippocampus and performed worse on spatial navigation tests. Injecting a class of natural opioids called endorphins directly into rat brains also made their memories worse. Injecting naloxone made their memory better. But this 1988 paper is the last one Barash finds. Like the small naltrexone clinical trials that petered out around the same time, the trail seems to go cold. The opioid antagonist solution to memory impairment reminds him of the impressionist paintings he saw during his medical school visits to a nearby museum for the Power of Observing class. If you look closely at each tiny dot—or data point—you can't see the beauty of the picture from afar. "I'm not sure if there's a Monet effect," Barash texts Butler, "where the further you get away from it, the more apparent it is." Or is he missing something? He e-mails Gallagher a link to the *NEJM* paper, describes the amnestic syndrome, asks her about the fate of her long-ago opioid and memory studies, and sits back to wait for a response. Unbeknownst to him, she is just a few months shy of enrolling the first patients in a Phase 3 clinical trial designed to tamp down hyperactivity in the hippocampus.

Gallagher is tall, thin, and intense, with a gravelly voice and a ready laugh. Today, she is an internationally respected Alzheimer's researcher and professor at Johns Hopkins University in Baltimore, Maryland, with a thirty-four-page curriculum vitae and grants totaling in the many millions. "The thing that's unusual about me," says Gallagher, "is that I do a lot of really basic neuroscience. One of my areas was studying aging as an experiment of nature to understand memory systems better. It wasn't because I wanted to understand Alzheimer's disease. It was just—I'm interested in memory."

In the late 1980s, many in the Alzheimer's field began investigating a neurotransmitter called acetylcholine, which has a role in learning and memory. Gallagher turned her attention to that neurotransmitter, leaving her opioid work behind. Acetylcholine is the basis for three of the four FDA-approved drugs for Alzheimer's. This includes Aricept, which can improve symptoms slightly for anywhere from a few months to a year over the entire course of the disease. In 2003, Gallagher's mother, a woman with a master's degree who never smoked or drank, died of Alzheimer's. Seeing the devastation of the disease firsthand sharpened Gallagher's resolve.

Around this time, while Kofke was beginning to wind down his work on hyperactivity in the hippocampus, the same phenomenon caught Gallagher's attention. The two areas affected are the dentate gyrus and the CA3—the place where similar memories are prevented from bleeding together. For many years, the reigning theory was that this hyperactivity was compensatory, making up for the loss of healthy neurons. But Gallagher didn't think so. Somewhat to her surprise, she found that this heightened activity began before neurons were dying. That aside, Gallagher didn't think extra activity would help the hippocampus make better memories. "It seems like it would, but if you understand how that system worked, you would say that's a big problem." Instead of turning up the volume on faltering communication, this activity acts like static in the background, interfering with the hippocampus's exquisitely tuned circuitry. Proving her point, she found that the higher the activity, the harder it was for rodents to navigate through a maze or find the escape platform in a tank of water. Gallagher decided to see if she could help by giving them a very low dose of an anti-seizure drug that tamps down runaway activity. It worked.

Gallagher's team then analyzed human brain scans to look for hyperactivity in the hippocampus. She discovered that older people, especially those with amnestic mild cognitive impairment, had this same signature, and so do *APOE4* carriers. "So, then I went into humans and said, okay, I'll just do the experiment. I'll see if this drug treatment will bring that overactivity down as we monitor it with brain imaging. If it's compensatory, their memory will get worse. If it's the same as in the rats, their memory will improve. And it turned out they were the same as rats."

Gallagher spent many years building her scientific case, even as so many other researchers had their sights set on amyloid beta. To keep her sanity, she started writing a mock TV series about her struggles. "I just keep myself going because there's some humor in it. We're in season three or four." On the day Gallagher finally finds time to speak with Barash, the research she did on naltrexone in the 1980s is the last thing on her mind, and she doesn't have the bandwidth to consider it carefully. Gallagher has much bigger fish to fry. After several smaller clinical trials, she now has the evidence to justify a Phase 3 clinical trial—the final step on the road to a new treatment. Although her proposed strategy is outside the mainstream, Gallagher's results and reputation were enough to secure multiple grants and launch a company called AgeneBio.

In January 2019, AgeneBio began enrolling patients in a trial slated to run at twenty-six sites across North America. The patients are between the ages of fifty-five and eighty-five and will take either a placebo or a very small, specially formulated dose of the anti-seizure drug leveti-racetam, called AGB101, for a year and a half and track whether the drug slows the progression of memory loss. Patients will also undergo PET imaging for tau and amyloid. If the drug works, Alzheimer's patients will have access to a once-a-day pill instead of the expensive infusions that would be required for anti-amyloid therapies. But it won't cure Alzheimer's. The goal is to stave off the progression of the disease. Even buying as little as one to two years could result in 10 to 20 percent fewer people with full-blown Alzheimer's. Gallagher has her sights set on a still bigger prize. "The idea here is that if our therapeutic works in MCI—and that's the study we can afford to do, barely—I think it will work earlier and truly be a preventative strategy."

—⁓—

The seeds of Li-Huei Tsai's interest in Alzheimer's were planted when she was three, on the day her grandmother forgot where they lived. On their way home from the market, the two took shelter from the rain at a bus stop. But when the rain ended and Tsai said it was time to go home, the old woman didn't know where home was. Tsai can still see the look on her grandmothers's face.

Today Tsai is the director of the Picower Institute for Learning and Memory at MIT in Cambridge. Rows of Veuve Clicquot champagne bottles, the souvenirs of decades of scientific achievements, line the shelves above her desk. Less conspicuous is a device you could mistake for a white-board, which Tsai turns on and sits in front of for as long as an hour a day, day in and day out. It emits flickering white light and a strange buzzing tone called pink noise. Both the light and the sound oscillate forty times a second in what's called a gamma wave. "I love to expose myself to gamma tone and gamma light," Tsai says.

The brain has a handful of innate brain rhythms—delta, theta, alpha, beta, and gamma—which range in frequency from about 1 to 150 times a second; scientists believe these waves make it possible for billions of neurons to coordinate the information they share. Imagine sitting by a lake watching four people in four rowboats, all rising and falling together on the waves. The people aren't secretly communicating with each other, nor are they physically connected. Instead, the same waves rock the boats. All four get the same information, so even though they're separated in space, they act as one.

Gamma waves orchestrate the activity of neurons throughout the brain. Stronger gamma, waves with higher peaks and lower troughs, are associated with paying attention, better working memory, sensory processing, and spatial navigation. Tsai is interested in gamma because it's weaker in people with Alzheimer's. What's more, gamma is weaker even before amyloid beta begins to accumulate. The possibility that weak gamma could be among the earlier things to go wrong in Alzheimer's patients is just one of many lines of evidence that lead her to believe that the hour each day she spends in front of her device is time well spent.

So, what sets up these waves in the first place? Neuroscientists will tell you that they don't completely understand it, like so much else about the brain. But it seems to work in a yin-yang kind of way. The firing rates of cells create brain waves, and brain waves, in turn, orchestrate cells' firing rates. One type of neuron, less plentiful but more diverse than the rest, is in charge of these rates. They're called inhibitory neurons, and without them, the brain would be chaotic. Most neurons are excitatory; they receive messages from other neurons, and if that message is loud enough, they'll fire and pass it on. Like bouncers at a bar, inhibitory neurons keep the excitatory neurons under control. As a general rule, when they fire, they turn those excitatory neurons off.

Inhibitory neurons come in many shapes and sizes throughout the brain. One is called a basket cell. Its axon splits into many filaments and wraps around an excitatory neuron's cell body, the point where the axon exerts maximum influence. A single basket cell can synchronize and control the output of hundreds or even thousands of excitatory neurons, switching them on and off with precise timing, setting up a rhythmic tug-of-war that creates the waves. When basket cells are activated and entrained at forty times a second—in theory, by any input that oscillates at 40 hertz, such as light, noise, smell, or even touch—the peaks and troughs of the gamma wave are strengthened.

Tsai's lab began to explore the effects of entrainment using light. They inserted fiberoptic wires into the brains of mice engineered to produce extra amyloid beta, and shone 40-hertz light down the wires directly into the hippocampus, where it activated only the inhibitory neurons. After one hour of treatment, scores of genes turned on. Microglia, the brain's immune cells, changed shape to prepare for their house-cleaning role and nearly doubled in number. Briefly, the levels of amyloid beta decreased. Tsai was floored. "I said, 'Oh my God, you've got to repeat it to see if this is real. This is too surprising.'" She also knew that even if her team could replicate the experiment, the results would only be useful for people if she could find a way to strengthen gamma without inserting wires into their brains.

In follow-up experiments, mice unencumbered by wires spent an hour a day hanging out in darkened rooms in small boxes lit by flashing LED spotlights. Light doesn't penetrate all the way to the hippocampus, but it can entrain brain waves in the visual cortex, where sight is processed. Treating the mice an hour a day for seven days reduced not just free-floating amyloid beta, but also amyloid plaques and toxic tau. But this was only in the visual cortex. Tsai needed to find out if the beneficial effects of flickering light could reach ground zero, the hippocampus. When her team upped the hour-a-day exposure times from one week to as much as six, gamma strengthened in the hippocampus, fewer neurons died, and the mice were better at remembering where to find the hidden escape platform in a water tank.

Tsai's team tried using 40-hertz pink noise instead of light and was able to achieve similar effects in only one week. The diameter of blood vessels also expanded, which may, in turn, help clear out amyloid beta and tau. And when mice were exposed to light and sound at the same time, there were fewer amyloid beta plaques throughout the brain, and microglia clustered around the ones that remained. Tsai's team is trying to work out why this non-invasive treatment has such profound effects, but the potential for Alzheimer's patients seems too great to wait for the answers.

In 2016, Tsai and a collaborator at MIT named Ed Boyden founded Cognito Therapeutics to explore whether gamma entrainment using light and sound works in people. Ninety participants are being tested at multiple locations in three trials using Cognito's device. One trial was slated to report results in October 2020, another in 2021, and the third in 2022. At MIT, Tsai is running her own small clinical trial to assess the effect of daily treatment for up to nine months, with results expected to be reported in 2025. "My dream is that perhaps one day we can try to create a 'gamma society,'" she testified before Congress in 2017. "We can try to change our lighting system(s) at home, or on the streets, the refresh rate of computer monitors or TV, or people can get exposure to the gamma flicker more readily, to create a healthy society." Her optimism is tempered with humility. "On the surface, we seem to know a lot about Alzheimer's disease, but when it comes to intervention, it has been a humbling process," Tsai told an audience of neuroscientists two years later. "We've been burned hundreds of times." Although neuroscientists agree that the function of brain waves remains enigmatic, manipulating this activity holds promise in other brain disorders, including Parkinson's, epilepsy, and mental illness.

—m—

Saul Villeda looks lovingly at his two mice, tracking their movements around the cage. He knows you should never name them, because you get too emotionally attached. He made that mistake once. One mouse is old, the other is young, and they're surgically connected along the length of their abdomens, moving as if they were born that way. Villeda is an assistant professor at the University of California San Francisco. He grew up poor in East Los Angeles, and he's the first in his family to go to college. He thought he wanted to be a mechanic, but a professor noticed that Villeda looked at things from a different angle and encouraged him to think again. Villeda, rather than tackling Alzheimer's head-on, looked at this disease of aging and asked, "Can I make something young again?"

That's where his stitched-together mice—whose blood flows freely from one to the other—come into the picture. These lab mice are inbred, so there's no tissue rejection, but they're otherwise natural. Typically, the young mouse is about three months, and the older one is between a year-and-a-half and two years old. That's like attaching a person in their midtwenties to someone in their late sixties or early seventies. Young mice run around a lot, their fur is glossy and thick, and they have an easier time remembering where to find a hidden platform in a large tank of water. Old mice are forgetful. They have dull, patchy fur. They're slow and frail.

Before, during, and after the operation, the mice are treated with as much care as human patients. The surgeons, one for each mouse, wear hairnets, booties, and smocks to keep conditions sterile. The mice are anesthetized before the surgeons cut through the skin and the membrane surrounding the abdominal cavity, then suture the two cavities together to create one big sac. They connect the leg joints where the two mice meet so the animals won't pull at the scar, then sew them up. Within three weeks, the mice have learned to walk as one. Within ten days, the stress hormones caused by surgery go down, just as they would in a person. Within two weeks, the same blood is flowing between them, and the transformation begins. After five weeks together, Villeda's old mice look healthier, and their memory improves. "We never get an old animal back to a full young animal," says Villeda. "It's probably bringing someone in their seventies back to around forty. But not twenty." But the effects go both ways, and the young mice get the short end of the stick, looking and acting like fifty-year-olds. Barbaric though parabiosis sounds, it's an efficient way to test the concept: infusing an old mouse with enough young blood plasma to make a difference would require sacrificing many more animals.

In 2014, when Villeda was a grad student at Stanford, *Nature Medicine* published his proof-of-concept results showing the rejuvenating power of young blood. Other scientists were also finding blood factors that reverse aging throughout the body, including the brain. "There's an incredible amount of plasticity left in the old body, including in places like the hippocampus," says Villeda. "And the best part about it is that you don't have to drill holes to get into the brain. It's screaming therapeutic potential." The results were so appealing that it wasn't long before a start-up California company named Ambrosia began charging people thirty-five years or older eight thousand dollars to be transfused with young blood plasma. After the FDA issued a warning that the treatment hadn't gone through rigorous testing to ensure that it was either safe or effective, the company shut down.

Essential safety concerns aside, there's not enough young blood in the world to treat every Alzheimer's patient today. Villeda's parabiosis experiments are just screening tools to help hone in on which of the hundreds of factors—proteins, antibodies, clotting factors, electrolytes, or hormones—circulating in blood plasma either turn back the clock or speed it up. Because exercise has such proven benefits on aging, Villeda decided to search there. After giving his aged mice continuous access to treadmills for six weeks, he sorted their blood plasma to see which factors went up. There were thirty—too many to test. So he winnowed it down to those involved in metabolism and, from there, zeroed in on an obscure enzyme that the liver pumps out after exercise. Why an obscure enzyme? "I've been pursuing factors that we don't know very much about," Villeda says, "because, in my mind, it's probably something we've overlooked."

To test whether his hunch about this enzyme was correct, Villeda created mice that pump out much more of it than usual. After three weeks, these genetically engineered mice had more neurogenesis and performed better on memory tests. Villeda partnered with Joel Kramer and Kaitlin Casaletto at the MAC, who found that this enzyme is also higher in older people after they exercise. But not everyone can exercise. If this liver enzyme could be distilled into a pill, it could help treat people too frail to get on the treadmill. Other labs such as Stanford, Columbia, and Harvard are also investigating the rejuvenating effects of young blood. Several clinical trials are already underway, including one at multiple locations across Spain and the United States.

Villeda, Sahay, Tanzi, Gallagher, and Tsai are just a handful among many researchers who want to intervene in the process of aging to slow down or prevent Alzheimer's. And while these approaches may seem vastly different, Villeda says that "there are enough commonalities between all this research that there's some fundamental truth there that we can tap into." Researchers are working on transplanting healthy inhibitory neurons into an aging mouse hippocampus, boosting the brain's ability to clear out plaques and tangles, or rejuvenating the neurons' metabolism to counteract the effects of aging. Like other chronic diseases—mental illness, HIV/AIDS, high blood pressure—multiple treatments may be most effective. "I think there's no silver bullet," says Gallagher. "Ultimately, when we have relegated Alzheimer's pretty much to the history books, it's going to look more like HIV/AIDS—not the biology of it, but there's going to be some combination of therapeutics."

Curing Alzheimer's will be hard, but no one in the field is giving up, despite past failures and all the reasons why Alzheimer's is a formidable disease. "Being a researcher is inherently an optimistic choice because it assumes that there's light at the end of some tunnel," says Howie Rosen at the MAC in San Francisco. "And that if you keep walking, you'll get there." The stakes are too high not to keep on walking. By some estimates, forty-six million people in the United States alone are on their way to developing Alzheimer's without even knowing it.

# PART FIVE

. . . . . . . . . . . . . . . .

# OUTSTANDING QUESTIONS

# Finding
# Lost Memories

· · · · · · · · · · · · · · ·

Memory has three stages: encoding, storing, and retrieving. If any one of them fails, an episode from three days or three decades ago will be missing from your personal story. In an obscure natural experiment that began on an operating table three quarters of a century ago, hints emerged that vanished memories could be found. Inside a turreted gray limestone building on a hilltop in Montreal, Dr. Wilder Graves Penfield gently lowers an electrode onto the glistening surface of the patient's right temporal lobe. She is a 32-year-old woman with severe epilepsy. Suddenly, she speaks.

"I hear singing."

Penfield lifts the electrode briefly, then lowers it again. "Yes, it is 'White Christmas,'" she says.

Penfield waits for his assistant to place a small white paper square with the number 15 to mark the spot before moving the tip of the electrode a quarter of an inch or so away.

"That is different, a voice—talking—a man." Penfield lifts the electrode, and the assistant marks the spot with a paper square numbered 16. Penfield discovers eight locations that each evoke different auditory memories—a radio program, a play, a violin. Twenty-six minutes after the patient hears the song "White Christmas," Penfield returns to location 15 and lowers the electrode again. Again, she tells him the orchestra is playing "White Christmas."

Penfield's plan was to find the source of this woman's seizures so he could remove the diseased brain tissue. He wasn't looking for memories—but he appeared to have found them. Between the 1930s and the early 1960s, he documented forty examples of what seemed like fragments of memories activated by brain stimulation. A 26-year-old woman says, "I hear voices. It's late at night, around the carnival somewhere, some sort of traveling circus. I just saw lots of big wagons they use to haul animals in."

"I hear my mother singing," says another young woman.

"My mother is telling my brother he has got his coat on backward," says a 12-year-old boy. A 21-year-old man who had recently traveled to Montreal from his home in South Africa describes hearing his cousins Bessie and Ann Wheliaw laughing, probably at some joke. Penfield wrote that "it was at least as clear to him as it would have been had he closed his eyes and ears thirty seconds after the event and rehearsed the whole scene 'from memory.' Sight and sound and personal interpretation—all were re-created for him by the electrode."

Penfield used a stenographer in the operating theater to document what the patients described. He and other surgeons would often ask clarifying questions. Some patients said they experienced the sights or sounds not so much as if they remembered a long-ago event but as if the event was happening in the present moment. And yet they were also aware that they were lying on an operating table, under bright lights, their brains exposed, sharing what they felt. Penfield was struck by how vivid these recollections were, which he took as evidence that the memories were real and unchanged. "It is a hearing-again and seeing-again—a living through moments of past time." Penfield called these phenomena recordings.

In the early 1950s, just a few years before a surgeon removed patient H.M.'s hippocampus, Penfield gave several lectures about the storehouse of memories in the brain. The temporal lobe, home to the hippocampus, appeared to contain what Penfield later called keys of access to these memories. The fact that the content of these experiential phenomena seemed trivial—standing on a street corner, hearing a mother call her child—supported Penfield's belief that the brain permanently stored every experience, no matter how inconsequential.

His interpretation may not be correct, but what he found in the operating room was no fluke. In the decades since, neurosurgeons from many countries have described what appeared to be fragments of past events elicited by brain stimulation. Some date back decades. A 34-year-old man hears the theme song from *The Flintstones*, a TV cartoon he'd last seen at the age of 15. Another hears "Wish You Were Here" by Pink Floyd. A woman smells burnt wood, which reminds her of an evening in Brittany sitting around a campfire when she was a teenager.

Penfield stumbled upon traces of memories, yet his accidental discovery could have been predicted by an evolutionary biologist named Richard Semon. Semon's ideas were shockingly before their time and are only now proving prescient. Back then, Semon's peers were focused mostly on how memories are encoded and stored. He was just as interested in the process of recall. Unlike Penfield, who believed that memories are permanently and immutably stored as if on a strip of film, Semon suspected that a memory changes every time it's replayed. He also thought that there might be multiple versions, or traces, of the original experience. Each one could be stored in slightly different connection patterns distributed across the brain, and these versions could interact with each other. He named these traces the "engram." Semon made no bones about the fact that he could only theorize about the nature of engrams, leaving it to future scientists and better tools to figure out if he was correct. He was also a century ahead of his time, and his ideas were ignored. At the age of fifty-nine, shortly after the end of World War I and his wife's death, Semon draped himself in a German flag, lay down on his bed, and shot himself in the heart.

An American zoologist named Karl Lashley was next to take up the search. He would devote more than thirty fruitless years to the hunt. Lashley had a practical, evidence-based method for finding the engram. He trained rats to navigate a maze and then systematically went about slicing through snippets of tissue in different parts of the brain. The rats' memory declined in proportion to how much tissue he damaged with his scalpel, but where he sliced at it didn't seem to matter much. Just as Semon had imagined, Lashley's evidence suggested that any given memory is distributed across many brain areas. But there was a problem. No matter how many sections of brain Lashley damaged, he could never fully abolish a memory. Memory seemed to be at once everywhere and nowhere. At the end of his career, he quipped, "I sometimes feel, in reviewing the evidence of the localization of the memory trace, that the necessary conclusion is that learning is just not possible." Lashley abandoned the search for the engram just as Penfield appeared to have brought one to life by electrically stimulating the temporal lobe, the home of the hippocampus.

"What Penfield did was proof of principle that this is not crazy. This can actually happen," says Steve Ramirez at Boston University's Center for Memory and Brain. He has decorated his light-filled office with a large blow-up T. Rex dinosaur, and his Twitter handle is @okaysteve, but despite his fun-loving, low-key attitude, Ramirez has the same lofty goal pursued by Semon and Lashley. "Can we go in and create maps for memories in the brain? Like, what does a memory even look like? Memory has sights and sounds and smells and emotions associated with it. Can we find those elements and say this is what the totality of one particular memory physically looks like?"

About a decade ago, inspired in part by a painful breakup with his girlfriend, Ramirez and his collaborator at MIT, Xi Liu, set out to answer this question by trying to find a fear memory in the hippocampus of a mouse. They seized on a newly invented tool called optogenetics, which lets scientists turn neurons on by shining light through a fiberoptic wire into the brain of a genetically engineered mouse. But which neurons? Out of millions of cells in the hippocampus, Ramirez and Liu only wanted to see the ones involved in making a specific memory. In a feat of precision engineering, they adapted the tool so the light-activated switch only worked in neurons that had recently fired—in other words, the neurons involved in learning something new. In this experiment, they put the mouse in a box with an electrified floor, where it quickly learned to be afraid—and to freeze. To see if they could artificially reactivate that memory, the scientists put the mouse in a different box without an electrified floor, sent light down an optical fiber into the hippocampus, and turned on the same collection of neurons that had fired in the first box. The mouse froze in fear even though there was no reason to freeze. This was a new environment; a space the animal should have assumed was safe. But by activating the tagged neurons, Ramirez and Liu had found an engram—or if not the totality of it, at least a sufficient number of neurons to turn it back on.

As if that hadn't been enough, they immediately took on what seemed like the next logical challenge; over three long days in the short life of a mouse, could they edit the memory of being in a safe space and turn it into a fear memory? Ramirez and Liu dubbed this experiment Project Inception. They put the same type of genetically engineered mouse in a small box that smelled of overripe fruit in a dimly lit room under warm red lights. The mouse had twelve minutes to roam about, learning that there was nothing to fear in this box. At the same time, Ramirez labeled only those neurons that were firing. He placed the same mouse into a different environment on day two, a well-lit, almond-scented box. The mouse shouldn't have mistaken it for the first box. But while it roamed around, Ramirez used light to reactivate the original memory. At the same time, he gave the mouse a gentle foot shock. On day three, he returned the mouse to box number one, the safe space. The familiar smell of overripe fruit and the warm red lights should have cued the mouse's brain to turn on the safe memory. But instead, the mouse instantly froze. It appeared that the scientists had edited the foot shock memory from box number two into the first memory. Ramirez and Liu proved what neuroscientists, psychologists, and anyone who's disagreed with a friend about what actually happened have long believed to be true. Memories change. They're updated—or reconsolidated—with new information that can be true or false. In 2012 and 2013, Ramirez and Liu announced the results of these experiments. Tragically, Liu passed away a few years later, but Ramirez continues his hunt for the engram.

So if a mouse's brain can't tell the difference between an edited engram and a real one, what about one created out of whole cloth? A few years ago, a team at the Hospital for Sick Children in Toronto led by Paul Frankland and Sheena Josselyn implanted an engram for a completely fake memory that linked the scent of oranges to either a rewarding or an aversive sensation. Mice have little bundles of neurons in the olfactory region that each code for specific smells. One of those smells is orange, an odor that the average mouse doesn't seem to care about one way or the other. In this experiment, scientists optogenetically stimulated that bundle of neurons simultaneously with another brain region that processes the aversive aspect of an experience. Afterward, when they put the mouse into a cage with a piece of filter paper infused with the orange scent, the mouse stayed away—just as it would have if the scientists had paired orange with a foot shock in the real world. Frankland and Josselyn's team tried it the other way around too, stimulating the orange-scent-detecting bundle of neurons at the same time as another brain region that processes a rewarding experience. When this mouse went into the cage, it approached the orange-scented paper as if there had been something good about it. Both memories were fake, created from scratch just by stimulating different brain regions at the same time. But from the mouse's perspective—or at least judging from its behavior—this engram was indistinguishable from reality.

"I'm really interested in understanding what's real, in a very, very deep way," says André Fenton from New York University. "I follow the line of reasoning that I am a biological object, and so for me to have experiences that I think are mine, I need to figure out the biology that gives rise to my experiences." If memories are contained in the collections of neurons, Fenton studies the molecules that hold those collections together. Or rather one particular molecule, a protein called PKMzeta, which is concentrated in the synapses between neurons. There's just one problem. Proteins only last about a week in the body before they get degraded and recycled. "So how can you build something permanent—or permanent-ish—with something that's very transient?" Fenton asks. As he and his collaborator, Todd Sacktor, discovered, PKMzeta has an unusual property that gets around that problem; once made, it can be replenished on site, generating a rotating cast of PKMzeta proteins that glue a particular synapse together indefinitely. There's at least one other leading candidate for a protein that holds synapses together, and likely many other molecular players that help turn a single experience into a lasting memory. But Fenton and his collaborators have shown that just by blocking the activity of PKMzeta, they can unglue synapses and remove a memory.

Ramirez's former professor at MIT, Susumu Tonegawa, looks at the question of what makes memories fail from the same perspective as

Richard Semon—it could just as easily have to do with faulty retrieval as faulty encoding and storage. In one test of this idea, Tonegawa's team discovered that even if they blocked the formation of proteins like PKMzeta that strengthen the connections between synapses, they could still artificially reactivate memories optogenetically. This also suggests that an even more transient, or "permanent-ish," event takes place before proteins can be made. A recent experiment in Li-Huei Tsai's lab revealed that the very first thing to happen after neurons are activated is that their DNA is primed—open for business—and can later begin to churn out proteins if the neurons are reactivated again during a very specific time window. Tonegawa later found evidence that two copies of the memory engram are created at once—one in the hippocampus and the second in other brain areas that store all the individual fragments. This experiment made headlines because it ran counter to one long-held theory that memories are laid down in the hippocampus and then gradually transferred and strengthened elsewhere for long-term storage.

Protein synthesis that strengthens connections made during the first few minutes after an experience, Tonegawa suggested, could be what makes it possible to retrieve a memory with natural cues, like a few notes from a familiar melody or the smell of burning wood. But that doesn't mean that that original engram has disappeared. It may be silent, capable of being reawakened only with a more powerful cue, like the stimulating tip of an electrode or a beam of colored light traveling down a fiberoptic wire into the hippocampus.

Neuroscientist György Buzsáki at the New York University School of Medicine prefers to steer clear of human constructs like engrams. He thinks they can get in the way of seeing things from the brain's perspective. "I don't know what an engram is, and I don't even try to understand it," he says. "Everything is a relationship between the brain and what happens outside." In his view, the brain can't possibly be a blank slate waiting for experiences to etch new pictures onto it. Instead, it comes equipped with a huge reserve of built-in patterns, each one created by a connected group of neurons. Memory formation is a game of matching those patterns with meaningful experiences so as to better predict the future and the consequences of its own actions.

This dynamic system is enormously flexible because neurons can be swapped in and out of any group and still generate very similar patterns. Buzsáki argues that you could get rid of every neuron involved in recognizing a specific person and still retain that memory. "You are welcome to erase all my Jennifer Aniston cells in my brain with a magic laser, and I guarantee you that my memory of Jennifer Aniston will stay. Because the runner-ups will occupy the space right away." Experiments by Ramirez and others support this dynamic view of memory, showing that the overlap between the neurons involved in the making of a memory and the neurons involved in reactivating it can be very small.

New frameworks for thinking about memory and new tools—like holographic optogenetics that can "see" in three dimensions, or noninvasive wireless optogenetics—may bring neuroscientists closer to understanding the fundamental underpinnings of this essential human faculty. But at the very least, such basic neuroscience research in rodents offers some distant vision of how to help people with memory disorders. Ramirez, for one, hopes there will be ways to translate memory-manipulation insights into therapies for people with mental illnesses like depression, anxiety, or PTSD. "More sensible and practical scientists work on how to do good things with this knowledge," says André Fenton. "I work on things like PKMzeta and mechanisms with the assumption that when we understand that sufficiently, we will figure out how to apply it usefully."

# 21

# The Case for Tau

. . . . . . . . . . . . . . . . . . .

## *December 2018*

In a darkened room at Biogen, Butler stands a few feet away from a six-foot-tall, three-dimensional projection of Owen's hippocampus. Two white C-shaped objects float in space before him, their outlines clearly defined. An engineer on Butler's team fed the data from the MRI scan taken before Owen's overdose into the system and rendered the images. It's a test run for a new visualization tool called VisCube. Joystick in hand, Butler manipulates Owen's hippocampus, flipping it upside down, spinning it around, zooming in and out, exploring the detailed anatomy of all the subregions, from the larger hot end of the hippocampus facing the front to the cold tip at the back.

Butler asks the technician to switch to the scan captured four months after Owen's overdose. From above, the sharpness of the images reveals a slight buckling along the ridge that runs from front to back. The damage looks even more apparent when Butler flips to the view from below. Both hippocampi have crumpled in on themselves, but the right side looks worse. It's normal to have some variation between the two sides, and Owen's right hippocampus was already smaller before the overdose. Now it actually looks very much like a seahorse. Before Thanksgiving, *JAMA Neurology* rejected their case report, but Butler and Barash plan to include these before-and-after 3D pictures in a new submission to another journal.

Owen's tests from his hospital stay in March ruled out traditional sources of oxygen deprivation, such as a heart attack, stroke, seizure, infection, or anemia. While it's possible that Owen's breathing was suppressed at some point during the overdose, the 3D images back up their long-standing suspicion that hypoxia alone cannot explain the syndrome. If that were the case, other brain areas would be affected, but Renaud La Joie's advanced imaging protocol shows that only Owen's hippocampus has atrophied. Advanced testing performed by colleagues at the University of Pennsylvania has ruled out rare autoimmune disorders that preferentially zero in on the hippocampus.

A direct toxic effect of fentanyl on inhibitory neurons remains the most plausible explanation for the amnestic syndrome. But neither the striking image nor the test results answer a glaring question: Why did this happen to Owen and not tens of thousands of other people who survived a fentanyl overdose? Butler has some hypotheses that he'll propose in the paper. There could be genetic differences between people that affect how blood flows through the hippocampus, how the neurons signal to each other, or how they're programmed to die. He also wonders if Owen's long history of opioid use followed by eighteen months of sobriety made his hippocampus somehow more susceptible.

Butler thanks the engineer as he hands him the controller and watches the luminous image of Owen's damaged hippocampus disappear. He has spent much more time on the case report than he anticipated, and now he's feeling the pressure to get it submitted and published. So this time around, when Barash texts him about whether tau PET scans could give more insight into the syndrome, Butler is all business.

Barash: Hey I was thinking of looking at the tau scans in ADNI to see what sample of opioid exposed was . . .
Butler: I just e-mailed you
Barash: Ok
Butler: Built 3D renderings of hippo. Patient pre and post
Barash: Oh cool

—m—

Tau is a shape-shifting protein with the freedom to take on different forms. Tau is concentrated mostly in the neurons' long and fragile axons, where it attaches to hollow cylinders called microtubules, which give cells their structure and shape. Microtubules also act like railways to help shuttle materials around. Traditionally, scientists believed that short lengths of tau served as railroad ties that stabilized the track. It's possible that, instead, tau gives microtubules the ability to flex and grow without coming apart. Either way, in a diseased brain, tau falls away from these microtubules and begins to accumulate inside the cell body. Some misfolded tau can form the toxic seed of a growing molecule by linking up with other copies of itself, lengthening into long sheets that fold back on each other like a hairpin, twisting together into pairs of long filaments, and clumping together into tangles. These tangles are thought to interfere with the transportation of proteins and other materials, ultimately killing the cells.

Tau tangles are a hallmark of other dementias and neurodegenerative diseases like frontotemporal dementia and chronic traumatic encephalopathy (CTE); they take on different shapes and accumulate in different places depending on the disease. In CTE, they build up around small blood vessels in the folds of the brain's outer layers. A concussion can pull an axon apart, while smaller blows may leave the axon intact but still damage the even more fragile microtubules. In frontotemporal dementia, tau tangles are shaped like a J and accumulate in the frontal lobes. But no matter where tangles accumulate, they spread in a pattern unique to each disease.

The spread of tau looks a lot like an infection. It begins in one place and invades connected areas, perhaps moving along the axons. Stanley Prusiner, the scientist who proposed that infectious proteins called prions cause Creutzfeld-Jakob disease, has expanded his research to encompass Alzheimer's, which he refers to as a double-prion disorder. In his view, which is shared by many scientists, distorted forms of both tau and amyloid beta spread through the brain by acting as templates, contorting other proteins into their virulent forms. In some patients, especially those who succumb to the disease at a young age, their version of tau may be more infectious or harder to clear.

In 1991, one year before scientists proposed the amyloid cascade hypothesis, German neuroscientists Eva and Heiko Braak showed that tau tangles spread in such an orderly and predictable way that researchers could use the pattern to define the stages of Alzheimer's. Amyloid beta plaques accumulate in the outermost layer of the brain, but the Braaks found that abnormal forms of tau tangles first appear in the entorhinal cortex, the gateway that funnels sensory information into the hippocampus. Where tau tangles accumulate corresponds more closely than amyloid beta plaques to brain damage and how cognitively impaired patients are. But despite the obvious connection between tau tangles and Alzheimer's disease, this enigmatic protein remained in the shadows for many years and has only recently come into the spotlight.

"When I started, in the nineties, I was always the last session, the same four people in the audience," says Karen Duff, the director of the United Kingdom Dementia Research Institute. "Everyone had heard the amyloid talks, and they couldn't care less about tau." Neither rogue protein on its own is enough to cause Alzheimer's. It's only when amyloid beta spreads down into the entorhinal cortex and meets up with tau that a deadly dance begins. Recent research shows that amyloid beta in the hippocampus makes neurons more hyperactive, which in turn releases more tau from the synapses. "And then it's unleashed," says Duff. "The tau pathology just spreads. It has its chosen routes and its chosen cell types, which are more vulnerable, and then you get the cognitive decline."

Because tau's presence and how much there is predicts how poorly patients fare and how quickly they decline, this protein is becoming a new center of gravity for Alzheimer's research and clinical trials. New clues that *APOE4* can make tau more toxic—especially in women—are helping put different pieces of the Alzheimer's puzzle together. To test drugs and understand how the disease unfolds, scientists are developing mouse models that accumulate both tau and amyloid beta. In recent years, therapies that target tau and prevent this abnormal protein's buildup have claimed a growing share of clinical trials. But the nature of the deadly dance, or whether directly targeting tau will be more effective than targeting amyloid beta, is unclear, especially since there are multiple forms of tau in Alzheimer's patients. Duff, for one, is agnostic. She has her sights set on improving the brain's ability to clear out both of these misfolded proteins before they become an unstoppable force. If successful, doctors might be able to use this therapy to treat other diseases that feature the accumulation of toxic proteins, like Parkinson's or Huntington's.

In Alzheimer's patients, tangles accumulate over many years, long before any symptoms show up. But the brains of young opioid abusers in Scotland also contained tangles, including one victim as young as seventeen. What about Owen? One catastrophic overdose was enough to destroy 10 percent of his hippocampus. Did that event also launch a relentless, self-seeding cascade of tau buildup in his hippocampus? Or is he young and healthy enough that his brain has cleared the debris left behind, especially now that he's no longer using? If there are tau tangles in the hippocampus of any of the amnestic syndrome victims, it would suggest that long-term use of opioids could lower the bar for eventually developing Alzheimer's. Barash and Butler have been circling around this possibility ever since the *MMWR* report was published. A relatively inexpensive spinal tap would be enough to reveal elevated levels of tau in the cerebrospinal fluid, which would signal a problem. But it couldn't show where in the brain the tau tangles, if any, were building up. And for now, there's only one proven way to do that: a PET scan costing thousands. One or two patients may not be enough to figure this out, but it would be a start. Barash will have to convince someone who has the time, the funding, and the expertise to pick up the baton and follow this line of research.

## 22

# The Road Ahead

· · · · · · · · · · · · · · ·

## *December 2018*

The drive from the Soldiers' Home in Chelsea to Mass General's Research Center in the former Navy Shipyard in Charlestown is barely five miles, but on a weekday afternoon, it can take forty-five minutes. A few days earlier, an e-mail to the head of anesthesiology at Mass General led Barash to Zhongcong Xie, an anesthesiologist who trained with Rudolph Tanzi and studies Alzheimer's. Xie, who was third in line on the e-mail chain, is investigating why older people often experience memory loss after surgery. As much as Barash prefers to do business from the comfort of his couch, Gillian has convinced him that this meeting is too important not to happen face-to-face. On the bridge over the Mystic River, with the Soldiers' Home water tower visible in the rearview mirror, Barash runs through what he plans to say. With rare exceptions—like his first phone call with Kofke not quite two years ago—it takes a lot of explaining before anyone understands what he's talking about. It probably doesn't help that he's just some medical director from a nursing home. But maybe this meeting will be different. Xie knows fentanyl because he uses it in almost every surgery, he understands Alzheimer's, and he answered Barash's e-mail right away. If anyone is going to care about what Barash has to say, it's probably Zhongcong Xie.

Xie is friendly and courteous but appears distracted as he walks Barash quickly through the Martinos Center's vaulted granite and glass lobby, past a water fountain surrounded by potted trees, and up to his lab, which takes up the better part of the fourth floor. Barash passes ten or so researchers peering through microscopes or bent over lab benches. He glances at the rooms containing mazes designed to test mice for memory loss and takes note of the tiny operating chambers that flank the hallways.

Xie has a lot of research projects going on, but they all revolve around the same question. Does anesthesia have long-lasting effects on the brain, and if so, how can it be made safer? Up to 45 percent of older people experience a brief period of delirium after surgery. An estimated 10 percent complain of trouble with thinking and memory for months afterward. One percent may never recover. Many factors likely play a role—especially the stress of surgery itself—which makes it hard to sort out what's behind POCD. Xie is interested in the anesthesia piece of the puzzle. Surgery is often unavoidable, but the practice of anesthesia can be changed.

The stakes for getting a clear picture of who may be at risk are high. The number of people around the world who have surgery each year is more than 300 million and rising. A third or more are over sixty-five. In people suffering from POCD, Xie and others have discovered higher than normal levels of tau protein and amyloid beta in the cerebrospinal fluid that bathes the brain. Cardiac patients, who appear especially vulnerable to POCD, can have increased levels of tau for as long as six months after surgery. When researchers analyzed men and women in the same group, they found no connection between the *APOE4* gene and the risk of POCD. But when they looked at each group separately, they discovered that men who carry the gene are more likely to develop memory loss.

Despite such worrying evidence, it's hard to know whether surgery merely unmasks cognitive decline that had previously gone unnoticed, accelerates it, or causes it. And many of the variables at play—like age, pain, sleep deprivation, stress, and inflammation following surgery—are themselves risk factors for Alzheimer's. The question for Xie is whether any of the ten or so drugs used during surgery should be avoided or administered with special care in older adults, especially anyone with mild cognitive impairment or at risk for Alzheimer's. Like others in this field, Xie has never studied fentanyl, even though it's used in most surgeries. One barrier to such research is strict regulations surrounding the control of opioids. Although Xie is an anesthesiologist, he would need to apply for a special license, store the drugs in a locked safety box, keep meticulous records, and allow inspectors into his lab to check on it regularly. The truth is, however, that Xie has never even considered fentanyl as a possible culprit.

When Barash and Xie arrive in the corner office at the far end of the lab, a radiologist and a research assistant are already waiting for the meeting to begin. Once everyone's seated, Barash sees that he has the room's full attention. So he starts at the beginning of a very long story. He describes the amnestic syndrome; the connection with the fentanyl epidemic; Kofke's research on fentanyl; the evidence for tau in the hippocampus of young opioid overdose victims; his fear that older people who take opioids long term for pain control could be gradually eroding their memories; his ambition to test naltrexone as a therapeutic for the early stages of Alzheimer's; and the possibility that mice with opioid-induced damage could be a new animal model for Alzheimer's. Heads are nodding. Xie's questions make it clear to Barash that everything he's saying makes sense.

"I never really thought about opioid use as an animal model for Alzheimer's." Xie looks out the window at the darkening sky. "No one else is looking at this. I'm interested."

"What about doing a pilot study with tau PET scans in a few amnesic patients? Just to get started," Barash says. "I can give you my write-up."

Without hesitating, Xie says he may be able to include those patients in his current project on tau and anesthesia. He'd need to amend it to include opioid abusers, who are automatically excluded from research projects, but that's just a clerical detail. And he's awaiting word on other funding. He'll think about whether he could use some of it to dig deeper into the relationship between opioids and Alzheimer's.

"So, what do you do over at the Soldiers' Home?" Xie asks as he escorts Barash to the elevator.

"I'm the medical director there. It's mostly an older population."

"You should come work with us instead!"

Barash smiles as he peels the visitor label off his shirt. "You know, I really like working with the staff over there and taking care of the veterans. It's a good job."

In the lobby, Xie asks, "By the way, how do you know Kofke?"

"My dad's an anesthesiologist."

A flash of recognition crosses Xie's face. "Your dad is Paul Barash?" he says as he shakes Barash's hand. "He's a legend! I saw him give a talk when I was just a fellow."

Back home, after the kids are in bed, Barash pulls out his to-do list. Somewhere along the line, he stopped writing lists by hand and stashing them in his pants pocket. Now they're on his phone, divided by projects. He annotates a few things as done and adds a few others:

SYNDROME
Go to DPH to review records for possible case
Create diagnostic criteria for amnestic syndrome
Monroe—ready to resubmit Owen's paper?
Speak with contact at National Center for Injury Prevention
   and Control (done)
E-mail CDC about nationwide search for cases (done)

NALTREXONE TRIAL
Request call with Alzheimer's Drug Discovery Foundation
Endogenous opioids rise with age?
Find academic hospital research partner—Beth Israel? Mass
   General?

TAU PET
Send Xie Tau PET proposal
Owen to Boston?
Chun Lim re other amnesic patients
Meeting with Xie (done)

CHRONIC USE
Monroe Project Seahorse status
CDC database overlap between opioid use and AD rates
Review Kofke's pain proposal (done)

—⁓—

Big changes are afoot at the Soldiers' Home in Chelsea on a late spring day in 2018. A helicopter buzzes a safe distance from the 145-foot-tall water tower. The red and white structure dominates the landscape but has stood empty for years. Now a long cable stretches from the top of one of the tower's six legs to earth-moving equipment. Neighbors, newscasters, Barash, and other Soldiers' Home staff line the street to watch. Once the tower begins to lean to one side, it takes fewer than five seconds before it lands facedown in the dirt. A rusty cloud of dust billows out the back through a panel blown off from the force of impact. "No way!" someone says. "Awesome!" says another spectator. "That's it. Back to work," says one of the staff.

The end of the water tower will make way for a new state-of-the-art facility with private rooms instead of wards. Barash will miss the charming unpretentiousness of the old Soldiers' Home building. But the wards hold about ten people to a room, which makes it hard to contain infectious diseases like the flu. In a few years, when they move into the new building, it will be easier to manage the veterans' health. "I hate to see it go," says one of the veterans with a sad smile, "but that's progress."

Ever since Barash's successful meeting with Zhongcong Xie at Mass General just before Christmas, progress on the amnestic syndrome research projects has taken one and a half steps backward for every two steps forward. One of Barash's former supervisors at Beth Israel Deaconess Medical Center told him that if he could come up with funding, they'd look into the feasibility of a naltrexone trial. A week later, Barash let him know that the Alzheimer's Drug Discovery Foundation had invited him to submit a grant proposal, but the neurologist admitted he already had too many trials in progress. Barash turned to a Mass General neurologist who also appeared interested, but he was overcommitted as well. Barash pitched the idea to an expert who oversees the portfolio of Alzheimer's clinical trials at the National Institute on Aging. Naltrexone is worth looking into, she told him. He should find a researcher who can take it on.

Barash takes comfort in the fact that whenever he gets his foot in the door to explain his ideas, people think he might be on to something, even if they're not in a position to help. "You excel in advancing knowledge without apparent support for your true skills," Butler responds via text when Barash gives him the latest update. Barash isn't surprised that he can't find a research partner. An established clinician or scientist at a major academic hospital like Beth Israel or Mass General is in the business of gathering data to win funding for the next grant. It's not easy to get off that track. And he or she is certainly not waiting around for a new project to surface.

With a full-time job and no lab of his own, Barash has four options in his weekend playbook to keep the momentum going: Sift through the scientific literature and online databases for more pieces of the puzzle; collaborate on an inexpensive pilot project with clinicians at a smaller hospital; ride the coattails of an established scientist whose research aligns with Barash's hypothesis; or e-mail experts at strategic intervals. He pursues every option.

—⁓—

Andrew Kofke is reading a letter from the Anesthesia Patient Safety Foundation, and he's mad as hell. A few months earlier, he sent an anesthesiologist named Ignacio Badiola the link to the HBO *VICE* documentary about the amnestic syndrome. Badiola works at the University of Pennsylvania's Pain Medicine Center. Kofke wanted to know if Badiola had noticed memory problems in patients who used opioids for many years. Badiola said he had. He'd always hoped the problem would go away when the patients stopped taking opioids, but he had no way to know if the drugs caused permanent damage. Kofke and Badiola decided to write a proposal for the Anesthesia Patient Safety Foundation, whose mission is "that no one shall be harmed by anesthesia care."

Their plan was to identify twenty to twenty-five pain patients who took long-term, high-dose opioids and compare them with patients whose pain was managed with drugs like aspirin, ibuprofen, or other nonsteroidal anti-inflammatory drugs. This approach addressed a weakness in the Project Seahorse study—the lack of enough patients in the opioid-free pain group. Kofke and Badiola wanted to measure hippocampus volume and do cognitive testing on both sets of patients to see if there is a detectable difference. The study wouldn't be able to determine cause and effect, but if the results showed a discrepancy that couldn't be readily explained by other variables, it would pave the way for a bigger research project on whether there's a link between prescription opioid use and insidious damage to the hippocampus.

Kofke figured their proposal was a good fit for a foundation dedicated to reducing harm, in part because the amnestic syndrome provides solid evidence that fentanyl can harm human memory. But the rejection letter explains that the proposal is not responsive to the topics the foundation cares about. Kofke suspects that the reviewers haven't even read the proposal. What makes him madder still is that some of the reviewers are the same experts who objected to his earlier research linking opioids to hippocampal damage. Kofke complains to Barash that there still seems to be resistance to the very possibility of harm. But Kofke is undaunted. He and Badiola will turn this project into a grant proposal for the National Institutes of Health (NIH), and maybe a government agency is more appropriate anyway.

A week later, over dinner in Kendall Square, Butler pulls the plug on Project Seahorse. Between Owen's case report and his full-time job at Biogen, it's been nine months since he's been able to focus on it. He tells Barash the data is too messy, and the documentation on dosing may not be completely reliable. He's worried about the small number of people in the non-opioid pain group. And while he could troll the MAC's database to beef up the numbers, he can't rule out the possibility that he'd be unwittingly selecting people that best support their hypothesis. Maybe Project Seahorse will find a signal in the noisy data, but it will be weak, possibly even meaningless. He doesn't want older adults to worry about losing their memory if they take a Percocet after getting a tooth pulled, or cancer patients to think they shouldn't take opioids to control their pain. Barash can see that Butler has made up his mind. Their only hope for figuring out whether long-term, high-dose opioid use damages memory is for the NIH to fund Kofke and Badiola's study.

—⁓—

*Dr. Butler sent me "my" case report. That was a heavy read. Turns out I'm like frickin' H.M. It used to be annoying to hear about him in every lecture on memory. Now hella weird to be memorialized that way. Could have prevented a lot of bad stuff if I'd taken better care of myself before the overdose. But everything that happened before made me who I am. Now just trying to be present around people I know and hugely grateful for their friendship. If I didn't have them and the foundation of everything I learned before, I would be destroyed in every way imaginable. I can't even fathom what life would be like today. Without some of these positive things that I just took for granted, I'd be living in a vacuum, in a void or something. My sense of appreciation has increased infinitely. That's how it feels.*

Owen has been working as a security guard at a senior living center for a few months. At first, the job was straightforward. All he had to do was sit in the gatehouse and wave people in. He kept meticulous notes about policies and procedures to make sure he didn't make any mistakes. Owen suspected that his coworkers noticed the note-taking and thought he was pretty strange. But because he compensates so well for his disability, his boss, who doesn't know about his memory loss, gives him new responsibilities, adding to Owen's anxiety.

He drives a security patrol vehicle to other gatehouses to give staff breaks or answer their questions. Another task is to motor around the grounds on a rover, collecting coins from unmarked laundry rooms, and keeping track of where he's already been so he doesn't waste time going to the same location twice. The task seems maximally designed to be a stress test for the spatial component of episodic memory. To make matters worse, there's a different route every day with a different sequence for collecting the coins. Owen solves the problem by creating a map and pinpointing the locations. Nevertheless, the growing responsibilities are exhausting.

Owen suspects his memory will never be much better than it is today, but there's good news too. He's on a waiting list for an intensive six-week-long OCD treatment that he believes might help. He's gone without abusing drugs for the longest period of his life since he took his first Vicodin pill at the age of eleven. Soon he'll start a new job working to get teens to avoid using drugs. It'll be a few hours a day, and it feels like a good fit. Owen signs up for a filmmaking class at the nearby community college. The memory loss is still hard. The depression and anxiety are still there, and he feels like he wakes up every day at some indeterminate time in the past. What memories he forms aren't time-stamped accurately—they could be from yesterday, or from long ago—and there's no rhythm. But despite these limitations, Owen is managing, and he knows his story isn't over.

—m—

In June, Barash finds an excuse to e-mail Zhongcong Xie about the tau PET study, which has been taken over by a medical imaging expert at Mass General. It's just one of a thousand or so e-mails about the amnestic syndrome that he's sent over the past few years. Xie's response is as simple and offhand as it is significant. He has the funding to systematically put to the test some of the ideas that Butler, Barash, and Kofke began spitballing over years earlier. Xie's lab could tease out the role of hypoxia versus fentanyl, look for genetic risk factors for hippocampal damage, or replicate Kofke's experiments showing that mice develop Alzheimer's-like brain damage after being given opioids. Xie's lab has the resources and staff to go even further than Kofke. They could also test the mice's ability to navigate through mazes to see if their memory is impaired. In short, they can explore whether damage to the hippocampus from opioids could be the basis for a new mouse model for Alzheimer's—and therefore a new way to study the disease.

Barash calls Gillian on his way home from Chelsea.

"Sorry, but traffic is brutal. We had a birthday party for one of the veterans, so I left late." He pauses while a driver nearby leans on the horn, then continues. "I'm standing still on the Mystic Valley parkway in Medford."

"No worries, the pizza won't go anywhere," Gillian says. "If you're not here by six, I'll feed the boys so they don't turn into monsters. How was your day?"

"Actually, pretty awesome for a self-loathing neurologist." The light turns green, Barash inches forward, and the frustrated driver leans on the horn again. "I'll tell you all about it when I get there."

Ten minutes later, Barash makes it through the bottleneck and cruises home. He's so busy thinking about the remarkable story of the past six years that he doesn't fully register the familiar landmarks along the way—the Mystic River, the gas station, the construction project on Route 2 that never seems to wrap up. It will mainly fall to other scientists to follow this research wherever it leads. If he can help, he will. For his part, Barash believes that if the only thing he and his colleagues have accomplished is to allow the next amnestic syndrome patient to go home with the dignity of a diagnosis, it will all have been worth it.

# Epilogue

· · · · · · · · · ·

Imagine waking up one morning to a world where tens of millions of people have lost their memories. Most victims are over the age of sixty-five, but some are younger. They don't recognize their family members. Many are so confused they're found wandering the streets, lost in their own neighborhoods. No one recovers, and there is no way to slow down or cure this disease. Scientists don't know what causes it or who will succumb. And the number of people afflicted grows with each passing day. This is the world we live in. But no CNN or Fox News anchor reports on the number of new cases, people hospitalized, the daily death toll, or the desperate effort to bring this memory pandemic to an end. Despite the human and economic cost, this devastating and all-too-common memory loss is invisible to most.

—·—

On February 1, 2020, the Massachusetts Department of Public Health announced the first confirmed case of COVID-19 in the state. Over the course of the previous year, Barash and his colleagues made slow but steady progress. They developed a formal case definition for a diagnosis of opioid-associated amnestic syndrome, OAS for short. All patients had to have amnesia lasting twenty-four hours or longer. To meet the criteria for a confirmed case, they would also need a scan showing damage to both hippocampi and an opioid-positive drug screen. To meet the criteria for probable, they would need the same imaging and a known history of opioid use. A possible case would be someone with a positive opioid screen, or a known history of opioid use, or damage to the hippocampus. Meanwhile, Barash and some of the doctors involved in identifying the first few patients sketched out rough plans for a small clinical trial for naltrexone at Lahey, back where the story began. Owen and two other patients agreed to come to Mass General for tau PET scans and cognitive testing. Kofke and Badiola resubmitted their opioids and memory grant to the NIH and received high enough scores from reviewers to feel optimistic about their prospects.

But as the number of COVID-19 cases in the state ticked up and evidence mounted of the virus's deadly and contagious nature, all this work was put on hold. DeMaria turned his attention to helping the DPH manage the crisis. Chelsea was the epicenter of the outbreak, and Barash and the Soldiers' Home staff went into full battle mode as they updated contingency plans daily and prepared for the inevitable onslaught of infections. The veterans on the long-term care side of the facility, who still lived ten people to a ward while awaiting the renovation, were especially vulnerable. By April 3, five staff members and thirteen residents had tested positive, two of whom had died. It was obvious to Barash that until the pandemic ended, there would be no late nights or spare weekend hours devoted to e-mailing about new cases of the amnestic syndrome or how to design a clinical trial for naltrexone.

Around the world, the COVID-19 pandemic disrupted memory research and Alzheimer's clinical trials both large and small. Participants could no longer come in for brain scans, treatments, or cognitive testing. Elaborate protocols years in the making had to be reconfigured overnight. Michela Gallagher was fortunate to have reached a point in her trial where she could minimize the fallout. Li-Huei Tsai salvaged what she could, although the pandemic delayed her trial at MIT by at least several years. Scientists like Zhongcong Xie, Steve Ramirez, and Amar Sahay had to put laboratory experiments on the back burner for three months or more.

But these are temporary setbacks, and although the quest to defeat Alzheimer's will be harder and longer than the battle against COVID-19, advances over the previous few years promise to bring the long stalemate to an end. With new animal models, blood tests, and imaging tools, new insights and perspectives, and the willingness to look outside the spotlight, researchers can finally see a way forward.

———✦———

In the fall of 2020, I asked Barash, Butler, DeMaria, Kofke, and Owen if they would each like to write a short essay for this book: looking back, how would they sum up what the amnestic syndrome meant to them? Happily, they all agreed. Here are their contributions:

## JED BARASH

Growing up in Connecticut as a fan of baseball's New York Yankees, I revered their Hall of Fame shortstop, Derek Jeter. It wasn't only his clutch performance as a member of five World Series champions, his grace on the field, or his kindness to me during a fortuitous meeting at an airport food court early in his career. It wasn't even that Jeter gave everything that he had in all the critical moments. No, it was the effort that he gave when the game *wasn't* on the line that made him all the more special.

Just hitting a routine ground ball at an infielder late in a regular-season game, with the Yankees way behind, there was Jeter running as hard as he could to first base, an out being all but certain. I can still picture his head down, arms pumping, legs churning, and then leaning forward like a sprinter through the finish line as he crossed the bag. Most fans found his hustle admirable, a sign of his respect for the sport, particularly in an age when many players might simply coast in a similar situation. But I always thought that Jeter ran hard for another reason: he truly believed there was a chance, even if highly improbable, that he might safely reach first. Maybe the fielder would briefly bobble the ball or make a throw that was slightly offline, but either way, hustling would at least give him an opportunity to get on base when merely jogging there wouldn't have. More than that, he seemed to believe that even an infield single could ignite a rally and, if everything went right, a win.

In medicine, doctors often look back fondly and recall the cases in which they made a particularly challenging diagnosis. And yet, in between those challenging situations, doctors make countless routine diagnoses, from high blood pressure to arthritis. These moments may be less memorable, but the discipline necessary to arrive at the "bread and butter" diagnoses makes it possible to rise to the occasion when the disorder is anything but ordinary. It would be difficult to estimate the number of CTs and MRIs our team had collectively reviewed for patients with memory loss in the years before identifying the unusual pattern seen in our first cases of the opioid-associated amnestic syndrome (OAS).

To be sure, the identification of the syndrome was, in part, the consequence of a team that was "in the right place at the right time," but the pattern recognition wouldn't have been possible without the group's dutiful review of image after image and sequence after sequence, day in and day out, for years. That the imaging in one of the first cases of OAS was initially interpreted as unrevealing by an outside radiologist only further highlights the importance of our examination of the primary data in a disciplined and detailed fashion. As it turned out, "hustling to first" also opened the door to new opportunities in understanding human memory.

Jed Barash, MD
Medical Director, Soldiers' Home
Chelsea, MA
October 2020

## MONROE BUTLER

A well-known argument in favor of the individual case report is that one talking pig is adequate enough to prove that pigs can talk. Before Jed involved me in the amnestic syndrome, I'd never had any reason to think about this adage too deeply, but as a young neurologist, I was nevertheless eager to get involved with the research. As we looked into the original Massachusetts cases of amnesia, Jed was drawn toward understanding the epidemiologic basis, and I gravitated toward working out the mechanism. My initial thinking was that there must be a common phenomenon—masquerading as something that only seemed novel—behind the syndrome. Specifically, I thought that depressed breathing caused by opioids created a global hypoxic state, which was just enough to injure susceptible regions in the hippocampus without leading to more widespread neuronal damage. However, when Jed and I studied the individual case thoroughly, my thinking shifted. I became convinced that this was indeed a novel mechanism and a syndrome worth exploring. The key finding from Owen Rivers's case report, where MRIs were taken before and after his overdose, revealed that there was focused damage to areas of the hippocampus that are not typically vulnerable to hypoxic injury.

Once I realized that the syndrome was truly rare, I grew intrigued by the implications of elucidating the cause. The hippocampus is plastic yet very vulnerable—susceptible to insults such as head trauma, stroke, seizures, severe stress, neurodegeneration, and dementia. Understanding risks to individual patients and the population at large from opioid exposure became critical in terms of public health and brain health awareness. With so many people exposed to opioids in controlled settings, like surgery, or uncontrolled environments, like with drug abuse, this individual case report's importance became apparent. Because thousands of individuals overdose on opioids in the United States each year, one might expect this amnestic syndrome to be more widespread. Yet, I suspect individual factors play a role in making only some people susceptible to the syndrome, such as genetic differences in how people metabolize opioids, how the cell recycles proteins, and variable effects in the central nervous system.

Postmortem analysis of brain tissue from chronic opioid abusers demonstrates that in a subset of overdose victims, these drugs can induce long-term injury to the hippocampus and incite an abnormal tau response, including the deposition of pathological neurofibrillary tangles, which are present in neurodegenerative diseases such as Alzheimer's. Whether these changes are compensatory and protective or initial provocateurs of a dangerous biochemical cascade leading to dementia is unknown. Regardless, the acute insults are inflaming a dangerous landscape—the hippocampus—which is a nexus of vulnerability for many neurodegenerative processes, such as pathological tau, beta amyloid, TDP-43, and argyrophilic grains disease. The pattern of focal brain injury and associated clinical deficits from opioid-associated amnesia has motivated us to explore this phenomenon in order to advance our understanding of the human mind.

P. Monroe Butler, MD, Ph.D.
Associate Medical Director
Biogen
Cambridge, MA

## AL DeMARIA

Public health resources are always limited. There are more public health issues to deal with than people with the time and resources to deal with them, and everyone agrees that health care costs too much. In this context, decisions are made and priorities set for public health investigations and interventions.

Emerging infections and concerns about terrorism have led public health agencies to encourage clinicians to report the unusual or unexpected. A clinician in Queens, New York, reported an unexpected cluster of encephalitis in 1999, leading to the recognition of West Nile virus's arrival into North America. In 2001, an infectious disease specialist in Florida immediately understood he had an unusual case of anthrax, which turned out to be the first signal of the anthrax attacks. More recently, a previously overlooked paralyzing disease in children led to the recognition of acute flaccid myelitis (AFM), perhaps associated with certain viral infections. Cases of a sepsis-like syndrome similar to Kawasaki disease were recognized by clinicians in Italy as associated with COVID-19 (a multisystem inflammatory syndrome in children, MIS-C). Public health depends on astute clinicians to recognize new diseases in order to respond to them in a timely way, and in some cases, even to recognize their existence at all.

Jed Barash contacted me in 2015 about cases of amnesia in people with substance use disorder associated with unusual findings on imaging the brain. It is not unusual for clinicians to report suspicious circumstances. After all, we encourage them to do so, and they are concerned about the implications of what they may be observing. Most of the time, nothing comes of these reports, but that is what is to be expected if such an informal surveillance system is to work. There must be many false alarms for every real event. So what was different here: (1) I knew Jed. He had worked with me when he was a Robert Wood Johnson Fellow at Yale, so I could judge his reliability; (2) he had already done research on what he was observing and was able to establish that it was unusual, putting the report in context; (3) there were objective findings on scanning the brain (something other clinicians can hone in on); and (4) we were in the middle of an epidemic of opioid use and overdose, so we were paying close attention to the impact of opioids. Without the knowledge about the reporter (Jed), the background research, and the context, this might not have been pursued by the department just based on resources and competing priorities. In February 2016, an e-mail was sent to physicians in Massachusetts to try to establish the extent of the cluster of cases; then in April 2017, the amnestic syndrome was made reportable under a regulation promulgated twenty years before to cover emerging infections and other conditions, making them reportable for a fixed time without a lengthy rule-making process. Ultimately, in Massachusetts and beyond, dozens of cases were identified, more cases were plausibly hypothesized as having gone unrecognized, and the evidence of association with opioids was strengthened.

Now, what next? A disabling amnesia is associated with exposure to opioids, perhaps more so with fentanyl and other synthetic opioids, which have flooded the illicit drug market. The opioid epidemic continues, but the COVID-19 pandemic's impact on whatever gains were being made still remains to be determined. While the association of hippocampal damage leading to amnesia and opioids is strong, the full impact is yet to be known. This syndrome represents a public health and clinical problem, but it affects a marginalized population at a time when myriad other public health and social problems must also be addressed. There are national surveillance and investigation programs for AFM and MIS-C, but opioid-associated amnesia is an orphan. AFM affects children, possibly due to enteroviral infection, MIS-C affects children and adults (MIS-A) with COVID-19, and the opioid amnesia syndrome affects predominantly young adults with opioid use disorder. While we have some capacity to recognize new threats, we do not have the capacity to prioritize action to address all the threats identified. It is appropriate to be especially concerned about protecting children. Still, it should not be to the exclusion of efforts to protect people who are *also* our children, have a medical condition of unclear prevalence—one that is difficult to treat and overcome—and that could have long-lasting consequences.

Alfred DeMaria, Jr., MD
Medical and Laboratory Consultant
Former State Epidemiologist and Medical Director
Bureau of Infectious Disease and Laboratory Sciences
Massachusetts Department of Public Health
November 2020

## ANDREW KOFKE

This is a story about connecting the dots. From 1990 to 2007, I published a series of reports of a reproducible effect of clinically used anesthetic opioid drugs that produce limbic system brain damage in physiologically controlled, non-hypoxic mammals. My colleagues and I found that this effect occurred with many clinically used mu opioids: fentanyl, alfentanil, sufentanil, and remifentanil—all opioids similar to morphine. We later determined this damage was due to limbic system hypermetabolism and seizure and duplicated these metabolic and seizure findings in humans. The limbic system includes areas of the brain responsible for emotion, executive function, and memory, most notably, the hippocampus. Other research my team did included:

1. Ascertaining drugs which could block or lessen opioid neurotoxicity.
2. Screening for responsible neurotransmitter systems.
3. Examining dose-response relationships.
4. Determining that forebrain ischemia is exacerbated in the context of opioid anesthesia.

5. Demonstrating the value of opioid anesthesia for successful electroconvulsive therapy in humans.

Moreover, in an effort to assess relevance to the human brain, we gave high-dose opioids to ventilated pharmacologically paralyzed sub-human primates and also to intubated pharmacologically paralyzed human volunteers in a PET scanner (which assesses brain metabolic rate), observing limbic activation in both situations. We also evaluated spontaneously breathing human volunteers with low sedative dose opioids, observing limbic activation in MRI blood flow studies, which varied according to genotype. Efforts to translate these observations to humans undergoing anesthesia or other contexts of opioid administration were not successfully funded. A common critique from referees was that the opioid neurotoxicity we reported in rodents could not possibly be clinically relevant. Indeed, an editor of a major journal approached me (a young assistant professor at the time) at a major national meeting, seemingly in jest, indicating that I was the one who was going to get us all sued!

It is notable that postoperative cognitive dysfunction (POCD) is now generally accepted and formally recognized with a system of nomenclature proposed for it. This is clearly a syndrome with multiple factors, including underlying medical conditions, age, genomics, type of surgery, and specific anesthetic drugs. Volatile anesthetics have undergone the most scrutiny in regards to POCD, while opioids remain relatively unstudied. This is not totally unreasonable given that most anesthetics administered with opioids during surgery have anticonvulsant properties, which would mitigate any harm. Opioids are often given on their own after surgery, although the dose is likely too low to be dangerous. Nevertheless, it remains unstudied. It does seem to me that prior criticisms of opioid neurotoxicity as being clinically unimportant should be reevaluated.

These events relevant to my own research were quite discouraging. POCD was yet to be described, and I eventually gave up on any efforts to further pursue opioid neurotoxicity as a fundable research area. Over twenty years passed. Then the opioid epidemic crisis hit, POCD became a hot research area, and Massachusetts neurologist Jed Barash and colleagues identified the amnestic syndrome. Dr. Barash found my work and contacted me to write a review with him, "connecting the dots" between my work over twenty years ago and his discovery.

The published data provide strong support for the potential for opioids to be neurotoxic and furthermore that what we observed in rodents translates to subhuman and human primates. This background research across multiple species in the context of the opioid epidemic led to NIH funding through the National Institute on Drug Abuse to evaluate whether pain patients receiving prescribed opioids develop evidence of the same or similar amnesia syndrome with supporting anatomical studies in MRI of areas of the brain responsible for memory. Thus, we are continuing to connect the dots from preclinical rodent studies to humans taking or being given relatively high doses of narcotics. It is increasingly apparent that prior critiques regarding clinical relevance are themselves becoming irrelevant.

W Andrew Kofke, MD MBA FNCS
Professor, Director Neuroscience in Anesthesiology
and Critical Care Program
Department of Anesthesiology and Critical Care
University of Pennsylvania
November 2020

## OWEN RIVERS

It's no exaggeration to say that I lived in near-complete darkness for the first months after the incident. At first, attempts at Google-searching various combinations of keywords like "overdose" + "memory loss" or "OxyContin" + "hippocampal damage" yielded nothing more than frustration. Eventually, I came across Lauren's article on the PBS website describing the amnestic syndrome, leading me to Dr. Jed Barash and Dr. Monroe Butler, in whose research I've been fortunate to participate.

As time pressed on, certain distant memories, with increasing regularity, would appear suddenly, as if jolted into recollection. One, in particular, hailed from years past: an undergraduate psychology course entitled Psychology of Human Learning and Memory taught by the late Arthur

Shimamura, whose recent passing I'd like to acknowledge. I first learned about my amnestic syndrome in his class, four years before it would happen to me. In fact, nearly every undergraduate psychology course I'd ever taken would at least briefly reference those famed cases. By the third or fourth mention, they'd completely lost their novelty and instead could rouse in me a fervent indignation, a conceited frustration, for having to be subjected yet again to the practically hackneyed subject. Patient H.M. and Clive Wearing—their names seemingly forever etched in my long-term memory. I wish now that I'd paid closer attention to each reference. No way this is happening to *me*. It's impossible. This is impossible. There must be a mistake.

My relationship with memory preceding the incident could, at best, be described as tenuous. A rare subtype of obsessive-compulsive disorder, comprising an irrational fear of forgetting, predates my hippocampal injury and subsequent amnestic syndrome by nearly a decade. That's right! No stranger to the bold fist of irony. What did that look like, or how did it express? I find the tag "memory hoarding" succinctly descriptive. Hundreds of notes, alarms, reminders, "to-do" lists, and daily (sometimes hourly) external hard-drive or email backups. More time spent scanning memory for forgotten thoughts, drafting and editing lists, tracing present lines of thinking to their origins, repeating incessantly in mind phrases and tasks deemed arbitrarily urgent, than actually living. My absurd relationship with memory not only contributed to the events preceding the overdose but also to my recovery. Ironically, the all-consuming, nightmarishly compulsive, thought checking and list-making, and real-time archiving, at once a debilitating source of misery, is now nothing short of life-preserving.

At the time of this writing, the memory impairment is still severe, and its subtle implications continue to surprise me. Without checking my Calendar, I don't know what day of the week today is. And the date? Forget about it. I do know the month, though, as we're approaching the celebration of a major holiday. Last year, I finally underwent UCLA's Intensive Outpatient OCD Treatment Program—long overdue. Despite the amnestic syndrome, their incredible team of providers was able to wring out of me a 30 percent reduction in OCD symptoms. I still rely heavily—and undeniably, obsessive-compulsively—on external memory aids, without the help of which I would likely remain largely incapacitated. Thankfully, I usually only have to imagine what life without a (fully charged) smartphone might be like. Without Calendar notifications, task organization apps (huge shout-out to Trello), alarms, and meticulous preplanning each day, navigating everyday life on my own would be unfeasible. Oh, not to mention literal navigation. Sense of direction? Gone. Or following the plot of a film? Who knew memory would be so integral? Who is that character? I know that one is significant, but *why* they're significant, not a chance.

How am I retelling this story, then? After any significant break (>10 seconds) from writing, or even spending too much time on a single sentence, I must retrace my steps with the hope that my original intention can be spurred by association. As for the literal retracing of steps, my spatial memory faculties—both consolidation and retrieval—are dismal. *Did I just walk by this building? I've never seen this car before in my life. Wait, is that the building I just exited, or . . . ?* Twenty-five minutes have passed, panic has set in, and I still can't find my car. Or Trello crashes while I scramble to record the instructions or directions given me by (insert: physician, professor, siblings, flight attendant, friend). Emotionally salient memories, however, particularly those memories whose affectation I'd consider negative or distressing—of all the memory faculties I could have lost—these remain impeccably intact. To relive, every day, with startling poignancy, last year's breakup, for instance, seems cruel and unusual. It still amazes me that the process of "moving on" too relies on memory; that is, to remember tomorrow the psychological progress I've made today. Where do all of yesterday's lessons go? Where did they disappear to?

Nevertheless, I've grown to believe that I'm incredibly lucky to have not lost more. I now have a profound respect for the intricacies and fragility of my biology, for its durability and adaptability, and of life. Whether there might have been an easier, softer way to impart such a sentiment and get it to stick, I do not know. My deepest gratitude to Lauren for writing this book and asking me to contribute. And to Dr. Monroe Butler, Dr. Jed Barash, and others researching this amnestic syndrome. And to my family and friends, without whose patience and care and love, I would not have stood a chance.

RO
November 2020

# Acknowledgments

· · · · · · · · · · · · · · · · · · · ·

The voices of everyone who helped bring this book to life kept me good company during the many hours I spent alone in a room writing. Above all I'm indebted to those whose lives were touched by the amnestic syndrome, and who spoke to me repeatedly and openly over several years, especially Barash and Owen, but also Butler, DeMaria, Haut, Kofke, and others.

Thank you to my enthusiastic agents at Aevitas Creative Management, Todd Shuster and Justin Brouckaert. They invested so much energy helping me develop my proposal that it served as a remarkably sound blueprint for the final book. Thank you to my editor at Pegasus Books, Jessica Case, who wholeheartedly embraced the nuance and messiness of this story without ever suggesting I simplify it for the sake of a better narrative. Copyeditor Judy Meyers's keen eyes saved me. I'm also grateful for generous financial support from the Alfred P. Sloan Foundation Program in Public Understanding of Science and Technology.

So many friends and former colleagues helped me editorially. Jon Palfreman started me off on the right foot, guiding me through the research and writing process and providing a sounding board for journalistic questions as they arose. Kelly Tyler-Lewis helped from beginning to end with insightful suggestions that helped shape the proposal, and later on, multiple iterations of the manuscript. It is a much better book for it. Tim De Chant, David Condon, Allison Eck, and Annie Valva, read rough drafts of the first few chapters and gave much-needed encouragement. So, too, did Ellie Baker and George Harrar.

Thank you to all the scientists and clinicians who took time out of their insanely busy lives and important work to describe their research and answer my questions—sometimes more than once! They did so, I think, not so much for their own benefit, but because they believe the public should have accurate information and understand how science works.

I owe a great debt of gratitude to two experts who read the entire manuscript and provided feedback: Bradford C. Dickerson, director of the Frontotemporal Disorders Unit at Massachusetts General Hospital, and Marc J. Kaufman, director of McLean Hospital's Translational Neuroimaging Laboratory. Other experts read short excerpts to help ensure accuracy. They are Marco Asaf, formerly at MIT's Picower Institute for Learning and Memory; Jonathan Curot at the Center for Brain and Cognitive Research at the University of Toulouse, in France; Javier De Felipe at the Cajal Institute in Madrid, Spain; Jonathan Epp at the Hotchkiss Brain Institute at the University of Calgary in Canada; David Epstein at the National Institute on Drug Abuse; Chris McBain and Kenneth Pelkey at the National Institutes of Health; David S. Knopman at the Mayo Clinic College of Medicine, Rochester, Minnesota; Jacob Vogel at the University of Pennsylvania; Barbara A. Wilson, founder of the Oliver Zangwill Centre for Neuropsychological Rehabilitation in the United Kingdom; and Eric Wish, director of the Center for Substance Abuse Research.

I also wish to thank the following scientists and clinicians: Chinnakkaruppan Adaikkan, Matthew P. Anderson, Marco Asaf, Ignacio Badiola, Erik Bloss, György Buzsáki, Gregory Carter, Kaitlin Casaletto, Elissa J. Chesler, Devyn Cotter, Ravi Das, Jonathan Dostrovsky, Karen Duff, Richard P. Dutton, Melissa K. Edler, Chris Evans, André Fenton, Gordon Fishell, Loren Frank, Paul Frankland, Max Gaitan, Michela Gallagher, Michael Ganetsky, Dan Geller, Jakub Hort, Adam Jasne, Catherine Cook Kaczorowski, Joshua P. Klein, James Knierim, Joel Kramer, Mara Kunst, Renaud La Joie, Michael Lev, Chun Lim, Donald G. MacKay, Eleanor Maguire, James J. Mahoney, Zach Malchano, Anthony Martorell, R. Kathryn McHugh, Nancy Merrill, Liv E. Miller, Kristen M. S. O'Connell, Mark Packard, Mercedes Paredes,

Jay Penney, Steve Ramirez, Howie Rosen, Shayna Rosenbaum, Jonathan Ross, Ueli Rutishauser, Amar Sahay, Bill Seeley, Juan Small, Nick Somerville, Shawn Sorrells, Hugo Spiers, Miriam Stoeber, Nanthia Suthana, Rudolph Tanzi, Ryan Taylor, Li-Huei Tsai, Justin Uzl, Mieke Verfaellie, Michael Yassa, and Yuval Zabar. Needless to say, I'm responsible for any inaccuracies.

And then there are the true experts in memory loss, the people who suffer from it and the friends and family who care for them. I thank them for sharing their experiences.

About a year after I started working on this book, my husband, Blaise Aguirre, read Chapter 1. He then very kindly stopped asking when the book would be done and he never read another word. But he listened to me worry, gave pep talks, sent related articles at all hours, and offered to introduce me to anyone he'd ever met who might possibly be helpful. My mother, Char Seeley, my aunt, Rita Gadel, and my son, Anthony Aguirre, slogged their way through very rough drafts. In their own way, each one helped me see this book through the eyes of someone who was neither a journalist, an agent, an editor, or an expert—but instead, just the type of curious reader I hope will enjoy and learn from this book.

# A Note About Sources

To cover the investigation into the amnestic syndrome, I interviewed clinicians, patients, and family members, in homes, offices, over the phone, Zoom, or via e-mail. They shared what they remembered about their experiences—although as I came to learn, what you remember depends on mood, previous experiences, what seems meaningful at the time, and the cues that bring a particular memory back to the surface. To supplement these interviews, I combed through hundreds of e-mails I requested through a Freedom of Information Act that were sent between clinicians, the Massachusetts DPH, the FDA, and the CDC. (Identifiable patient information was removed by the DPH.) I also used text messages between doctors, discharge notes, drawings, photos, site visits, and information from more than a dozen published papers on the syndrome.

E-mails and text messages are excerpted verbatim from the originals. I was not present for any of the scenes, with or without dialogue, that appear in this book. All scenes are based in part on interviews with people who were present. In rare instances when an event seemed so routine at the time that the participants couldn't remember whether it took place over the phone or in person, I took the liberty of setting it in a location. To protect patient privacy, I changed names as well as other identifying information.

No one involved in the amnestic syndrome has reviewed any portion of this manuscript. For them, it was a leap of faith to believe that I could tell it appropriately. Inevitably I bring my own lens to this story, but my goal was to capture the substance and import of what happened in the most faithful version that I could attain.

# Notes

· · · · · · · ·

### Prologue

The prologue is based on interviews with Owen Rivers in July and August 2018; January, April, and August 2019; and September 2020; his mother in August 2018 and January 2019; his father in April 2019; his siblings in April 2019; Monroe Butler in May and July 2018 and September 2019; hospital discharge note; e-mail between Owen and his doctor; and P. Monroe Butler et al., text messages between Jed Barash and Monroe Butler; "An Opioid-Related Amnestic Syndrome with Persistent Effects on Hippocampal Structure and Function," *The Journal of Neuropsychiatry and Clinical Neurosciences* 31, no. 4 (March 2019): 392–96, https://doi.org/10.1176/appi.neuropsych.19010017.

### Chapter 1

3    *includes an attention-grabbing title:* Franz H. Messerli, "Chocolate Consumption, Cognitive Function, and Nobel Laureates," *The New England Journal of Medicine* 367, no. 16 (2012): 1562–64.

5    *He has to see the patient for himself:* Azeen Ghorayshi, "14 People in Massachusetts Suddenly Lost Their Memories. Could Heroin Be the Culprit?" BuzzFeed, April 15, 2017; Juan E. Small et al., "Complete, Bilateral Hippocampal Ischemia: A Case Series," *Neurocase* 22, no. 5 (October 2016): 411–12, https://doi.org/10.1080/13554794 .2016.1213299.

11   *is called Creutzfeldt-Jakob disease, or CJD:* "Creutzfeldt-Jakob Disease Fact Sheet," NIH, National Institute of Neurological Disorders and Stroke, https://www .ninds.nih.gov/Disorders/Patient-Caregiver-Education/Fact-sheets/Creutzfeldt -Jakob-Disease-Fact-Sheet.

12   *Prusiner's discovery showed:* Stanley B. Prusiner, "Prions Are Novel Infectious Pathogens Causing Scrapie and Creutzfeldt-Jakob Disease," *BioEssays: News and Reviews in Molecular, Cellular and Developmental Biology* 5, no. 6 (December 1986): 281, https://doi.org/10.1002/bies.950050612.

**Additional Sources:** Interviews: Jed Barash: July and December 2017, January and March 2018, March and April 2019, August 2020; Max's grandmother: August 2019; Max's mother: November 2017 and November 2019; Juan Small: November 2017 and August 2020; Yuval Zabar: November 2017 and November 2019.

### Chapter 2

15   *the hippocampus, which makes up less than one one-hundredth of the volume of the brain\*:* Michio Suzuki et al., "Male-Specific Volume Expansion of the Human Hippocampus during Adolescence," *Cerebral Cortex* 15, no. 2 (February 2005): 191, https://doi .org/10.1093/cercor/bhh121.
     *brain is defined as gray matter

15–16    *Henry Molaison was a perfectly healthy boy. . . . and listening to the radio:* Suzanne
         Corkin, *Permanent Present Tense: The Unforgettable Life of the Amnesic Patient H. M.*
         (New York, NY: Basic Books, 2014).

16–17    *By the time Henry was twenty-seven . . . removing both of them had obliterated it:*
         Luke Dittrich, *Patient H.M., A Story of Memory, Madness and Family Secrets* (New
         York: Random House, 2017); Wilder Penfield, *No Man Alone, A Neurosurgeon's Life*
         (Boston/Toronto: Little, Brown, and Company, 1977); W. B. Scoville and B. Milner,
         "Loss of Recent Memory after Bilateral Hippocampal Lesions," *Journal of Neurology,
         Neurosurgery, and Psychiatry* 20, no. 1 (1957), https://doi.org/10.1136/jnnp.20.1.11;
         Heinrich Klüver et al., " 'Psychic blindness' and other symptoms following bilateral
         temporal lobectomy in Rhesus monkeys," *American Journal of Physiology* 119 (September
         1937): 33; and Corkin, *Permanent Present Tense.*

17–20    *By one measure, the operation was successful . . . sophisticated vocabulary words like
         espionage:* Dittrich, *Patient H.M.*; Corkin, *Permanent Present Tense*; Donald G.
         MacKay, *Remembering: What 50 Years of Research with Famous Amnesia Patient H.M.
         Can Teach Us about Memory and How It Works* (Amherst, New York: Prometheus
         Books, 2019); Suzanne Corkin, "Tactually-Guided Maze Learning in Man: Effects
         of Unilateral Cortical Excisions and Bilateral Hippocampal Lesions," *Neuropsychologia*
         3, no. 4 (August 1965), https://doi.org/10.1016/0028-3932(65)90006-0; and NOVA,
         "The Man Who Couldn't Remember" (Interview with Suzanne Corkin), June 1, 2009,
         https://www.pbs.org/wgbh/nova/article/corkin-hm-memory/.

20       *An interview at MIT in 1970 with a visiting researcher:* [this interview has been
         copyedited to remove repeated words, ellipses, and "uhs"] W. Marslen-Wilson,
         Biographical interviews with H.M. Unpublished transcript. M.I.T., Cambridge, Mass.:
         1970; Time 53:30, p. 27, http://mackay.bol.ucla.edu/1970%20HM%20transcript%20
         word%20document%20Final%20Revised.pdf. Digitized and re-checked by Lori James
         and Don MacKay, for publication on //www.mackay.bol.ucla.edu/.

20–21    *The answers research on Henry provided . . . has stood the test of time:* Corkin, *Permanent
         Present Tense;* NOVA, "The Man Who Couldn't Remember"; Jacopo Annese et al.,
         "Postmortem Examination of Patient H.M.'s Brain Based on Histological Sectioning
         and Digital 3D Reconstruction," *Nature Communications* 5, no. 1 (January 2014): 1–9,
         https://doi.org/10.1038/ncomms4122.

22       *When Dr. Alois Alzheimer met 51-year-old August Deter in 1901:* Konrad Maurer,
         "Auguste D and Alzheimer's Disease," *Lancet* 349, no. 9064 (May 1997): 1546–49,
         https://doi.org/10.1016/S0140-6736(96)10203-8.

## Chapter 3

26       *His seventh paper on the prion disease CJD:* Jed A. Barash et al., "Accuracy of
         Administrative Diagnostic Data for Pathologically Confirmed Cases of Creutzfeldt-
         Jakob Disease in Massachusetts, 2000–2008," *American Journal of Infection Control* 42,
         no. 6 (June 2014): 659–64, https://doi.org/10.1016/j.ajic.2014.02.002.

29–31    *When a 41-year-old construction worker . . . "come back and see me in eight weeks if you
         can":* Juan E. Small et al., "Complete, Bilateral Hippocampal Ischemia: A Case Series,"
         *Neurocase* 22, no. 5 (October 2016): 2–3, https://doi.org/10.1080/13554794.2016.1213
         299; Ghorayshi, "14 People in Massachusetts Suddenly Lost Their Memories. Could
         Heroin Be The Culprit?"

33       *Butler's fourth paper, coauthored with an anthropologist:* Luke J. Matthews et al.,
         "Novelty-seeking DRD4 polymorphisms are associated with human migration

distance out-of-Africa after controlling for neutral population gene structure," *American journal of physical anthropology* 145, no. 3 (April 2011): 382–389, https://doi .org/10.1002/ajpa.21507.

34  *a 42-year-old man from San Jose named George Carillo ... treat the cause, not the symptoms:* J. William Langston, M.D., and Jon Palfreman, *The Case of the Frozen Addicts* (Amsterdam: IOS Press, 2014), 3–6, 11–18, 34–35, 120; J. William Langston, "The MPTP Story," *Journal of Parkinson's Disease* 7, no. s1 (March 2017): S11–S19, https://doi.org/10.3233/JPD-179006; and J. William Langston et al., "Parkinsonism Induced By 1-Methyl-4-Phenyl-1, 2, 3, 6-Tetrahydropyridine (MPTP): Implications for Treatment and the Pathogenesis of Parkinson's Disease," *Canadian Journal of Neurological Sciences*, 11, no. S1 (February 1984): 160–165, https://doi.org/10.1017 /S0317167100046333.

37  *Butler comes across the case of a 73-year-old man in Korea:* Jihoon Kim et al., "Isolated Bilateral Hippocampal Lesions Following Carbon Monoxide Poisoning," *European Neurology* 66, no. 1 (August 2011): 64, https://doi.org/10.1159/000329271.

37  *A 33-year-old man who had inhaled heroin:* Aurélien Benoilid et al., "Heroin Inhalation-Induced Unilateral Complete Hippocampal Stroke," *Neurocase* 19, no. 4 (August 2013): 313–15, https://doi.org/10.1080/13554794.2012.667125.

37  *three cases of amnesia and hippocampal damage after cocaine use:* Mohammad Reza Bolouri et al., "Neuroimaging of Hypoxia and Cocaine-Induced Hippocampal Stroke," *Journal of Neuroimaging: Official Journal of the American Society of Neuroimaging* 14, no. 3 (July 2004): 290–91, https://doi.org/10.1111/j.1552 -6569.2004.tb00254.x; Morales Vidal et al., "Cocaine Induced Hippocampi Infarction," *BMJ Case Reports* (July 2012): 1–2, https://doi.org/10.1136 /bcr.03.2012.5998; and Kathryn L. Connelly et al., "Bilateral Hippocampal Stroke Secondary to Acute Cocaine Intoxication," *Oxford Medical Case Reports* 2015, no. 3 (2015): 215–17, https://doi.org/10.1093/omcr/omv016.

**Additional Sources:** Interviews: Jed Barash: December 2017, April and October 2018; Monroe Butler: August 2017, May 2018. Massachusetts Department of Public Health e-mails: October 2015. Text messages: Jed Barash and Monroe Butler: May 2016.

## Chapter 4

42  *the first known examples of West Nile Virus in the United States:* Sam Roberts, "Dr. Deborah Asnis, Who Sounded Alert on West Nile Virus Outbreak, Dies at 59," *New York Times*, September 16, 2015.

42  *the AIDS epidemic, which began with a handful of patients with a rare fungal pneumonia:* Center for Disease Control, "People living with HIV/AIDS," https://www .cdc.gov/fungal/infections/hiv-aids.html.

43–44  *Butler and radiologist Juan Small unearth the third case. . . . She is case 4, the evidence he needs to go back to DeMaria:* Juan E. Small et al., "Complete, Bilateral Hippocampal Ischemia: A Case Series," *Neurocase* 22, no. 5 (October 2016): 3–4, https://doi.org/10.1 080/13554794.2016.121329.

**Additional Sources:** Interviews: Jed Barash: October, December, 2017, April 2018; Monroe Butler: May 2018, August 2020; Al DeMaria: October 2017, August 2018, October 2020; Juan Small: August 2020; Nicholas Somerville: December 2017. Massachusetts Department of Public Health e-mails: October, November, December 2015, January and February 2016.

## Chapter 5

This chapter is based largely on the following sources: Eric A. Newman, Alfonso Araque, Janet M. Dubinsky, Larry W. Swanson, Lyndel King, and Eric Himmel, *The Beautiful Brain, The Drawings of Santiago Ramon y Cajal* (New York: Abrams, 2017); Javier DeFelipe, "The dendritic spine story: an intriguing process of discovery," *Frontiers in Neuroanatomy* 9, no. 14 (March 2015): 1–13, https://doi.org/10.3389/fnana.2015.00014; and Interview with Javier DeFelipe, Cajal Institute, April 2020.

## Chapter 6

56        ***For the last thirty years, Lim has wanted to piece together a human memory circuit:*** M. Ferguson et al., "A Human Memory Circuit Derived From Brain Lesions Causing Amnesia," *Nature Communications* 10, no. 3497 (2019): 1–9, https://doi.org/10.1038/s41467-019-11353-z.

60        ***seeing an uptick in patients with substance use disorders:*** Renata R. Almeida et al., "Temporal Trends in Imaging Utilization for Suspected Substance Use Disorder in an Academic Emergency Radiology Department," *Journal of the American College of Radiology* 16, no. 10 (October 2019): 1440–46, https://doi.org/10.1016/j.jacr.2019.03.013.

65        ***"Complete, Bilateral Hippocampal Ischemia: A Case Series":*** Juan E. Small et al., "Complete, Bilateral Hippocampal Ischemia: A Case Series," *Neurocase* 22, no. 5 (October 2016): 411–15, https://doi.org/10.1080/13554794.2016.1213299.

65        ***Barash reads a report from the Centers for Disease Control and Prevention:*** Alexis B. Peterson et al., "Increases in Fentanyl-Related Overdose Deaths – Florida and Ohio, 2013-2015," *Morbidity and Mortality Weekly Report* 65, no. 33 (August 2016): 844–49, https://doi.org/10.15585/mmwr.mm6533a3.

67        ***Klein has collected sixteen cases:*** Shamik Bhattacharyya et al., "Bilateral Hippocampal Restricted Diffusion: Same Picture Many Causes," *Journal of Neuroimaging: Official Journal of the American Society of Neuroimaging* 27, no. 3 (May 2017): 300–05, https://doi.org/10.1111/jon.12420.

67        ***The CDC experts are sufficiently convinced by the worrisome health implications:*** Jed A. Barash et al., "Cluster of an Unusual Amnestic Syndrome—Massachusetts, 2012–2016," *Morbidity and Mortality Weekly Report* 66, no. 3 (January 2017): 76–79, https://doi.org/10.15585/mmwr.mm6603a2.

**Additional Sources:** Interviews: Jed Barash: October, November, December 2017, October 2018, February and October 2019; Al DeMaria: October 2017, March and August 2018; Joshua P. Klein: November 2017, October 2019; Monroe Butler: May 2018, Mara Kunst: January 2018; Michael Lev: December 2017, October 2019; Chun Lim: November 2017, December 2018; Nick Somerville: October 2019. Massachusetts Department of Public Health e-mails: February, March, July, August, September, November, December 2016. Text messages: Jed Barash and Monroe Butler: January 2017.

## Chapter 7

69–72     Biographical and medical information regarding Clive Wearing is drawn from Deborah Wearing, *Forever Today: A Memoir of Love and Amnesia* (New York: Doubleday, 2011), 17, 75–76, 80, 82, 107, 112, 117, 121, 157, 200, 228; Barbara A. Wilson et al., "Dense Amnesia in a Professional Musician Following Herpes Simplex

Virus Encephalitis," *Journal of Clinical and Experimental Neuropsychology* 17, no. 5 (October 1995): 668–81, https://doi.org/10.1080/01688639508405157; and Oliver Sacks, "The Abyss," *New Yorker*, September 17, 2007.

71    ***When Deborah asked Wearing if he would like to conduct:*** "Life without Memory: The Case of Clive Wearing," May 16, 2015, Youtube, video, 10:25, https://www.youtube.com/watch?v=nFoUvF9PiWo.

73    ***Maguire asked ten normal test subjects and five amnestic patients to imagine:*** Demis Hassabis et al., "Patients with Hippocampal Amnesia Cannot Imagine New Experiences," *Proceedings of the National Academy of Sciences of the United States of America* 104, no. 5 (January 2007): 1727, https://doi.org/10.1073/pnas.0610561104.

74    ***unusual things you could do with a cardboard box:*** Melissa C. Duff et al., "Hippocampal Amnesia Disrupts Creative Thinking: Creative Thinking and the Hippocampus," *Hippocampus* 23, no. 12 (December 2013): 4, 12, https://doi.org/10.1002/hipo.22208.

**Additional Sources:** Interviews: Jed Barash: July and November 2017; Mieke Verfaellie: September 2017, August 2020.

## Chapter 8

77    ***The anesthetized rat lies facedown:*** Jonathan Dostrovsky et al., "The hippocampus as a spatial map. Preliminary evidence from unit activity in the freely moving rat," *Brain Research* 34, no. 1 (November 1971): 171–75, https://doi.org/10.1016/0006-8993(71)90358-1.

77    ***In an earlier experiment:*** John O'Keefe, "Spatial Cells in the Hippocampal Formation," Nobel Prize Lecture, December 7, 2014, University College London, United Kingdom, transcript, https://www.nobelprize.org/uploads/2018/06/okeefe-lecture.pdf.

77    ***Experiments with rodents suggest the same:*** Robert J. Douglas, "The Hippocampus and Behavior," *Psychological Bulletin*, 67, no. 6 (1967): 429, https://doi.org/10.1037/h0024599; Dostrovosky et al., "The hippocampus as a spatial map," 171.

79    ***O'Keefe and collaborator Lynn Nadel went further:*** John O'Keefe and Lynn Nadel, *The Hippocampus as a Cognitive Map* (Oxford, United Kingdom: Oxford University Press, 1978).

79    ***In 1984, James Rank Jr. discovers head-direction cells:*** Jeffrey S. Taube et al., "Head-Direction Cells Recorded from the Postsubiculum in Freely Moving Rats. I. Description and Quantitative Analysis," *The Journal of Neuroscience: The Official Journal of the Society for Neuroscience* 10, no. 2 (February 1990): 420–35, https://doi.org/10.1523/JNEUROSCI.10-02-00420; Eun Hye Park et al., "How the Internally Organized Direction Sense Is Used to Navigate," *Neuron* 101, no. 2 (January 2019): 285–93, https://doi.org/10.1016/j.neuron.2018.11.019.

79    ***The Mosers named them grid cells:*** Torkel Hafting et al., "Microstructure of a Spatial Map in the Entorhinal Cortex," *Nature* 436, no. 7052 (June 2005): 801–06, https://doi.org/10.1038/nature03721.

79    ***which show the rodent how far it is from the edge of an enclosure:*** Trygve Solstad et al., "Representation of Geometric Borders in the Entorhinal Cortex," *Science* 322, no. 5909 (December 2008): 1865–68, https://doi.org/10.1126/science.1166466.

79    ***they identify speed cells:*** Emilio Kropff et al., "Speed Cells in the Medial Entorhinal Cortex," *Nature* 523, no. 7561 (July 2015): 419–24, https://doi.org/10.1038/nature14622.

79        *it can remap to new places:* Patrick Latuske et al., "Hippocampal Remapping and its
         Entorhinal Origin," *Frontiers in Behavioral Neuroscience* 11, no. 253 (January 2018):
         1–13, https://doi.org/10.3389/fnbeh.2017.00253.

79        *the idea that our brains attach memories to places isn't new:* Lynne Kelly, *Memory Craft:
         Improve Your Memory with the Most Powerful Methods in History* (New York: Pegasus
         Books, 2020).

79        *Known as the "method of loci":* Frances A. Yates, *The Art of Memory* (London: Pimlico, 1966).

80        *may be the spaces where ancient people encoded memories:* Duane W. Hamacher, "The
         Memory Code: How Oral Cultures Memorise So Much Information," *The Conversation*,
         September 26, 2016.

80        *to look for cells that track the flow of time:* Christopher J MacDonald et al.,
         "Hippocampal 'Time Cells' Bridge the Gap in Memory For Discontiguous Events,"
         *Neuron* 71, no. 4 (August 2011): 737–49, https://doi.org/10.1016/j.neuron.2011.07.012.

81        *To see how time cells operate on their own:* Christopher J. MacDonald et al., "Distinct
         Hippocampal Time Cell Sequences Represent Odor Memories in Immobilized Rats,"
         *The Journal of Neuroscience: The Official Journal of the Society for Neuroscience* 33, no. 36
         (September 2013): 14607–16, https://doi.org/10.1523/JNEUROSCI.1537-13.2013.

83        *cells in people that act like place and time cells in other animals:* Gray Umbach et al.,
         "Time Cells in the Human Hippocampus and Entorhinal Cortex Support Episodic
         Memory," *Proceedings of the National Academy of Sciences* 117, no. 45 (October 2020):
         28463–74, https://doi.org/10.1073/pnas.2013250117.

84        *Sam wears a form-fitting black bodysuit:* "This neuroscientist is using VR to learn more
         about the brain," Mashable, August 30, 2017, https://mashable.com/2017/08/30
         /how-she-works-neuroscientist/.

85        *what she found was remarkably similar to what neuroscientists observed in rodents:*
         Zahra M. Aghajan et al., "Theta Oscillations in the Human Medial Temporal Lobe
         during Real World Ambulatory Movement," *Current Biology,* 27, no. 24 (December
         2017): 3743–51, https://doi.org/10.1016/j.cub.2017.10.062.

85        *when the pattern of the theta rhythms during the learning phase matched:* Zahra M.
         Aghajan et al., "Modulation of Human Intracranial Theta Oscillations during Freely
         Moving Spatial Navigation and Memory," *BioRxiv* (August 2019): 1, https://doi
         .org/10.1101/738807.

86        *bringing the research that much closer to the real world:* Uros Topalovic et al., "Wireless
         Programmable Recording and Stimulation of Deep Brain Activity in Freely Moving
         Humans," *Neuron* 108, (October 2020): 322, https://doi.org/10.1016/j.neuron.2020.08.021.

86        *translate experimental results into memory-saving therapies:* Nanthia Suthana et al., "Memory
         Enhancement and Deep-Brain Stimulation of the Entorhinal Area," *The New England Journal
         of Medicine* 366, no. 5 (February 2012): 502–10, https://doi.org/10.1056/NEJMoa1107212; Ali
         S. Titiz et al., "Theta-burst microsimulation in the human entorhinal area improves memory
         specificity," *Elife* 6 (October 2017): 1–18, https://doi.org/10.7554/eLife.29515.

## Chapter 9

88        *The elevator pitch for why spending money on basic Alzheimer's research is so essential\*:*
         2020 Alzheimer's Disease Facts and Figures, Alzheimer's Association, 17-18; 2018
         Alzheimer's Disease Facts and Figures, Alzheimer's Association, 24; World Alzheimer
         Report 2015, Alzheimer's Disease International
         *The following recent study finds that the incidence of Alzheimer's is decreasing in
         Europe and the United States. This suggests that the number of people in those regions

who will have the disease in the coming decades may be fewer than projected: Martin Prince et al., "Recent global trends in the prevalence and incidence of dementia, and survival with dementia," *Alzheimer's Research & Therapy* 8, no. 1 (July 2016): 1–13, https://doi.org/10.1186/s13195-016-0188-8.

88    *Sahay made a name for himself in 2011:* Amar Sahay et al., "Increasing adult hippocampal neurogenesis is sufficient to improve pattern separation," *Nature* 472, no. 7344 (April 2011): 466–70, https://doi.org/10.1038/nature09817.

88    *The theory, called the amyloid cascade hypothesis:* John A. Hardy et al., "Alzheimer's Disease: The Amyloid Cascade Hypothesis," *Science* 256, no. 5054 (April 1992): 184–85, https://doi.org/10.1126/science.1566067.

89    *The first mutation affecting APP was definitively identified in 1991:* Alison Goate et al., "Segregation of a missense mutation in the amyloid precursor protein gene with familial Alzheimer's disease," *Nature* 349, no. 6311 (February 1991): 704–06, https://doi.org/10.1038/349704a0; Dmitry Goldgaber et al., "Characterization and chromosomal localization of a cDNA encoding brain amyloid of Alzheimer's disease," *Science* 235, no. 4791 (1987): 877–80, https://doi.org/10.1126/science.3810169; and Rudolph E. Tanzi et al., "Amyloid beta protein gene: cDNA, mRNA distribution, and genetic linkage near the Alzheimer locus," *Science* 235, no. 4791 (February 1987): 880–84, https://doi.org/10.1126/science.2949367.

89    *In 1995, scientists in California added a gene into a mouse's DNA:* Dora Games et al., "Alzheimer-type neuropathology in transgenic mice overexpressing V717F β-amyloid precursor protein," *Nature* 373 (February 1995): 523–27, https://doi.org/10.1038/373523a0.

90    *and the mice had trouble learning:* Paula M. Moran et al., "Age-related learning deficits in transgenic mice expressing the 751-amino acid isoform of human beta-amyloid precursor protein," *Proceedings of the National Academy of Sciences* 92, no. 12 (June 1995): 5341–45, https://doi.org/10.1073/pnas.92.12.5341; Karen Hsiao et al., "Correlative memory deficits, Aβ elevation, and amyloid plaques in transgenic mice," *Science* 274, no. 5284 (October 1996): 99–103, https://doi.org/10.1126/science.274.5284.99.

90    *Hints that the story was more complicated and targeting amyloid beta might not work:* Heiko Braak et al., "Frequency of stages of Alzheimer-related lesions in different age categories," *Neurobiology of Aging* 18, no. 4 (July–August 1997): 351–57, https://doi.org/10.1016/s0197-4580(97)00056-0.

90    *In scores of costly trials, experimental drugs that seemed so promising:* Eric Perakslis et al., "A call for a global 'bigger' data approach to Alzheimer disease," *Nature Reviews Drug Discovery* 18 (July 2018): 319, https://doi.org/10.1038/nrd.2018.86.

90    *Some drugs worsened cognitive decline and accelerated brain shrinkage:* Francesco Panza, "Do BACE inhibitor failures in Alzheimer patients challenge the amyloid hypothesis of the disease?," *Expert review of neurotherapeutics* 19, no. 7 (2019): 599–602, https://doi.org/10.1080/14737175.2019.1621751; Reisa Sperling et al., "Findings of Efficacy, Safety, and Biomarker Outcomes of Atabecestat in Preclinical Alzheimer Disease: A Truncated Randomized Phase 2b/3 Clinical Trial," *JAMA neurology* (January 2021): E1-E9, https://doi.org/10.1001/jamaneurol.2020.4857.

90    *For years, more than half the therapies being tested:* Jeffrey Cummings et al., "Alzheimer's disease drug development pipeline: 2018," *Alzheimer's & Dementia: Translational Research & Clinical Interventions* 4, no. 1 (May 2018): 195–214, https://doi.org/10.1016/j.trci.2018.03.009.

91    *They may even start out as a way to protect the brain from infection:* Brian J. Balin and Alan P. Hudson, "Herpes viruses and Alzheimer's disease: new evidence in the debate,"

*The Lancet Neurology* 17, no. 10 (October 2018): 839–41, https://doi.org/10.1016 /S1474-4422(18)30316-8.

91     ***the late onset–variant, which accounts for about 95 percent or more of cases:*** Denham Harman, "Alzheimer's disease pathogenesis: role of aging," *Annals of the New York Academy of Sciences* 1067, no. 1 (May 2006): 454–60, https://doi.org/10.1196 /annals.1354.065; Mario F. Mendez, "Early-onset Alzheimer disease and its variants," *Continuum* 25, no. 1 (February 2019): 34, https://doi.org/10.1212/CON.0000000000 000687.

91     ***A list of suspected risk factors:*** Gill Livingston et al., "Dementia prevention, intervention, and care: 2020 report of the Lancet Commission," *The Lancet* 396, no. 10248 (August 2020): 413–46, https://doi.org/10.1016/S0140-6736(20)30367-6; Dana M. Cairns et al., "A 3D human brain–like tissue model of herpes-induced Alzheimer's disease," *Science Advances* 6, no. 19 (May 2020): eaay8828, https://doi.org/10.1126 /sciadv.aay8828; Keith A. Vossel et al., "Epileptic activity in Alzheimer's disease: causes and clinical relevance," *The Lancet Neurology* 16, no. 4 (April 2017): 311–22, https://doi.org/10.1016/S1474-4422(17)30044-3; and Ophir Keret et al., "Association of late-onset unprovoked seizures of unknown etiology with the risk of developing dementia in older veterans," *JAMA neurology* 77, no. 6 (March 2020): 710–15, https ://doi.org/10.1001/jamaneurol.2020.0187.

91     ***About two-thirds of all cases are women:*** Dena Dubal, "Chapter 16: Sex difference in Alzheimer's disease: An updated, balanced and emerging perspective on differing vulnerabilities," in *Handbook of Clinical Neurology, 175 (3rd series) Sex Differences in Neurology and Psychiatry*, eds. R. Lanzenberger, G.S. Kranz and I. Savic (Amsterdam: Elsevier B.V., 2020), 261–73, https://doi.org/10.1016/B978-0-444-64123-6.00018-7.

92     ***a protective factor on the X chromosome:*** Emily J. Davis et al., "A second X chromosome contributes to resilience in a mouse model of Alzheimer's disease," *Science Translational Medicine* 12, no. 558 (August 2020): 1–16, https://doi.org/10.1126/scitranslmed.aaz5677.

93     ***One, like a simple blood test for tau:*** Sebastian Palmqvist et al., "Discriminative Accuracy of Plasma Phospho-tau217 for Alzheimer Disease vs Other Neurodegenerative Disorders," *JAMA* 324, no. 8 (July 2020): 772–81, https://doi .org/10.1001/jama.2020.12134; Nicolas R. Barthélemy et al., "Blood plasma phosphorylated-tau isoforms track CNS change in Alzheimer's disease," *Journal of Experimental Medicine* 217, no. 11 (July 2020): 1–11, https://doi.org/10.1084 /jem.20200861.

93     ***may protect people from Alzheimer's even when the deck is stacked against them:*** Michael E. Belloy et al., "Association of Klotho-VS Heterozygosity With Risk of Alzheimer Disease in Individuals Who Carry APOE4," *JAMA neurology* 77, no. 7 (April 2020): 849–62, https://doi.org/10.1001/jamaneurol.2020.0414.

94     ***Altman didn't seek out controversy, but his formative years:*** Joseph Altman, "Memoir, The discovery of adult mammalian Neurogenesis," in *Neurogenesis in the Adult Brain I: Neurobiology*, ed. Tatsunori Seki ( New York: Springer, 2011), 3. This can be accessed online: https://neurondevelopment.org/sites/default/files/memoir.pdf; Shirley A. Bayer, "Joseph Altman (1925–2016): A life in neurodevelopment," *The Journal of Comparative Neurology* 524, no. 15 (June 2016): 2933–43, https://doi.org/10.1002/cne.24058.

95     ***But one result surprised and confused Altman:*** Joseph Altman, "Are New Neurons Formed in the Brains of Adult Mammals?," *Science* 135, no. 3509 (March 1962): 1127– 28, https://doi.org/10.1126/science.135.3509.1127.

95     ***In a brief and sometimes bitter memoir written in 2008:*** Altman, "Memoir, The discovery of adult mammalian Neurogenesis," 30.

95      *songbirds produce as many as a thousand new neurons a day:* Fernando Nottebohm, "Neuronal Replacement in Adulthood," *Annals of the New York Academy of Sciences* 457, no. 1 (December 1985): 143–61, https://doi.org/10.1111/j.1749-6632.1985.tb20803.x.

95      *evidence for neurogenesis in macaque monkeys, tree shrews, and marmoset monkeys:* Elizabeth Gould et al., "Hippocampal neurogenesis in adult Old World primates," *Proceedings of the National Academy of Sciences* 96, no. 9 (April 1999): 5263–67, https://doi.org/10.1073/pnas.96.9.5263; Elizabeth Gould et al., "Neurogenesis in the Dentate Gyrus of the Adult Tree Shrew Is Regulated by Psychosocial Stress and NMDA Receptor Activation," *The Journal of Neuroscience* 17, no. 7 (April 1997): 2492–98, https://doi.org/10.1523/JNEUROSCI.17-07-02492.1997; Elizabeth Gould et al., "Proliferation of granule cell precursors in the dentate gyrus of adult monkeys is diminished by stress," *Proceedings of the National Academy of Sciences* 95, no. 6 (March 1998): 3168–71, https://doi.org/10.1073/pnas.95.6.3168.

95      *They analyzed the brain tissue of five cancer patients:* Peter S. Eriksson et al., "Neurogenesis in the adult human hippocampus," *Nature Medicine* 4, no. 11 (November 1998): 1313–17, https://doi.org/10.1038/3305.

95      *international team of scientists found a way to quantify:* Kirsty L. Spalding et al., "Dynamics of hippocampal neurogenesis in adult humans," *Cell* 153, no. 6 (June 2013): 1219–27, https://doi.org/10.1016/j.cell.2013.05.002.

96      *Sahay's goal is to harness this neuroplasticity:* Nannan Guo et al., "Dentate granule cell recruitment of feedforward inhibition governs engram maintenance and remote memory generalization," *Nature Medicine* 24, no. 4 (May 2018): 438, https://doi.org/10.1038/nm.4491.

**Additional Sources:** Interviews: Jed Barash: October 2017, January 2018; Amar Sahay: September 2017. Text messages: Jed Barash and Amar Sahay: February 2017.

## Chapter 10

103     *the CDC reported that there were 67,367 drug overdose deaths in 2018:* National Center for Health Statistics, "Drug Overdose Deaths in the United States, 1999–2018," NCHS Data Brief No. 356, January 2020, https://www.cdc.gov/nchs/products/databriefs/db356.htm, figures 1 and 3.

103     *and West Virginia were particularly hard hit:* National Institute on Drug Abuse, "Opioid Summaries by State," April 2020, https://www.drugabuse.gov/drug-topics/opioids/opioid-summaries-by-state.

104–105 *Fifty years ago, a unique set of circumstances . . . "results that differ so much from clinical experience":* Lee N. Robins, "Vietnam veterans' rapid recovery from heroin addiction: a fluke or normal expectation?," *Addiction* 88, no. 8 (August 1993): 1041–54, https://doi.org/10.1111/j.1360-0443.1993.tb02123.x; Lee N. Robins et al., "How Permanent Was Vietnam Drug Addiction?," *American Journal of Public Health* 64, no.12_Suppl (December 1974): 38–43, https://doi.org/10.2105/ajph.64.12_suppl.38; Lee N. Robins et al., "Vietnam veterans three years after Vietnam: how our study changed our view of heroin," *The American Journal on Addictions* 19, no. 3 (April 2010): 203–11, https://doi.org/10.1111/j.1521-0391.2010.00046.x.

106     *by one estimate, twenty-three million people in long-term recovery:* Cori Kautz Sheedy et al., "Guiding principles and elements of recovery-oriented systems of care: what do we know from the research?," *Journal of Drug Addiction, Education and Eradication* 9, no. 4 (August 2009): 225–86.

106     *a little more than 3 percent of people prescribed opioids ever become addicted:* David H.
Epstein et al., "Science-Based Actions Can Help Address the Opioid Crisis," *Trends in
Pharmacological Sciences* 39, no. 11 (November 2018): 912, https://doi.org/10.1016
/j.tips.2018.06.002.

106     *a slow decrease in the number of opioid prescriptions:* Centers for Disease Control and
Prevention, "U.S. Opioid Dispensing Rate Maps," https://www.cdc.gov/drugoverdose
/maps/rxrate-maps.html.

106     *has probably cut the number:* American Medical Association Opioid Task Force 2020
Progress Report

106     *in 2019 the combined number of overdose deaths:* Centers for Disease Control and Prevention,
National Center for Health Statistics, Provisional Drug Overdose Death Counts

106     *By one recent estimate, close to a quarter of men:* Dana A. Glei et al., "Estimating the
impact of drug use on US mortality, 1999-2016," *PLOS One* 15, no. 1 (January 2020):
1, https://doi.org/10.1371/journal.pone.0226732.

107     *Neurons that respond to dopamine are exquisitely sensitive to the difference:* Wolfram
Schultz, "Dopamine Reward Prediction Error Coding," *Dialogues in clinical neuroscience*
18, no. 1 (March 2016): 23–32, https://doi.org/10.31887/DCNS.2016.18.1/wschultz.

107     *Drugs create a surge of dopamine that is by some estimates\*:* National Institute on Drug
Abuse, "Drugs, Brains, and Behavior: The Science of Addiction," NIH, originally
printed in April 2007, https://www.drugabuse.gov/sites/default/files/soa.pdf, 22.
\*It is not clear why this is the case, but in 2018 scientists reported that opioid drugs
activate receptors inside the neuron, whereas the body's natural opioids only act on
receptors at the cell surface. This surprising discovery could lead to the development of
better drugs for pain control, or to new ways to treat addiction: Miriam Stoeber et al.,
"A genetically encoded biosensor reveals location bias of opioid drug action." *Neuron*
98, no. 5 (June 2018): 963–76, https://doi.org/10.1016/j.neuron.2018.04.021.

107     *effective technique of rewarding abstinence with small prizes:* Franco De Crescenzo
et al., "Comparative efficacy and acceptability of psychosocial interventions for
individuals with cocaine and amphetamine addiction: A systematic review and network
meta-analysis," *PLOS Medicine* 15, no. 12 (December 2018): e1002715, https://doi
.org/10.1371/journal.pmed.1002715.

108     *Das's team recruited ninety people in the United Kingdom:* Ravi K. Das and Sunjeev K.
Kamboj et al., "Ketamine can reduce harmful drinking by pharmacologically rewriting
drinking memories," *Nature Communications* 10, no. 1 (November 2019): 1–10, https
://doi.org/10.1038/s41467-019-13162-w.

109     *In follow-up research with 120 participants:* Grace Gale et al., "Long-term behavioural
rewriting of maladaptive drinking memories via reconsolidation-update mechanisms,"
*Psychological Medicine* (June 2020): 1–11, https://doi.org/10.1017/S0033291720001531.

109     *"The questions are," he writes, "when, for whom, and to what extent":* David H. Epstein,
"Let's agree to agree: a comment on Hogarth (2020), with a plea for not-so-competing
theories of addiction," *Neuropsychopharmacology* 45, no. 5 (January 2020): 716, https
://doi.org/10.1038/s41386-020-0618-y.

110     *The Drug Enforcement Agency (DEA) classifies fentanyl:* U.S. Department of Justice, DEA,
Diversion Control Division, "Controlled Substances - Alphabetical Order," December 21,
2020, https://www.deadiversion.usdoj.gov/schedules/orangebook/c_cs_alpha.pdf.

110     *Even the NIH, a government-run research facility:* Kelly Burch, "DEA Restrictions
Delaying Fentanyl Research That Could Save Live," *The Fix*, April 24, 2018.

**Additional Sources:**  Interviews: Jed Barash: November 2018; Owen Rivers: August 2018. E-mail:
Jed Barash and Bertha Madras: March 2017.

## Chapter 11

111    *Medical students and practicing anesthesiologists across the country:* Paul G. Barash, Michael K. Cahalan, Bruce F. Cullen, Christine M. Stock, Robert K. Stoelting, Rafael Ortega, Sam R. Sharar, and Natalie Holt, *Clinical Anesthesia*, 8th edition (Philadelphia: Wolters Kluwer/Lippincott Williams & Wilkins, 2017).

112    *Paul Janssen began the search for a better synthetic opioid:* Theodore H. Stanley, "The Fentanyl Story," *The Journal of Pain* 15, no. 12 (October 2014): 1215–26, https://doi.org/10.1016/j.jpain.2014.08.010.

112    *Their concerns turned out to be well-founded:* Harry D. Silsby et al., "Fentanyl Citrate Abuse Among Health Care Professionals," *Military Medicine* 149, no. 4 (April 1984): 227–28, https://doi.org/10.1093/milmed/149.4.227.

113    *This seemed like a problem worth fixing:* W. Andrew Kofke et al., "Neuropathologic effects of anesthetics used to stop status epilepticus in rats," *The Journal of the American Society of Anesthesiologists* 67, no. 3 (September 1987): A390.

113    *Kofke's first major experiment examined the effect of alfentanil:* W. Andrew Kofke et al., "Alfentanil-induced hypermetabolism, seizure, and histopathology in rat brain," *Anesthesia and Analgesia* 75, no. 6 (December 1992): 953–64, https://doi.org/10.1213/00000539-199212000-00014.

114    *he presented a poster at an annual anesthesiology meeting:* W. Andrew Kofke et al., "Opioid neurotoxicity: fentanyl-induced exacerbation of forebrain ischemia in rats," *Journal of Neurosurgical Anesthesiology* 6, no. 4 (October 1994): 323, https://doi.org/10.1016/s0006-8993(98)01228-1.

114    *On rats he tested a range of doses of fentanyl comparable to what humans receive:* W. Kofke et al., "Opioid neurotoxicity: fentanyl dose-response effects in rats," *Anesthesia and Analgesia* 83, no. 6 (December 1996): 1298–1306, https://doi.org/10.1097/00000539-199612000-00029.

114    *He tested multiple types of fentanyl:* W. Andrew Kofke et al., "Opioid neurotoxicity: neuropathologic effects in rats of different fentanyl congeners and the effects of hexamethonium-induced normotension," *Anesthesia and Analgesia* 83, no. 1 (July 1996): 141–46, https://doi.org/10.1097/00000539-199607000-00025.

114    *Kofke also found drugs that protected rats:* Elizabeth H. Sinz et al., "Phenytoin, midazolam, and naloxone protect against fentanyl-induced brain damage in rats," *Anesthesia and Analgesia* 91, no. 6 (December 2000): 1443–49, https://doi.org/10.1097/00000539-200012000-00027.

114    *Four healthy volunteers agreed to take a brief, FDA-approved dose of remifentanil:* W. Andrew Kofke et al., "The neuropathologic effects in rats and neurometabolic effects in humans of large-dose remifentanil," *Anesthesia and Analgesia* 94, no. 5 (May 2002): 1229–36, https://doi.org/10.1097/00000539-200205000-00033.

115    *Ran a larger experiment:* W. Andrew Kofke et al., "The neuropathologic effects in rats," 1234.

115    *would become his final study:* W. Andrew Kofke et al., "Remifentanil-induced cerebral blood flow effects in normal humans: dose and ApoE genotype," *Anesthesia and Analgesia* 105, no. 1 (July 2007): 167–75, https://doi.org/10.1213/01.ane.0000266490.64814.ff.

116    *he wrote a paper with recommendations:* W. Andrew Kofke, "Anesthetic management of the patient with epilepsy or prior seizures," *Current Opinion in Anesthesiology* 23, no. 9 (June 2010): 391–99, https://doi.org/10.1097/ACO.0b013e328339250b.

**Additional Sources:** Interviews: Jed Butler: October 2017, October 2018, August 2020; Andrew Kofke: August 2017, June 2019.

## Chapter 12

119     ***"This paper was published right after our MMWR paper came out":*** Marc W. Haut
        et al., "Amnesia Associated with Bilateral Hippocampal and Bilateral Basal Ganglia
        Lesions in Anoxia with Stimulant Use," *Frontiers in Neurology* 8, no. 27 (February
        2017): 1–5, https://doi.org/10.3389/fneur.2017.00027.

120     ***West Virginia ranks first in the nation for overdose deaths, diabetes, and cigarette
        smoking:*** Centers for Disease Control and Prevention, "Map of Current Cigarette Use
        Among Adults," https://www.cdc.gov/statesystem/cigaretteuseadult.html.

120     ***And has the third-highest percentage of people living in poverty:*** The Kaiser Family
        Foundation State Health Facts. Data Source: Kaiser Family Foundation estimates
        based on the Census Bureau's March Current Population Survey (CPS: Annual Social
        and Economic Supplements), 2017, "Distribution of Total Population by Federal
        Poverty Level."

120     ***a man who shot himself in the chest after being diagnosed with Alzheimer's:*** Jennifer
        Wiener Hartzell et al., "Completed suicide in an autopsy-confirmed case of early onset
        Alzheimer's disease," *Neurodegenerative disease management* 8, no. 2 (January 2018):
        81–88, https://doi.org/0.2217/nmt-2017-0045.

120     ***He studied railroad workers whose exposure to solvents:*** Marc W. Haut, "Corpus
        callosum volume in railroad workers with chronic exposure to solvents," *Journal of
        occupational and environmental medicine* 48, no. 6 (June 2006): 615–24, https://doi.org
        /10.1097/01.jom.0000205211.67120.23.

120     ***another case report about a man with a rare and incurable brain disorder:*** Jennifer
        Wiener et al., "Completed suicide in a case of clinically diagnosed progressive
        supranuclear palsy," *Neurodegenerative disease management* 5, no. 4 (August 2015):
        289–92, https://doi.org/10.2217/nmt.15.24.

120     ***which is what happened in 2015:*** Uzoma B. Duru et al., "An Unusual Amnestic
        Syndrome Associated With Combined Fentanyl and Cocaine Use," *Annals of internal
        medicine* 169, no. 9 (November 2018): 662–63, https://doi.org/10.7326/L18-0411.

121     ***He'd seen a strange MRI just like it a year earlier:*** Blair Suter et al., "'Found down':
        Patient with lesions impacting globus pallidus and hippocampus following suspected
        drug overdose," *West Virginia Medical Journal* 113, no. 3 (May–June 2017): 40–44.

122     ***some songbirds must cache their food:*** Anat Barnea et al., "Seasonal recruitment of
        hippocampal neurons in adult free-ranging black-capped chickadees," *Proceedings of the
        National Academy of Sciences* 91, no. 23 (November 1994): 11217–21, https://doi.org
        /10.1073/pnas.91.23.11217.

122     ***Maguire had recently discovered that the hippocampus is more active:*** Eleanor A.
        Maguire et al., "Human spatial navigation: cognitive maps, sexual dimorphism, and
        neural substrates," *Current opinion in neurobiology* 9, no. 2 (April 1999): 171–77, https
        ://doi.org/10.1016/S0959-4388(99)80023-3.

122     ***more activity in their hippocampus than ordinary people:*** Eleanor A. Maguire et al.,
        "Routes to remembering: the brains behind superior memory," *Nature neuroscience* 6,
        no. 1 (January 2003): 90–95, https://doi.org/10.1038/nn988.

122     ***The learning process takes between two and four years:*** Katherine Woollett et al.,
        "Talent in the taxi: a model system for exploring expertise," *Philosophical Transactions of
        the Royal Society B: Biological Sciences* 364, no. 1522 (May 2009): 1407–16, https://doi
        .org/10.1098/rstb.2008.0288.

123     ***she had shown that experience could measurably remodel the brain:*** Eleanor A. Maguire
        et al., "Navigation-related structural change in the hippocampi of taxi drivers,"

*Proceedings of the National Academy of Sciences* 97, no. 8 (April 2000): 4398–4403, https://doi.org/10.1073/pnas.070039597.

123  **made them better equipped from the outset:** Katherine Woolett et al., "Acquiring 'the Knowledge' of London's Layout Drives Structural Brain Changes," *Current Biology* 21, no 24 (December 2011): 2109–14, https://doi.org/10.1016/j.cub.2011.11.018.

124  **and the connect-the-dots test proved him correct:** Haut et al., "Amnesia Associated with Bilateral Hippocampal," 2.

124  **talking about his work for the BuzzFeed News story:** "I Lost My Memory After a Heroin Overdose," *Buzzfeed News*, April 22, 2017, YouTube, video, https://www.youtube.com/watch?v=aHkh19RZe9w.

**Additional Sources:** Interviews: Jed Barash: September 2018, August 2020; Marc Haut: August 2017, September 2018, August 2020; Liv Miller: September 2018. Massachusetts Department of Public Health e-mails: March 2017. Text messages: Jed Barash and Monroe Butler: April and August 2017.

## Chapter 13

128  **Within seventy-two hours, most of the drug:** Mayo Clinic Laboratories, "Fentanyl with Metabolite Confirmation, Random, Urine," accessed February 11, 2021, https://www.mayocliniclabs.com/test-catalog/Clinical+and+Interpretive/89655.

128  **victims of cardiac arrest who had a history of opioid use:** Jed A. Barash et al., "Opioid-associated Acute Hippocampal Injury with Cardiac Arrest," *Radiology* 289, no. 2 (2018): 315, https://doi.org/10.1148/radiol.2018181379.

129  **They'd analyzed autopsy records for six months:** Nicholas J. Somerville et al., "Characteristics of Fentanyl Overdose — Massachusetts, 2014–2016," *Morbidity and Mortality Weekly Report* 66, no. 4 (April 2017): 382–86, https://doi.org/10.15585/mmwr.mm6614a2.

133  **"Liv, can you please go see this patient?":** Duru et al., "An Unusual Amnestic Syndrome," 895–96.

**Additional Sources:** Interviews: Jed Barash, Monroe Butler: May and December 2018, September 2019; Alfred DeMaria, Marc Haut: August 2017, September 2018; Liv Miller: September 2018. Massachusetts Department of Public Health e-mails: April and May 2017.

## Chapter 14

141  **It typically strikes people between their late forties and early sixties:** Katrina M. Moore et al., "Age at symptom onset and death and disease duration in genetic frontotemporal dementia: an international retrospective cohort study," *The Lancet Neurology* 19, no. 2 (February 2020): 145–56, https://doi.org/10.1016/S1474-4422(19)30394-1.

143  **Even in healthy people, the hippocampus begins to shrink:** Mark A. Fraser et al., "A systematic review and meta-analysis of longitudinal hippocampal atrophy in healthy human ageing," *Neuroimage* 112 (May 2015): 364–74, https://doi.org/10.1016/j.neuroimage.2015.03.035; Lisa Nobis, Sanjay G. Manohar, Stephen M. Smith, Fidel Alfaro-Almagro, Mark Jenkinson, Clare E. Mackay, and Masud Husain, "Hippocampal volume across age: Nomograms derived from over 19,700 people in UK Biobank," *NeuroImage: Clinical* 23, no. 101904 (June 2019): 1–13, https://doi.org/10.1016/j.nicl.2019.101904.

143     *and is significantly faster in people with Alzheimer's disease:* Josephine Barnes et al.,
        "A meta-analysis of hippocampal atrophy rates in Alzheimer's disease," *Neurobiology
        of Aging* 30, no. 11 (November 2009): 1711–23, https://doi.org/10.1016/j.neurobiolaging
        .2008.01.010.

144     *cognitive trajectory climbs rapidly from birth until about age thirty:* UCSF Weill
        Institute for Neurosciences, "Healthy Aging vs. Diagnosis," accessed February 11,
        2021, https://memory.ucsf.edu/symptoms/healthy-aging.

144     *exercise is one of the most effective ways to change how your brain ages:* Joseph Michael
        Northey et al., "Exercise interventions for cognitive function in adults older than 50: a
        systematic review with meta-analysis," *British journal of sports medicine* 52, no. 3 (2018):
        154–60, https://doi.org/10.1136/bjsports-2016-096587; Alana M. Horowitz et al.,
        "Therapeutic potential of systemic brain rejuvenation strategies for neurodegenerative
        disease," *F1000Research* 6, no 1291 (August 2017): https://doi.org/10.12688/f1000
        research.11437.1; and Kirk I. Erickson et al., "Physical activity, cognition, and brain
        outcomes: a review of the 2018 physical activity guidelines," *Medicine and science in sports
        and exercise* 51, no. 6 (June 2019): 1242.

144     *when rodents run on treadmills more than their peers:* Henriette Van Praag et al.,
        "Exercise enhances learning and hippocampal neurogenesis in aged mice," *Journal of
        Neuroscience* 25, no. 38 (September 2005): 8680–85, https://doi.org/10.1523
        /JNEUROSCI.1731-05.2005; Carla M. Yuede et al., "Effects of voluntary and forced
        exercise on plaque deposition, hippocampal volume, and behavior in the Tg2576 mouse
        model of Alzheimer's disease," *Neurobiology of disease* 35, no. 3 (September 2009):
        426–32, https://doi.org/10.1016/j.nbd.2009.06.002.

144     *dig into these details to find out if the results in mice apply to people:* Kaitlin B. Casaletto
        et al., "Late-life physical and cognitive activities independently contribute to brain and
        cognitive resilience," *Journal of Alzheimer's Disease* 74, no. 1 (March 2020): 363–76,
        https://doi.org/10.3233/JAD-191114.

144     *Yassa's research with young adults:* Kazuya Suwabe et al., "Rapid stimulation of human
        dentate gyrus function with acute mild exercise," *Proceedings of the National Academy of
        Sciences* 115, no. 41 (October 2018): 10487–92, https://doi.org/10.1073/pnas.1805
        668115.

145     *Okonkwo's previous research showed that people who reported exercising more:* Ozioma
        C. Okonkwo et al., "Physical activity attenuates age-related biomarker alterations in
        preclinical AD," *Neurology* 83, no. 19 (November 2014): 1753–60, https://doi
        .org/10.1212/WNL.0000000000000964.

145     *For his pilot study, Okonkwo's team:* Julian M. Gaitán et al., "Protocol of Aerobic
        Exercise and Cognitive Health (REACH): A Pilot Study," *Journal of Alzheimer's
        Disease Reports* 4, no. 1 (March 2020): 107–21, https://doi.org/10.3233/ADR-200180.

145     *A woman from Colombia with a genetic mutation:* Joseph F. Arboleda-Velasquez et al.,
        "Resistance to autosomal dominant Alzheimer's disease in an APOE3 Christchurch
        homozygote: a case report," *Nature Medicine* 35, no. 11 (November 2019): 1680–83,
        https://doi.org/10.1038/s41591-019-0611-3.

146     *Another patient followed for years at Johns Hopkins University's:* Greg Rienzi, "Banking
        on Brain Science," Johns Hopkins Medicine News and Publications, August 2, 2019,
        https://www.hopkinsmedicine.org/news/articles/banking-on-brain-science.

146     *increasingly less likely to be the case that education:* Andrea M. Rawlings et al.,
        "Cognitive Reserve in Midlife is not Associated with Amyloid- Deposition in Late-
        Life," *Journal of Alzheimer's Disease* 68, no. 2 (March 2019): 517–21, https://doi.org
        /10.3233/JAD-180785.

146    *Women, who tend to have better verbal skills than men:* Erin E. Sundermann et al., "Female advantage in verbal memory: Evidence of sex-specific cognitive reserve," *Neurology* 87, no. 18 (2016): 1916–24, https://doi.org/10.1212/WNL.0000000000003288.

146    *a third or more of cases of Alzheimer's are due to lifestyle factors:* Tiia Ngandu et al., "A 2 year multidomain intervention of diet, exercise, cognitive training, and vascular risk monitoring versus control to prevent cognitive decline in at-risk elderly people (FINGER): a randomised controlled trial," *The Lancet* 385, no. 9984 (June 2015): 2255–63, https://doi.org/10.1016/S0140-6736(15)60461-5; Sam Norton et al., "Potential for primary prevention of Alzheimer's disease: an analysis of population-based data," *The Lancet Neurology* 13, no. 8 (August 2014): 788–94, https://doi.org/10.1016/S1474-4422(14)70136-X.

146    *Some programs aim to tailor the treatment:* Richard S. Isaacson et al., "Individualized clinical management of patients at risk for Alzheimer's dementia," *Alzheimer's & Dementia* 15, no. 12 (December 2019): 1588–602, https://doi.org/10.1016/j.jalz.2019.08.198.

146    *Perhaps as many as forty-six million people in the United States who have no symptoms:* Ron Brookmeyer, Nada Abdalla, Claudia H. Kawas, and María M. Corrada, "Forecasting the prevalence of preclinical and clinical Alzheimer's disease in the United States," *Alzheimer's & Dementia* 14, no. 2 (February 2019): 126, https://doi.org/10.1016/j.jalz.2017.10.009.

146    *their memory can be on par with someone thirty or more years younger:* Felicia W. Sun et al., "Youthful Brains in Older Adults: Preserved Neuroanatomy in the Default Mode and Salience Networks Contributes to Youthful Memory in Superaging," *Journal of Neuroscience* 36, no. 37 (2016): 9659–68, https://doi.org/10.1523/JNEUROSCI.1492-16.2016.

147    *"your Letter to the Editor entitled":* Jed A. Barash et al., "Acute amnestic syndrome associated with fentanyl overdose," *New England Journal of Medicine* 378, no. 12 (March 2018): 1157–58, https://doi.org/10.1056/NEJMc1716355.

**Additional Sources:** Interviews: Jed Barash: September 2018, January 2020; Monroe Butler: May and December 2018; Devyn Cotter: May 2018, April and December 2019; Suzy Kwok: May 2018. Massachusetts Department of Public Health e-mails: August and October 2017. Text messages: Jed Barash and Monroe Butler: April and August 2017.

## Chapter 15

152    *yesterday morning, he e-mailed a journalist:* Lauren Aguirre, "A Mysterious Amnesia, Related to Opioid Overdose, Creeps Beyond New England," *Undark*, January 29, 2018.

154    *the practice led to significantly more strokes and more deaths:* Marc I. Chimowitz et al., "Stenting versus Aggressive Medical Therapy for Intracranial Arterial Stenosis," *New England Journal of Medicine* 365, no. 11 (September 2011): 993–1003, https://doi.org/10.1056/NEJMoa1105335; Gina Kolata, "Study is Ended as a Stent Fails to Stop Strokes," *New York Times*, September 7, 2011.

155    *a computer program called FreeSurfer has automated the process:* Mike F. Schmidt et al., "A comparison of manual tracing and FreeSurfer for estimating hippocampal volume over the adult lifespan," *Human brain mapping* 39, no. 6 (June 2018): 2500–13, https://doi.org/10.1002/hbm.24017.

156    *A paper by forensic scientists who examined autopsy reports:* Iain C. Anthony et al., "Pre-disposition to accelerated Alzheimer-related changes in the brains of HIV negative opiate abusers," *Brain* 133, no. 12 (December 2010): 3685–98, https://doi.org/10.1093/brain/awq263.

157 *A similar study of Norwegian heroin users with brain damage:* Solveig Norheim Andersen et al., "Hypoxic/ischaemic brain damage, especially pallidal lesions, in heroin addicts," *Forensic Science International* 102, no. 1 (May 1999): 51–9, https://doi .org/10.1016/S0379-0738(99)00040-7.

157 *an earlier paper from these same Scottish scientists:* Jeanne E. Bell et al., "Hyperphosphorylated tau and amyloid precursor protein deposition is increased in the brains of young drug abusers," *Neuropathology and applied neurobiology* 31, no. 4 (August 2005): 444, https://doi.org/10.1111/j.1365-2990.2005.00670.x.

158 *Barash and his co-authors' response to the criticism:* Richard P. Dutton, "More on Acute Amnestic Syndrome Associated with Fentanyl Overdose," *The New England Journal of Medicine* 378, no. 23 (June 2018): 2247-48, https://doi.org/10.1056/nejmc1805681.

158–159 *A middle-aged woman in Ohio developed brain swelling:* Adam S. Jasne et al., "Cerebellar Hippocampal and Basal Nuclei Transient Edema With Restricted diffusion (CHANTER) Syndrome," *Neurocritical care* 31, no. 2 (February 2019): 290, https://doi.org/10.1007/s12028-018-00666-4.

159 *A 63-year-old woman in Canada had used fentanyl patches prescribed for pain:* Ryan G. Taylor et al., "Opioid-associated amnestic syndrome observed with fentanyl patch use," *The Canadian Medical Association Journal* 191, no. 12 (March 2019): E337–E339, https://doi.org/10.1503/cmaj.181291; correction: (November 2019) 191: E1228. https ://doi.org/10.1503/cmaj.191311.

159 *"Attention to this syndrome may stimulate further investigation":* Dutton, "More on Acute Amnestic Syndrome," 2247–48.

**Additional Sources:** Interviews: Jed Barash: October 2018; Monroe Butler: May 2018; Richard Dutton: April 2019; Adam Jasne: April 2019; Owen Rivers: August 2018; Ryan Taylor: March 2019. Massachusetts Department of Public Health e-mails: May 2018.

## Chapter 16

161 *she'll administer more than a dozen tests to probe his memory:* P. Monroe Butler et al., "An opioid-related amnestic syndrome with persistent effects on hippocampal structure and function," *The Journal of neuropsychiatry and clinical neurosciences* 31, no. 4 (March 2019): 392–96, https://doi.org/10.1176/appi.neuropsych.19010017 (see also online-only supplement at appi.neuropsych.19010017.ds001.pdf).

166 *A young neuropsychologist named Renaud La Joie has designed:* Butler et al., "An opioid-related amnestic syndrome," online-only supplement.

166 *La Joie developed an MRI scanning technique that measures:* Renaud La Joie et al., "Differential effect of age on hippocampal subfields assessed using a new high-resolution 3T MR sequence," *Neuroimage* 53, no. 2 (November 2010): 506–514, https://doi.org/10.1016 /j.neuroimage.2010.06.024

**Additional Sources:** Interviews: Jed Barash: August 2018; Monroe Butler: July, December 2018; September 2019; Kaitlin Casaletto: April 2019; Devyn Cotter: Renaud La Joie: April 2019; Owen Rivers: April, August 2019.

## Chapter 17

170 *The human brain contains about eighty-six billion neurons:* Frederico AC Azevedo et al., "Equal numbers of neuronal and nonneuronal cells make the human brain an

isometrically scaled-up primate brain," *Journal of Comparative Neurology* 513, no. 5 (February 2009): 532–541, https://doi.org/10.1002/cne.21974.

170    **the hippocampus gives birth to dentate gyrus cells throughout our lives in a process\*:** Elena P. Moreno-Jiménez et al., "Adult hippocampal neurogenesis is abundant in neurologically healthy subjects and drops sharply in patients with Alzheimer's disease," *Nature medicine* 25, no. 4 (March 2019): 554–60, https://doi.org/10.1038/s41591-019-0375-9.

\*A 2018 paper by Shawn F. Sorrells et al., "Human hippocampal neurogenesis drops sharply in children to undetectable levels in adults," *Nature*, 555 no. 7696 (March 2018): 377-81, https://doi.org/10.1038/nature25975, created a stir because the researchers found no evidence of adult neurogenesis. In a blog post, "WTF! No neurogenesis in humans??," http://snyderlab.com/2018/03/07/wtf-no-neurogenesis-in-humans/, Jason Snyder analyzed the strengths and weaknesses of the research and placed it the context of previous work. Subsequent studies, like the one referenced above, continue to find evidence for neurogenesis. Research has also found evidence that neurogenesis may occur in the human olfactory bulb, but this area has not been studied as extensively. For example, Michael A. Durante et al., "Single-cell analysis of olfactory neurogenesis and differentiation in adult humans." *Nature neuroscience* 23, no. 3 (February 2020): 323–26, https://doi.org/10.1038/s41593-020-0587-9.

170    **The phenomenon is called pattern separation:** Joshua P. Neunuebel et al., "CA3 retrieves coherent representations from degraded input: direct evidence for CA3 pattern completion and dentate gyrus pattern separation," *Neuron* 81, no. 2 (January 2014): 416–27, https://doi.org/10.1016/j.neuron.2013.11.017.

171    **A decade ago, when Sahay began researching neurogenesis:** Amar Sahay et al., "Increasing adult hippocampal neurogenesis is sufficient to improve pattern separation," *Nature* 472, no. 7344 (April 2011): 466–70, https://doi.org/10.1038/nature09817.

171    **new neurons in the dentate gyrus were evening out the balance:** Kathleen M. McAvoy et al., "Adult hippocampal neurogenesis and pattern separation in DG: a role for feedback inhibition in modulating sparseness to govern population-based coding," *Frontiers in Systems Neuroscience* 9, no. 120 (August 2015): 1–7, https://doi.org/10.3389/fnsys.2015.00120.

172    **Allowing some details to escape one's memory is equally important:** Blake A. Richards et al., "The Persistence and Transience of Memory," *Neuron* 94, No. 6 (June 2017): 1071–84, figure 2, https://doi.org/10.1016/j.neuron.2017.04.037.

173    **researchers like Sahay aim to tweak neurogenesis:** Kathleen M. McAvoy et al., "Modulating Neuronal Competition Dynamics in the Dentate Gyrus to Rejuvenate Aging Memory Circuits," *Neuron* 91, no. 6 (September 2016): 1356–73, https://doi.org/10.1016/j.neuron.2016.08.009.

173    **genetically engineered Alzheimer's mouse models have less neurogenesis:** Oliver Wirths, "Altered neurogenesis in mouse models of Alzheimer disease," *Neurogenesis (Austin)* 4, no. 1 (May 2017): e1327002, https://doi.org/10.1080/23262133.2017.1327002.

173    **artificially boosting neurogenesis and adding a growth factor called BDNF:** Se Hoon Cho et al., "Combined adult neurogenesis and BDNF mimic exercise effects on cognition in an Alzheimer's mouse model," *Science* 361, no. 6406 (September 2018): 1–15, https://doi.org/10.1126/science.aan8821.

174    **Owen's hippocampus has shrunk by at least as much as the average sixty-year-old's:** Ruth Peters, "Ageing and the brain," *Postgraduate Medical Journal* 82, no. 964 (February 2006): 84–88, https://doi.org/10.1136/pgmj.2005.036665; "Hippocampal volume across age: Nomograms derived from over 19,700 people in UK Biobank," *NeuroImage: Clinical* 23, no. 101904 (June 2019): 1–13, https://doi.org/10.1016/j.nicl.2019.101904.

175    *A typical Alzheimer's drug development program:* Jeffrey Cummings et al., "The price
       of progress: Funding and financing Alzheimer's disease drug development," *Alzheimer's
       & Dementia: Translational Research & Clinical Interventions* 4 (June 2018): 331, https
       ://doi.org/10.1016/j.trci.2018.04.008; Alzheimer's Association, "How Clinical Trials
       Work," accessed February 11, 2021, https://www.alz.org/alzheimers-dementia
       /research_progress/clinical-trials/how-trials-work.

175    *Memantine in 2003. This drug, along with a handful of others:* Robert J. van Marum,
       "Update on the use of memantine in Alzheimer's disease," *Neuropsychiatric Disease
       Treatment* no. 5 (May 2009): 237–47, https://doi.org/10.2147/ndt.s4048; Alzheimer's
       Association, "FDA-approved treatments for Alzheimer's," 2019, https://www.alz.org
       /media/documents/fda-approved-treatments-alzheimers-ts.pdf.

175    *In 2018, Pfizer shut down its Alzheimer's division:* Colin Dwyer, "Pfizer Halts
       Research into Alzheimer's Treatments," National Public Radio, January 8, 2018,
       https://www.npr.org/sections/thetwo-way/2018/01/08/576443442
       /pfizer-halts-research-efforts-into-alzheimers-and-parkinsons-treatments.

**Additional Sources:** Interviews: Monroe Butler: July 2018, August 2020, January 2021; Owen
Rivers: January and September 2020.

## Chapter 18

180    *the process can easily take more than ten years:* Gail A. Van Norman, "Drugs, Devices,
       and the FDA: Part 1: An Overview of Approval Processes for Drugs," *JACC: Basic to
       Translational Science* 1, no. 3 (April 2016): 170–79, https://doi.org/10.1016/j.jacbts
       .2016.03.002.

180    *But it's not uncommon for safety concerns to crop up long after approval:* Syed
       Rizwanuddin Ahmad, "Adverse Drug Event Monitoring at the Food and Drug
       Administration Your Report Can Make a Difference," *Journal of General Internal
       Medicine,* 18 no. 1 (January 2003): 57–60, https://doi.org/10.1046/j.1525-
       1497.2003.20130.x; Ellen Pinnow et al., "Postmarket Safety Outcomes for New
       Molecular Entity Drugs Approved by the Food and Drug Administration Between
       2002 and 2014," *Clinical Pharmacology Therapy,* 104 no. 2 (August 2018): 390–400,
       https://doi.org/10.1002/cpt.944.

181    *called the FDA Adverse Event Reporting System:* U.S. Food & Drug Administration,
       "FDA Adverse Event Reporting System (FAERS) Public Dashboard," accessed
       February 11, 2021, https://www.fda.gov/drugs/questions-and-answers-fdas-adverse
       -event-reporting-system-faers/fda-adverse-event-reporting-system-faers-public-dashboard.

183–184 *a small clinical trial published in 1985:* Bradley T. Hyman et al., "Effect of naltrexone on
       senile dementia of the Alzheimer type," *Journal of Neurology, Neurosurgery & Psychiatry*
       48, no. 11 (November 1985): 1169–71, https://doi.org/10.1136/jnnp.48.11.1169.

184    *for the longest period—two weeks—seems most promising:* David S. Knopman et al.,
       "Cognitive effects of high-dose naltrexone in patients with probably Alzheimer's
       disease," *Journal of Neurology, Neurosurgery, and Psychiatry* 49, no. 11 (November 1986):
       1321–26, https://doi.org/10.1136/jnnp.49.11.1321-a.

**Additional Sources:** Interviews: Jed Barash: June, November, and December 2018, June 2019; Erica:
January 2019; Monroe Butler: December 2018; Alfred DeMaria: August 2018; David Knopman:
August 2020; Owen Rivers: August 2018. Massachusetts Department of Public Health e-mails:
March 2017, June, July, August, October, and November 2018.

# Chapter 19

187      *publishing her first paper on the topic:* Michela Gallagher et al., "Manipulation of Opiate Activity in the Hippocampus Alters Memory Processes," *Life Sciences* 23, no. 19 (November 1978): 1973–77, https://doi.org/10.1016/0024-3205(78)90565-9.

187      *improved the animals' ability to remember the location of foot shocks:* Michela Gallagher, "Naloxone Enhancement of Memory Processes: effects of Other Opiate Antagonists," *Behavioral and Neural Biology* 35, no. 4 (August 1982): 375–82, https://doi.org/10.1016/S0163-1047(82)91020-2.

187      *where to find food pellets in a maze:* Michela Gallagher et al., "Opiate Antagonists Improve Spatial Memory," *Science* 221 no. 4614 (September 1983): 975–76, https://doi.org/10.1126/science.6879198; Michela Gallagher, "Effects of Opiate Antagonists on Spatial Memory in Young and Aged Rats," *Behavioral and Neural Biology* 44, no. 3 (November 1985): 374–85, https://doi.org/10.1016/S0163-1047(85)90688-0.

187–188      *she found that old rats naturally have more of their own opioids:* Hann-Kuang Jiang et al., "Elevated dynorphin in the hippocampal formation of aged rats: relation to cognitive impairment on a spatial learning task," *Proceedings of the National Academy of Sciences of the United States of America* 86, no. 8 (April 1989): 2948–51, https://doi.org/10.1073/pnas.86.8.2948.

188      *Injecting more of these natural opioids . . . directly into rat brains:* Elizabeth Bostock et al., "Effects of opioid microinjections into the medial septal area on spatial memory in rats," *Behavioral Neuroscience* 102, no. 5 (October 1988): 643–52, https://doi.org/10.1037/0735-7044.102.5.643.

188      *Gallagher turned her attention to that neurotransmitter:* Michela Gallagher et al., "Ageing: the cholinergic hypothesis of cognitive decline," *Current Opinion in Neurobiology* 5, no. 2 (April 1995): 161–68, https://doi.org/10.1016/0959-4388(95)80022-0.

188      *the four FDA-approved drugs:* Justin M. Long et al., "Alzheimer Disease: An Update on Pathobiology and Treatment Strategies," *Cell* 179, no. 2 (October 2019): 11, https://doi.org/10.1016/j.cell.2019.09.001.

188      *This includes Aricept, which can improve symptoms slightly for anywhere from a few months:* Joanne Knowles, "Donepezil in Alzheimer's disease: an evidence-based review of its impact on clinical and economic outcomes," *Core Evidence* 1, no. 3 (March 2006): 205–206.

189      *the higher the activity, the harder it was for rodents to navigate:* Ming Teng Koh et al., "Treatment Strategies Targeting Excess Hippocampal Activity Benefit Aged Rats with Cognitive Impairment," *Neuropsychopharmacology* 35, no. 4 (March 2010): 1016–25, https://doi.org/10.1038/npp.2009.207.

189      *especially those with amnestic mild cognitive impairment:* Michael A. Yassa et al., "Pattern separation deficits associated with increased hippocampal CA3 and dentate gyrus activity in nondemented older adults," *Hippocampus* 21, no. 9 (August 2011): 968–79, https://doi.org/10.1002/hipo.20808.

189      *and so do APOE4 carriers:* Ramsey Najm et al., "Apolipoprotein E4, inhibitory network dysfunction, and Alzheimer's disease," *Molecular neurodegeneration* 14, no. 1 (June 2019): 1–13, https://doi.org/10.1186/s13024-019-0324-6.

189      *"I'll just do the experiment":* Rebecca P. Haberman et al., "Targeting Neural Hyperactivity as a Treatment to Stem Progression of Late-Onset Alzheimer's Disease," *Neurotherapeutics* 14, no. 3 (May 2017): 662–76, https://doi.org/10.1007/s13311-017-0541-z.

190     ***AgeneBio began enrolling patients in a trial:*** "Study of AGB101 in Mild Cognitive
        Impairment Due to Alzheimer's Disease," Clinicaltrials.gov, accessed February 11,
        2021, https://clinicaltrials.gov/ct2/show/NCT03486938.

190     ***buying as little as one to two years could result in 10 to 20 percent fewer:*** Julie
        Zissimopoulos et al., "The Value of Delaying Alzheimer's Disease Onset," *Forum for
        health economics & policy* 18, no. 11 (November 2014): 6, figure 1, https://doi.org
        /10.1515/fhep-2014-0013.

191     ***Are associated with paying attention:*** Jessica A. Cardin et al., "Driving fast-spiking cells
        induces gamma rhythm and controls sensory responses," *Nature* 459, no. 7247 (April
        2009): 663–67, https://doi.org/10.1038/nature08002.

192     ***and shone 40-hertz light down the wires directly into the hippocampus:***
        Chinnakkaruppan Adaikkan et al., "Gamma Entrainment: Impact on Neurocircuits,
        Glia, and Therapeutic Opportunities," *Trends in Neurosciences* 43, no. 1 (January 2020):
        24–41, https://doi.org/10.1016/j.tins.2019.11.001.

192     ***In follow-up experiments, mice unencumbered by wires:*** Chinnakkaruppan Adaikkan
        et al., "Gamma Entrainment Binds Higher-Order Brain Regions and Offers
        Neuroprotection," *Neuron* 102 no. 5 (June 2019): 929–43, https://doi.org/10.1016/j
        .neuron.2019.04.011; Anthony J. Martorell et al., "Multi-sensory gamma stimulation
        ameliorates alzheimer's - associated pathology and improves cognition," *Cell* 177 no. 2
        (April 2019): 256–71, https://doi.org/10.1016/j.cell.2019.02.014.

193     ***at multiple locations in three trials using Cognito's device:*** "Study of Tolerability,
        Safety and Efficacy of Sensory Stimulation at Multiple Dose Levels to Improve Brain
        Function (Etude Study)," Clinicaltrials.gov, accessed February 11, 2021, https
        ://clinicaltrials.gov/ct2/show/NCT03661034; "Stimulating Neural Activity to Improve
        Blood Flow and Reduce Amyloid: Path to Clinical Trials," Clinicaltrials.gov, accessed
        February 11, 2021, https://clinicaltrials.gov/ct2/show/NCT03543878; "Multi-Center
        Study of Sensory Stimulation to Improve Brain Function," Clinicaltrials.gov, accessed
        February 11, 2021, https://clinicaltrials.gov/ct2/show/NCT03556280.

193     ***At MIT, Tsai is running her own small clinical trial:*** "Daily Light and Sound
        Stimulation to Improve Brain Functions in Alzheimer's Disease," Clinicaltrials.gov,
        accessed February 11, 2021, https://clinicaltrials.gov/ct2/show/NCT04055376.

194     ***proof-of-concept results showing the rejuvenating power of young blood:*** Saul A. Villeda
        et al., "Young blood reverses age-related impairments in cognitive function and
        synaptic plasticity in mice," *Nature Medicine* 20, no. 6 (May 2014): 659–63, https://doi
        .org/10.1038/nm.3569.

195     ***began charging people thirty-five years or older eight thousand dollars:*** Shanley Pierce,
        "High risks and high costs for young blood donations to older people," *Texas Medical
        Center,* September 3, 2019.

195     ***after the FDA issued a warning that the treatment:*** Statement from FDA
        Commissioner Scott Gottlieb, M.D., and Director of FDA's Center for Biologics
        Evaluation and Research Peter Marks, M.D., Ph.D., cautioning consumers against
        receiving young donor plasma infusions that are promoted as unproven treatment for
        varying conditions, February 19, 2019, https://www.fda.gov/news-events/press
        -announcements/statement-fda-commissioner-scott-gottlieb-md-and-director-fdas
        -center-biologics-evaluation-and-0.

195     ***To test whether his hunch about this enzyme was correct:*** Alana M. Horowitz et al.,
        "Blood factors transfer beneficial effects of exercise on neurogenesis and cognition to
        the aged brain," *Science* 369, no. 6500 (July 2020): 167–73, https://doi.org/10.1126
        /science.aaw2622.

195     *Several clinical trials are already underway:* "A Study to Evaluate Albumin and Immunoglobulin in Alzheimer's Disease (AMBAR)," Clinicaltrials.gov, updated July 31, 2019, https://clinicaltrials.gov/ct2/show/NCT01561053; "Young Donor Plasma Transfusion and Age-Related Biomarkers," Clinicaltrials.gov, accessed February 11, 2021, https://clinicaltrials.gov/ct2/show/NCT02803554; "The PLasma for Alzheimer SymptoM Amelioration (PLASMA) Study (PLASMA)," Clinicaltrials.gov, accessed February 11, 2021, https://clinicaltrials.gov/ct2/show/NCT02256306.

196     *transplanting healthy inhibitory neurons into an aging mouse hippocampus:* Magdalena Martinez-Losa et al., "Nav1.1-Overexpressing Interneuron Transplants Restore Brain Rhythms and Cognition in a Mouse Model of Alzheimer's Disease," *Neuron* 98, no. 1 (April 2018): 75–89, https://doi.org/10.1016/j.neuron.2018.02.029.

196     *boosting the mouse brain's ability to clear out plaques and tangles:* Gustavo A. Rodriguez et al., "Chemogenetic attenuation of neuronal activity in the entorhinal cortex reduces Aβ and tau pathology in the hippocampus," *PLoS biology* 18, no. 8 (August 2020): e3000851, https://doi.org/10.1371/journal.pbio.3000851.

196     *or rejuvenating neurons' metabolism:* Joshua Goldberg et al., "The mitochondrial ATP synthase is a shared drug target for aging and dementia," *Aging Cell* 17, no. 2 (January 2018): e12714, https://doi.org/10.1111/acel.12715.

196     *forty-six million people:* Ron Brookmeyer, Nada Abdalla, Claudia H. Kawas, and María M. Corrada, "Forecasting the prevalence of preclinical and clinical Alzheimer's disease in the United States," *Alzheimer's & Dementia* 14, no. 2 (February 2019): 126, https://doi.org/10.1016/j.jalz.2017.10.009.

**Additional Sources:** Interviews: Jed Barash: November 2018; Michela Gallagher: August 2020. Text messages: Jed Barash and Monroe Butler: December 2018.

## Chapter 20

199     *Dr. Wilder Graves Penfield gently lowers an electrode:* Wilder Penfield et al., "The Brain's Record of Auditory and Visual Experiences," *Brain* 86, no. 4 (December 1963): 596, 617–619, https://doi.org/10.1093/brain/86.4.595.

200     *Between the 1930s and the early 1960s:* Wilder Penfield, *Mystery of the Mind, A Critical Study of Consciousness and the Human Brain* (New Jersey: Princeton University Press, 1975), 66, 71.

200     *Penfield wrote that "it was at least as clear to him":* Wilder Penfield and Lamar Roberts, *Speech and Brain Mechanisms* (New Jersey: Princeton University Press, 1959), 50.

200     *"It is a hearing-again and seeing-again":* Wilder Penfield and Herbert Jasper, *Epilepsy and the functional anatomy of the human brain* (Boston: Little, Brown & Co., 1954), 137.

200     *Penfield called these phenomena recordings:* Penfield et al., "The Brain's Record of Auditory and Visual Experiences"; Presidential address at the seventy-sixth Annual Meeting of the American Neurological Association, Atlantic City, N.J., June 18, 1951, "Memory Mechanisms," Wilder Penfield, 185; Alison Winter, *Memory: Fragments of a Modern History* (Chicago, The University of Chicago Press Books: 2013), 92–96.

200     *appeared to contain what Penfield later called keys of access:* Penfield, *Mystery of the Mind*, 66, 71. Penfield's sketch of this concept can be found here: Robert Nitsch et al., "Neuronal Mechanisms Recording the Stream of Consciousness–A Reappraisal of Wilder Penfield's (1891–1976) Concept of Experiential Phenomena Elicited by Electrical Stimulation of the Human Cortex," *Cerebral Cortex* 28, no. 9 (July 2018): 3349, Figure 1, https://doi.org/10.1093/cercor/bhy085.

201      *similar fragmentary memories elicited by brain stimulation:* Jonathan Curot et al.,
         "Awake Craniotomy and Memory Induction Through Electrical Stimulation: Why
         Are Penfield's Findings Not Replicated in the Modern Era?," *Neurosurgery* 87 no. 2
         (August 2020): E130–E137, https://doi.org/10.1093/neuros/nyz553; Jonathan Curot
         et al., "Memory scrutinized through electrical brain stimulation: A review of 80 years
         of experiential phenomena," *Neuroscience & Biobehavioral Reviews* 78 (July 2017):
         161–77, https://doi.org/10.1016/j.neubiorev.2017.04.018.

201–202  *his accidental discovery could have been predicted by an evolutionary biologist. . . . "that
         the necessary conclusion is that learning is just not possible":* Stefan Köhler et al., "Heroes
         of the Engram," *Journal of Neuroscience* 37, no. 18 (May 2017): 4648–50, https://doi
         .org/10.1523/JNEUROSCI.0056-17.2017.

203      *Ramirez and Liu had found an engram:* Xu Liu et al., "Optogenetic stimulation of
         a hippocampal engram activates fear memory recall," *Nature* 484, no. 7394 (March
         2012): 381–85, https://doi.org/10.1038/nature11028.

203      *Ramirez and Liu dubbed this experiment Project Inception:* Steve Ramirez et al.,
         "Creating a False Memory in the Hippocampus," *Science* 341, no. 6144 (July 2013):
         387–91, https://doi.org/10.1126/science.1239073.

203      *a team . . . led by Paul Frankland and Sheena Josselyn implanted an engram:* Gisella
         Vetere et al., "Memory Formation in the Absence of Experience," *Nature Neuroscience*
         22, no. 6 (April 2019): 933–40, https://doi.org/10.1038/s41593-019-0389-0.

204      *one particular molecule, a protein called PKMZeta:* Eva Pastalkova et al., "Storage
         of Spatial Information by the Maintenance Mechanism of LTP," *Science* 313 no. 5790
         (August 2006): 1141–44, https://doi.org/10.1126/science.1128657; Panayiotis Tsokas et al.,
         "Compensation for PKMZeta in long-term potentiation and spatial long-term memory in
         mutant mice," *eLife* (May 2016): 5:e14846, https://doi.org/10.7554/eLife.14846.

205      *Tonegawa's team found that even if they blocked the formation of proteins:* Tomás J. Ryan
         et al., "Engram cells retain memory under retrograde amnesia," *Science* 348, no. 6238
         (2015): 1007–13, https://doi.org/10.1126/science.aaa5542.

205      *Tonegawa later found evidence that two copies:* Takashi Kitamura et al., "Engrams and
         circuits crucial for systems consolidation of a memory," *Science* 356, no. 6333 (April
         2017): 73–78, https://doi.org/10.1126/science.aam6808.

206      *Experiments by Ramirez and others support this dynamic view of memory:* Asaf Marco
         et al., "Mapping the epigenomic and transcriptomic interplay during memory formation
         and recall in the hippocampal engram ensemble," *Nature Neuroscience* 23, no. 12 (October
         2020): 1606–17, https://doi.org/10.1038/s41593-020-00717-0; Christine A. Denny et al.,
         "Hippocampal Memory Traces Are Differentially Modulated by Experience, Time, and
         Adult Neurogenesis," *Neuron* 83, no. 1 (July 2014): 189–201, https://doi.org/10.1016/j
         .neuron.2014.05.018; and Casey J. Guenthner et al., "Permanent genetic access to
         transiently active neurons via TRAP: targeted recombination in active populations,"
         *Neuron* 78, no. 5 (June 2013): 773–84, https://doi.org/10.1016/j.neuron.2013.03.025.

## Chapter 21

207      *The right side looks worse:* P. Monroe Butler et al., "An Opioid-Related Amnestic
         Syndrome with Persistent Effects on Hippocampal Structure and Function," *The
         Journal of Neuropsychiatry and Clinical Neurosciences* 31, no. 4 (March 2019): 394, Figure 4,
         https://doi.org/10.1176/appi.neuropsych.19010017.

209      *mostly in the neurons' long and fragile axons:* Goran Šimić et al., "Tau protein
         hyperphosphorylation and aggregation in Alzheimer's disease and other tauopathies,

and possible neuroprotective strategies," *Biomolecules* 6, no. 1 (January 2016): 6, https://doi.org/10.3390/biom6010006.

209   ***The spread of tau looks a lot like an infection:*** Jacob W. Vogel et al., "Spread of pathological tau proteins through communicating neurons in human Alzheimer's disease," *Nature Communications* 11, no. 2612 (May 2020): 1–15, https://doi.org/10.1038/s41467-020-15701-2.

209   ***which he refers to as a double-prion disorder:*** Atsushi Aoyagi et al., "Amyloid beta and tau prion-like activities decline with longevity in the Alzheimer's disease human brain," *Science Translational Medicine* 11, no. 490 (May 2019): 1–13, https://doi.org/10.1126/scitranslmed.aat8462.

210   ***German neuroscientists Eva and Heiko Braak showed that tau tangles:*** Heiko Braak et al., "Neuropathological stageing of Alzheimer-related changes," *Acta Neuropathologica* 82, no. 4 (September 1991): 239–59, https://doi.org/10.1007/BF00308809.

210   ***meets up with tau that a deadly dance begins:*** Bernard J. Hanseeuw et al., "Association of Amyloid and Tau with Cognition in Preclinical Alzheimer Disease," *JAMA Neurology* 76, no. 8 (August 2019): 915–24, https://doi.org/10.1001/jamaneurol.2019.1424; Michael J. Pontecorvo et al., "A multicentre longitudinal study of flortaucipir (18F) in normal ageing, mild cognitive impairment and Alzheimer's disease dementia," *Brain* 142, no. 6 (June 2019): 1723–35, https://doi.org/10.1093/brain/awz090; and Renaud La Joie et al., "Prospective longitudinal atrophy in Alzheimer's disease correlates with the intensity and topography of baseline tau-PET," *Science Translational Medicine* 12, no. 524 (January 2020): 1–12, https://doi.org/10.1126/scitranslmed.aau5732.

210   ***amyloid beta in the hippocampus makes neurons more hyperactive:*** Gustavo A. Rodriguez et al., "Chemogenetic attenuation of neuronal activity in the entorhinal cortex reduces amyloid beta and tau pathology in the hippocampus," *PLOS Biology* 18 no. 1 (August 2020): e3000851, https://doi.org/10.1371/journal.pbio.3000851.

210   ***New clues that*** APOE4 ***can make tau more toxic:*** "ApoE4 and Tau in Alzheimer's: Worse Than We Thought? Especially in Women," Alzforum, Networking for a Cure, November 1, 2019, https://www.alzforum.org/news/research-news/apoe4-and-tau-alzheimers-worse-we-thought-especially-women.

210   ***therapies that target . . . this abnormal protein's buildup have claimed a growing share of clinical trials:*** Jeffrey Cummings et al., "Alzheimer's disease drug development pipeline: 2020," *Alzheimer's and Dementia Translational Research and Clinical Intervention* 6, no. 1 (July 2020): 272–93, https://doi.org/10.1002/trc2.12050.

210–211   ***there are multiple forms of tau in Alzheimer's patients:*** Simon Dujardin et al., "Tau diversity contributes to clinical heterogeneity in Alzheimer's disease," *Nature Medicine* 26, no 8 (June 2020): 1256–63, https://doi.org/10.1038/s41591-020-0938-9.

211   ***But the brains of young opioid abusers in Scotland also contained tangles:*** Jeanne E. Bell et al., "Hyperphosphorylated tau and amyloid precursor protein deposition is increased in the brains of young drug abusers," *Neuropathology and applied neurobiology* 31, no. 4 (August 2005): 444, https://doi.org/10.1111/j.1365-2990.2005.00670.x; G. G. Kovacs et al., "Heroin abuse exaggerates age-related deposition of hyperphosphorylated tau and p62-positive inclusions," *Neurobiology of Aging* 36, no. 11 (August 2015): 3100–07.

**Additional Sources:** Interviews: Monroe Butler: December 2018. Massachusetts Department of Public Health e-mail: December 2018. Text messages: Jed Barash and Monroe Butler: December 2018.

## Chapter 22

213     *As many as 45 percent of older people experience a brief period of delirium after surgery:*
        Laszlo Vutskits and Zhongcong Xie, "Lasting impact of general anaesthesia on the
        brain: mechanisms and relevance," *Nature Reviews Neuroscience* 17, no. 11 (November
        2016): 707, https://doi.org/10.1038/nrn.2016.128.

213     *who have surgery each year is more than 300 million and rising:* Thomas G. Weiser,
        Alex B. Haynes, George Molina, Stuart R. Lipsitz, Micaela M. Esquivel, Tarsicio
        Uribe-Leitz, Rui Fu, Tej Azad, Tiffany E. Chao, William R. Berry, and Atul A.
        Gawande, "Size and distribution of the global volume of surgery in 2012," *Bulletin of
        the World Health Organization,* November 2015, https://www.who.int/bulletin
        /volumes/94/3/15-159293/en/.

213     *A third or more are over sixty-five:* Katie J. Schenning et al., "Sex and genetic
        differences in postoperative cognitive dysfunction: a longitudinal cohort analysis,"
        *Biology of Sex Differences* 10, no. 14 (March 2019): 1, https://doi.org/10.1186
        /s13293-019-0228-8.

213     *higher than normal levels of tau protein and amyloid beta in the cerebrospinal fluid:*
        Mattias Danielson et al., "Association between cerebrospinal fluid biomarkers of
        neuronal injury or amyloidosis and cognitive decline after major surgery," *British
        Journal of Anaesthesia* 126, no. 2 (February 2020): 467–76, https://doi.org/10.1016/j
        .bja.2020.09.043.

213     *increased levels of tau for as long as six months after surgery:* Matthew DiMeglio et al.,
        "Observational study of long-term persistent elevation of neurodegeneration markers
        after cardiac surgery," *Scientific reports* 9, no. 1 (2019): 5, https://doi.org/10.1038
        /s41598-019-42351-2.

213     *When researchers analyzed men and women in the same group*: Schenning et al.,
        "Sex and genetic differences in postoperative cognitive dysfunction," *Biology of Sex
        Differences* 10, no. 14 (March 2019): 1, https://doi.org/10.1186/s13293-019-0228-8.

217     *the link to the HBO* VICE *documentary about the amnestic syndrome:* Seth Dalton,
        "'My Brain Was Fucked up': This Guy Could Be the First Person to Get Amnesia from
        a Fentanyl Overdose," August 13, 2018, vice.com, https://www.vice.com/en/article
        /kzykzn/my-brain-was-fucked-up-this-guy-could-be-the-first-person-to-get-amnesia
        -from-a-fentanyl-overdose.

220     *the ideas that Butler, Barash, and Kofke began spitballing over:* Jed A. Barash et al.,
        "Connecting the dots: an association between opioids and acute hippocampal injury,"
        *Neurocase, The Neural Basis of Cognition* 24, no. 2 (May 2018): 124–31, https://doi.org
        /10.1080/13554794.2018.1475572.

**Additional Sources:** Interviews: Ignacio Badiola: June 2019; Jed Barash: December 2018, June 2019;
Andrew Kofke: June 2019; Monroe Butler: December 2018, September 2019; Owen Rivers: April
2019; Zhongcong Xie: February 2019, September 2020. Text messages: Jed Barash and Monroe
Butler: January 2019.

## Epilogue

223     *On February 1, 2020, the Massachusetts Department of Public Health announced:* Mass.
        gov press release, "Man returning from Wuhan, China is first case of 2019 Novel
        Coronavirus confirmed in Massachusetts," February 1, 2020, https://www.mass.gov

/news/man-returning-from-wuhan-china-is-first-case-of-2019-novel-coronavirus
-confirmed-in.

223    *a diagnosis of opioid-associated amnestic syndrome:* Jed A. Barash et al., "Opioid-associated amnestic syndrome: Description of the syndrome and validation of a proposed definition," *Journal of the Neurological Sciences* 417 (October 2020): 117048, https://doi.org/10.1016/j.jns.2020.117048.

224    *By April 3, five staff members and thirteen residents:* Matt Stout, "Amid Pandemic, state moving infected and non-infected alike from Chelsea Soldiers' Home," *Boston Globe*, April 3, 2020.

224    *the COVID-19 pandemic disrupted memory research and Alzheimer's clinical trials:* Kelly Servick, Adrian Cho, Giorgia Guglielmi, Gretchen Vogel, Jennifer Couzin-Frankel, "Labs go quiet as researchers brace for long-term coronavirus disruptions," *Science Magazine*, March 16, 2020; Aaron van Dorn, "COVID-19 and readjusting clinical trials," *The Lancet* 396, no. 10250 (August 2020): P523-24, https://doi.org/10.1016/S0140-6736(20)31787-6.

224    *National Institute of Health, Project Title:* Prescribed opioid induced brain damage in chronic pain patients, Application Number 1R21DA051737-01A1.

# Glossary

. . . . . . . . . . .

It seemed like a good idea to provide a glossary of memory and memory disorders—until I started writing. Where one type of memory leaves off and the other begins is unclear in part because the brain doesn't respect the tidy borders of our classification schemes. The same can be said for memory disorders. But perhaps that difficulty is the point. People have been trying to pin down the nature of memory for centuries, and it's still a work in progress. (Note that the partial list of terms and definitions below is organized conceptually rather than alphabetically.)

**Memory:** The ability to encode, store, and then recall, consciously or unconsciously, information, skills, and past experiences. Learning is intertwined with memory and is defined as acquiring new information or skills such that knowledge or behavior can change. Without memory, learning would be impossible.

**Explicit:** A memory that can be consciously recalled; in other words, something you know that you know. Explicit memory is sometimes called declarative memory. It's divided into two categories, episodic and semantic.

**Episodic:** Memory for a past event. In its richest form, this is a type of story with a who, what, where, when, and why. It could be something that happened to you, or the plot of a movie. Episodic memories rely heavily on encoding in the hippocampus and are made and remembered in the context of prior experience and knowledge about the world. Often, "you" are at the center of such memories.

**Semantic:** Knowledge about yourself and the world that can be learned through personal experience or from others. An example would be knowing that the sky is blue, why it is blue, and that azure is a shade of blue. Some knowledge is extracted from multiple personal episodic memories, and some can be learned from books or teachers. In its purest form, semantic memory is unrelated to a specific episodic memory, but the two types likely lie on a continuum. The formation of new semantic memories requires more effort with a damaged hippocampus, and the recall of old ones is also impaired.

**Implicit:** A memory you're not consciously aware of.

**Procedural:** A memory of how to do something, like riding a bike.

**Pavlovian:** Also known as classical conditioning. Pavlovian memories are created by repeatedly pairing a stimulus, like a bell ringing, with the appearance of food. Eventually, the animal learns to salivate when a bell rings, even in the absence of food.

**Priming:** Taps into the association between previously learned memories. Seeing an orange might make it easier for you to remember the word citrus.

**Memory Systems:** Forming a new memory often requires a combination of systems. For example, the entire experience of using a drug involves anticipating the result (reward), knowing where to get it (semantic), knowing how to do it (procedural), and the story of the event itself (episodic).

**Short-term memory:** A small amount of information kept for a brief period. The precise point where short-term memory ends and long-term memory begins has never been clarified. According to most experts, short-term memories last for seconds to at most a minute or two, but some clinicians and most people characterize short-term memory in terms of minutes, hours, or even a few days.

**Long-term memory:** Memories that are more durable and can last for decades.

**Working memory:** The ability to hold information in mind temporarily while solving a problem. It's often referred to as the brain's Post-It notes and relies on short-term memory, concentration, and attention.

**Amnesia:** Loss of memory. Anterograde amnesia is the inability to form new memories. Retrograde amnesia refers to the loss of previous memories and can be limited to memories that were only recently acquired or extend to those that stretch back decades.

**Dementia:** Dementia refers to a state in which persons who previously were independent in all activities of daily living are impaired because of cognitive dysfunction. Dementia is a "syndrome" and does not signify any one particular cause. The four most common are Alzheimer's, frontotemporal dementia, Lewy body, and vascular. It's not unusual for people to have more than one type of dementia.

**Alzheimer's disease:** Represents about two thirds of all cases of dementia. Alzheimer's disease is a specific biological constellation of the abnormal accumulation of two proteins, amyloid beta peptide and tau, which ultimately leads to plaques and tangles. There are different symptomatic variants due to Alzheimer's disease, but, in the most common, memory loss is one of the earliest problems to surface. Some experts argue there are two ways to define and diagnose Alzheimer's. In the past, "Alzheimer's disease" was used synonymously with dementia with prominent memory loss. But research and clinical experience in the past twenty years has shown that diseases other than Alzheimer's can cause memory problems and that Alzheimer's disease itself can present with cognitive problems that do not initially involve memory dysfunction. New tools that capture what's happening inside the brain while a person is still alive and can be tested may bridge the gap.

**Frontotemporal dementia:** A less common form of dementia due to a group of disorders called frontotemporal degenerations that afflict between fifty and sixty thousand Americans. It typically strikes people between their late forties and early sixties and can cause distressing personality changes, like lack of empathy and inhibition. Memory is usually unaffected. In contrast to Alzheimer's disease, frontotemporal degenerations affect the frontal and temporal lobes, regions of the brain that are involved in the control of personality, interpersonal relations, and social skills.

**Lewy body disease:** Lewy body disease is caused by the build-up of a unique protein called alpha-synuclein in many parts of the brain. Lewy body dementia is diagnosed in a person with dementia who also has two of the following three symptoms: prominent well-formed visual hallucinations, parkinsonism (slowness, difficulty walking, resting tremor), and disordered nighttime sleep or excessive daytime sleepiness. Parkinson's disease itself is a type of Lewy body disease.

**Vascular cognitive impairment (previously known as vascular dementia):** Due to a disease of blood vessels. Vascular cognitive impairment is very common but rarely occurs in isolation and is often accompanied by Alzheimer's disease or Lewy body disease.

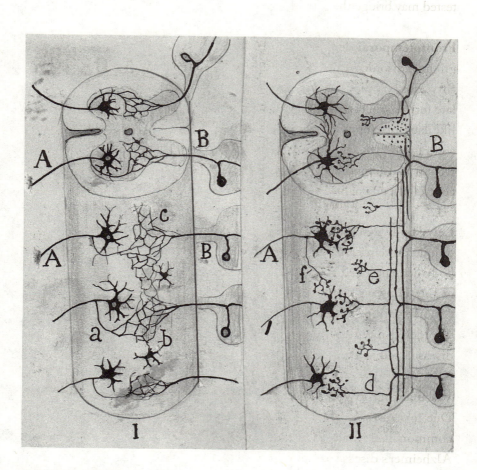

**Cajal's Neuron Doctrine.**

These two nearly identical schematics of the spinal cord illustrate vastly different theories of how information moves through the brain. In drawing I, Cajal depicts the popular theory of the day, in which an unbroken web connects one neuron to the next. In drawing II, Cajal correctly deduces that there are tiny gaps between neurons, which we now know as synapses. This revelation is a cornerstone of modern neuroscience.

# Image Credits

· · · · · · · · · · · · · · · · · · · · · · · ·

# Index

· · · · · · ·

## A

acute bilateral ischemia, 66
acute flaccid myelitis (AFM), 228–229
addiction, 101–110. *See also* opioids
AgeneBio, 190
aging process, slowing, 193–196
Aguirre, Lauren, 152, 232, 234
AIDS, 41–42, 45, 130, 196
alcoholism, 108–109, 131
Altman, Joseph, 94–95
Alzheimer, Alois, 22, 88, 157
Alzheimer's disease
   *APOE4* gene and, 91, 115, 139–143,
    155, 166, 189, 210, 213
   definition of, 269–270
   description of, 4, 22, 25–26
   diagnosis of, 89, 92, 139, 144
   early symptoms of, 82, 88–93, 145–146
   exercise and, 144–146
   familial Alzheimer's, 89
   misfolded proteins and, 22, 90, 209, 211
   opioids and, 115–116
   plaques and, 22, 88–93, 145–146,
    189–193, 209–213
   preventing, 146, 193–196
   research on, x, 87–93, 115–116,
    142–147, 154–157, 170–176, 183–196,
    208–217, 220, 224, 227–228
   risk factors for, 91–93, 115–116, 139–
    146, 155, 161, 189, 210–214, 220
   signs of, 82, 88–93, 145–146
   tau tangles and, 22, 89, 92–93, 145–
    146, 157, 190–192, 207–215, 220,
    224, 227–228
   treating, 183–196
Alzheimer's Disease Neuroimaging
   Initiative (ADNI), 142–143, 154, 208
amnesia. *See also* memory
   anterograde amnesia, 7, 18, 30, 70, 75,
    269
   definition of, 269
   degrees of severity, 69
   diagnosis of, 30, 46, 66–67, 223–224
   environment for, 72–74
   fentanyl and, 131–134, 151–153, 177–
    178, 208
   interventions and, 149–196
   memory aids for, 72–74, 233–234
   opioids and, ix–xi, 23–24, 37, 121–122,
    151–153, 177–178, 208, 226–232
   oxygen deprivation and, 42–43, 46, 72,
    208
   recognizing, 57–58
   retrograde amnesia, 19, 70, 75, 269
   transient global amnesia, 138–139
amnestic mild cognitive impairment
   (aMCI), 91–92, 144, 185, 189–190
amnestic syndrome. *See also* amnesia
   cases of, ix–xi, 9–10, 33, 56–59, 139–
    140, 232–234

classic example of, 9–10

diagnosis of, 223–224

opioid-associated amnestic syndrome, 223–224, 226

research on, x–xi, 65–68, 74–75, 119, 126–133, 139–140, 147, 152–153, 157–159, 175, 178–188, 208–218, 220–234

amygdala, 16, 21

amyloid beta plaques, 22, 88–93, 145–146, 189–193, 209–213

amyloid cascade hypothesis, 88–93, 210

amyloid precursor protein (APP), 89

Anesthesia Patient Safety Foundation, 115, 217

Aniston, Jennifer, 206

anoxia, 124, 126–127

anterograde amnesia, 7, 18, 30, 70, 75, 269. *See also* amnesia

Anthony (patient), 29–32, 35–37, 40, 42, 148

*APOE4* gene, 91, 115, 139–143, 155, 161, 189, 210, 213

Associated Press, ix

*Atlantic*, ix

auras, x

axons, 52–55, 170, 191–192, 209


**B**

Badiola, Ignacio, 217–219, 224

Barash, Jed, x, 3–13, 21–47, 58–68, 75–76, 87–88, 96–97, 102–103, 106, 110–120, 124–143, 147–148, 152–159, 163–165, 176–189, 207–221, 223–226, 228–229, 231–234

Barash, Paul G., 111, 215

Baumann, Michael, 102–103

BDNF growth factor, 173–174

Ben (patient), 178

Berra, Yogi, 62, 148, 179

Beth Israel Deaconess Medical Center, 56, 75, 131–132, 136, 147, 215–217

bilateral hippocampal injury, 33, 57, 153–154

bilateral hippocampal ischemia, 33, 41–47, 57–59, 65–66, 124–125, 131–136, 153–154. *See also* hippocampus

Biogen, 175–176, 186, 207, 218, 228

border cells, 79–80

Boston City Hospital, 41

*Boston Globe*, ix, 67

Boston Medical Center, 128–129, 137

Boston University, 72, 80

Boyden, Ed, 193

Braak, Eva, 210

Braak, Heiko, 210

brain

"broken" brains, 1–47

description of, 14–15, 51–55

electricity in, 15–16, 53–57, 78, 83–86, 168

electrodes in, 16–17, 77–85, 113–114, 199–200, 205

examination of, 21–22

exercise and, 144–146

forming new memory, 18–19

injury to, 34, 46, 61, 72, 75, 91, 115, 136, 174, 228

memory center of, ix, 5, 17–19, 21–22, 55, 167

plasticity of, 96, 121–123, 163, 194–195

samples of, 21, 51–52

stimulation of, 86, 200–201

study of, 21, 51–52

three-dimensional view of, 21

traumatic brain injury, 72, 91

waves in, 83, 85–86, 190–193

brain banks, 142, 157

brain lesions, x, 21, 58, 124, 134, 178

brain scans

experience with, x, xv, 4–5

of hippocampus, xv, xvii, 4–11

viewing, 21–22, 31–32, 44, 57–58, 70

brain tumors, xv, 5, 8, 166

brainwaves, 190–193

Brigham and Women's Hospital, 58, 130

Butler, P. Monroe, 33–39, 43, 64–68, 97, 126–127, 137–143, 153–169, 174–178, 183–188, 207–208, 211, 217–220, 225–228, 232, 234
Buzsáki, György, 205–206
BuzzFeed, ix, 119, 124

C
CADASIL, 140
Cajal, Santiago Ramón, 51–55, 94–96, 170
Cajal's neuron doctrine, 54–55, 271
Camus, Albert, 152
Carillo, George, 34–36
Casaletto, Kaitlin, 161–166, 195
Castle for the Insane, 22
CAT scans, 66, 211
Center for Regenerative Medicine, 87, 170
Center for the Neurobiology of Learning and Memory, 144
Centers for Disease Control and Prevention (CDC), x, 42–43, 45–47, 65–67, 103, 215–216
cerebellum, 21, 52
CESAResearch Network, 103, 105–106
Charles, Prince, 69
Christopher (patient), 133–136
chronic traumatic encephalopathy (CTE), 209
"Cluster of an Unusual Amnestic Syndrome—Massachusetts, 2012–2016," 67, 74–75, 152
CNN, 67, 223
cocaine, using, 8, 28–29, 37, 57, 67, 103–105, 109, 121, 128, 132–136, 165, 178
Cognito Therapeutics, 193
Columbia University, 88, 194, 195
Commission on Combating Drug Addiction and the Opioid Crisis, 106–107
"Complete, Bilateral Hippocampal Ischemia: A Case Series," 65

Corkin, Suzanne, 19, 21, 77
cortex, 22
Cotter, Devyn, 141–143, 154, 160–168
Cotter, Samantha, 180–181
COVID-19, 223–225, 228–229
creativity, 74
Creutzfeldt-Jakob disease (CJD), 11–12, 23, 26, 40–41, 126, 130, 209. See also prion disease

D
Damon, Matt, 185
Das, Ravi, 108–109
David (patient), 178
deep brain stimulation, 86, 200–201
DeMaria, Alfred, 26, 40–47, 57–61, 65–67, 127–131, 134, 137–138, 148, 152, 155–158, 178–181, 185, 224–225, 228–230
dementia. See also Alzheimer's disease
    causes of, 140–141
    definition of, 269
    diagnosis of, 25–26, 92, 144
    exercise and, 146
    frontotemporal dementia, 26, 64, 126, 141, 175, 209, 269, 270
    preventing, 146, 161
    research on, 59, 146, 161–163, 175–176, 183–184, 209–210, 227–228
    risk factors for, 146, 161
    senile dementia, 183–184
Dementia Research Institute, 210
dendrites, 28, 52–55, 170
Denison, Dr., xiv
Department of Defense, 104
Department of Justice, 103
Deter, Auguste, 22, 157
Dostrovsky, Jonathan, 77–79
Down's syndrome, 89
Drug Enforcement Administration (DEA), 97, 103, 110, 113
drug screenings, 104–105, 112, 121, 125, 135, 177–178, 223–224
Duff, Karen, 210–211
Duke University Medical Center, 115

**E**

ecstasy, using, 8, 121, 128

Eichenbaum, Howard, 80–81

electrical impulses, 15–16, 53–57, 78, 83–86, 168

electrodes, 16–17, 77–85, 113–114, 199–200, 205

encephalitis, 69–70

engrams, 201–205

environment, creating, 72–74

Epidemic Intelligence Service, 60, 62

epilepsy, 15–16, 193

episodic memory, 18, 71–72, 75, 79–81, 86, 121, 135, 162, 170–172, 187, 219, 267–269

Epstein, David, 108–109

Erica (patient), 177

excitatory neurons, 115, 191–192

exercise, benefits of, 144–146

explicit memory, 267. *See also* memory

**F**

Facebook, xviii, 141, 178

fake memories, 203–204

familial Alzheimer's, 89. *See also* Alzheimer's disease

FDA Adverse Event Reporting System (FAERS), 181

fentanyl. *See also* opioids
    amnesia and, 131–134, 151–153, 177–178, 208
    classifying, 110
    research on, 114–116, 126–136, 177–178, 214
    screening for, 131–134
    using, x, xv–xvi, 32, 36, 40, 101–102, 110–120, 126–136, 177–178

Fenton, André, 204–206

*Flintstones, The*, 201

Flores, Efren, 60–61

Food and Drug Administration (FDA), 112, 114, 175, 180–181, 184, 188, 195

Fox News, 223

Frankland, Paul, 171–172, 203–204

FreeSurfer program, 155–156

frontotemporal dementia, 26, 64, 126, 141, 175, 209, 269, 270. *See also* dementia

**G**

Galen, Gillian, 8–9, 27, 39–40, 43, 62, 110–111, 116–117, 119, 127, 133, 136, 139–140, 147, 158, 179, 184, 212, 220–221

Gallagher, Michela, 187–190, 195–196, 224

gamma waves, 191–193

Glenn, John, 21

glossary, 267–270

Golgi, Camillo, 52–54

grid cells, 79–80

Guevara, Che, 41

**H**

Hamilton College, 11

harm reduction strategies, 107

Harrelson, Woody, xvii, xviii

Hartford Hospital, 16

Harvard Medical School, 40

Harvard University, 40–41, 106, 195

Haut, Marc W., 120–125, 133–136, 178

HBO, 158, 217

Healthy Aging Study, 141–145

heroin, using, 6, 8, 23, 29, 31–37, 40–42, 45, 59, 62, 101–104, 109, 112, 129, 133–134, 157, 165, 177–179

herpes simplex viral encephalitis, 69–70

hippocampus
    creativity and, 74
    damage to, ix–xi, xvii, 4–10, 13, 22–24, 31–44, 56–57, 70, 77, 113–117, 121–127, 139–143, 158–160, 226–233
    description of, 14–15, 52–55
    encephalitis and, 69–70
    exercise and, 144–146
    hyperactivity in, 114–116, 188–189, 210
    imagination and, 74
    map building, 78–79, 122–123
    as memory center, ix, 5, 17–19, 21–22, 55, 167
    opioid overdose and, ix–xi, 5–10, 37, 40, 44, 121–127

removal of, 16–18, 21
scans of, xv, xvii, 4–11
scenes in, 74
vulnerability of, 22
*Hippocampus as a Cognitive Map, The,* 79
HIV/AIDS, 45, 196
H.M. (patient). *See* Molaison, Henry
Holy Family Hospital, 57
Hospital for Sick Children, 171, 203
*House,* 25, 36
Hull, Mindy, 138
Huntington's disease, 211
hyperactivity, 114–116, 188–189, 210
hypoxia, 43, 126–127, 208, 220

**I**

imagination, 74
implicit memory, 71, 268
inhibition, 115
inhibitory neurons, 55, 115, 191-192, 196, 208
intelligence quotient (IQ), 20, 74, 122, 146, 165
Intensive Outpatient OCD Treatment Program, 233
interventions, 149–196

**J**

*JAMA,* 37, 47, 61, 64
*JAMA Neurology,* 175, 184, 207
Janssen, Paul, 112
Jeter, Derek, 225
Johns Hopkins University, 146, 188
Joshua (patient), 177–178
Josselyn, Sheena, 203–204

**K**

Kaczorowski, Catherine, 93
Kafka, Franz, 151
Kaling, Mindy, 185
Kant, Immanuel, 28, 152
Kennedy, John F., 21
Klein, Joshua P., 58, 66–67
Knopman, D. S., 184–185

Kofke, Andrew, 111–118, 126–136, 140, 143, 158–159, 178, 183–185, 189, 212–220, 224–225, 230–232
Kramer, Joel, 143–147, 161, 195
Kunst, Mara, 57–58
Kwok, Suzy, 141–142
Kyle (patient), 178

**L**

Lafleche, Ginette, 73–75
Lahey Hospital & Medical Center, 4, 6, 9–10, 23–25, 29–32, 36–37, 39, 43–44, 62, 64, 119, 121, 127, 139, 157, 165, 176, 224
La Joie, Renaud, 166–168, 208
Langston, Bill, 34
Lashley, Karl, 201–202
Lawrence General Hospital, 131
learning process, 122, 267
Lev, Michael, 58–62, 66, 127–128
Lewy body disease, 26, 269, 270
Lim, Chun, 56–57, 75–76, 215
Limm, Wendy, 28–29
LinkedIn, 87
Liu, Xi, 202–203
lobotomies, 16
Locke, Dr., xvii–xviii
long-term memory, 19, 233, 269

**M**

MacDonald, Christopher, 80–81
mad cow disease, 11–12, 64, 130. *See also* Creutzfeldt-Jakob disease
Madras, Bertha, 106–107
Maguire, Eleanor, 73–74, 122–123
Mahoney, James, 109
maps, building, 78–79, 122–123
Marslen-Wilson, William, 20
Massachusetts Department of Public Health (DPH), xi, 26, 40–46, 56–61, 67, 127–132, 158, 215, 224, 230
Massachusetts General Hospital, 58–62, 87–88, 97, 113, 128, 131, 138, 147, 170, 173, 212, 215–217, 220, 224

Massachusetts Institute of Technology (MIT), 20, 94–95, 185, 190, 193, 202, 204

Mayo Clinic, 184

McLean Hospital, 27, 106

MDMA, using, 121, 128

MedWatch, 180–181

Meehan, Max, 5–13, 18, 21–25, 30–32, 35–37, 40, 42, 44, 46, 58, 66, 148, 157, 162, 182–183

memory. *See also* amnesia; hippocampus
aids for, 2, 233–234
catching, 77–78
center of, ix, 5, 17–19, 21–22, 55, 167
changing, 203–206
encoding, 28, 79–80, 199–201, 205, 267
episodic memory, 18, 71–72, 75, 79–81, 86, 121, 135, 162, 170–172, 187, 219, 267–269
explicit memory, 267
fake memories, 203–204
finding, 199–206
implicit memory, 71, 268
improving, 27–28, 187–196, 199–206
interaction of, 21
interventions and, 149–196
IQ and, 20, 74, 122, 146, 165
long-term memory, 19, 233, 269
loss of, ix–xi, xv–xix, 7–10, 18–44, 68–75, 82–89, 96, 120–140, 156–163, 170–182, 190, 199–213, 219–226, 232–234
making, 77–81, 199–206
matching patterns, 205–206
new memories, ix, 7, 18–19, 77–79, 205–206
Pavlovian memories, 108, 268
place cells and, 79–80
procedural memory, 18–19, 27, 71, 121–123, 134–135, 187, 268
reactivating, 28, 108, 202–206
reconsolidation, 108–109, 203
retrieving, 199, 205, 234
reward memories, 108–109
semantic memory, 19, 172, 267, 268
short-term memory, 269

smells and, 27–28
spatial memory, 83, 122, 234
stages of, 199–206, 267
storing, 85, 94, 123, 199–201, 205
strengthening, 27–28
systems, 21, 125, 188, 268
temporary loss of, ix–x
testing, 18–20, 24, 85
time and, 80–81
working memory, 85, 164, 191, 269

memory aids, 72–74, 233–234

Memory and Aging Center (MAC), 64, 90, 126, 137–144, 147, 153–157, 160, 165, 174–175, 195–196, 218

Memory Disorders Research Center, 72

Memory Palace, 79–80, 122

memory systems, 21, 125, 188, 268

*Metamorphosis*, 151

methadone, 107, 112

"Methadone Mile," 41, 129

"method of loci," 79–80

Miller, Bruce, 64–65, 90–91, 161, 175

Miller, Liv, 121, 133–135

Milner, Brenda, 18–19, 21, 77

misfolded proteins, 22, 90, 209, 211

Molaison, Henry (H.M.), 13, 15–22, 28, 55, 73, 77, 82, 121, 135, 160, 200, 219, 233

Montefiore Medical Center, 40

Montreal Neurological Institute and Hospital, 16

*Morbidity and Mortality Weekly Report (MMWR)*, 65–67, 74–75, 87, 96, 102, 119, 129, 132, 180, 182, 211

Moser, Edvard, 79

Moser, May-Britt, 79

MPTP, 34–36, 102

MRI scans
experience with, x, xv, 4–5
of hippocampus, xv, xvii, 4–11
machine, 166–168, 207–208
viewing, 21–22, 31–32, 44, 57–58, 70

multisystem inflammatory syndrome (MIS-C), 228–229

Murray, Bill, 185

"Music the Gift of God," 69, 71

**N**

Nadel, Lynn, 79
naloxone, 44, 114, 120, 185–188
naltrexone, 107, 152, 183–189, 214–217, 224
National Drug Early Warning System
  (NDEWS) Network, 103–107
National Institute on Aging, 217
National Institute on Drug Abuse
  (NIDA), 97, 102, 106–111, 232
National Institutes of Health (NIH), 103,
  110, 218–219, 224, 232
*Nature Medicine*, 194
navigation process, 77–86, 122–124, 170–
  171, 187–191, 201–202, 220, 234
*Neurocase*, 65, 139
neurogenesis, 94–96, 123, 169–174, 195
neurological exams, x, 165
neuron doctrine, 54–55, 271
neurons
  bundle of, 35, 204
  changes to, 28–29
  dead neurons, 10–11, 22–24, 35, 89–90,
    170, 173
  drawings of, 54–55, 271
  excitatory, 115, 191–192
  firing of, x, 78–83
  inhibitory, 55, 115, 191–192, 196, 208
  new neurons, 54, 94–96, 169–174
  number of, 170
  study of, 52–55, 78–82, 89–94
  types of, 170
NeuroPace, 83–85
neuroplasticity, 96, 121–123, 163, 194–195
New Britain Museum of American Art, 12
*New England Journal of Medicine (NEJM)*,
  3–4, 147–148, 158–159, 188
New York City Department of Public
  Health, 42
*New York Times*, 67
New York University, 204–205
New York University School of Medicine,
  205
Nields, Henry, 129–133, 138
Nietzsche, Friedrich, 28, 33, 140, 152,
  175, 182

Nixon, Richard, 104
NOVA, ix, 152
NPR, 88

**O**

obsessive-compulsive disorder (OCD),
  102, 153–154, 220, 233–234
O'Keefe, John, 77–79, 81–83
Okonkwo, Ozioma, 145
opioid-associated amnestic syndrome
  (OAS), 223–224, 226. *See also*
  amnestic syndrome
opioid crisis, 40, 45, 103–107, 129, 141–142
opioids. *See also* fentanyl
  addiction and, 101–110
  Alzheimer's disease and, 115–116
  amnesia and, ix–xi, 23–24, 37, 121–122,
    151–153, 177–178, 208, 226–232
  damage from, ix–xi, 5–10, 37–44,
    56–57, 67, 113–117, 121–127, 139–
    143, 158–160, 226–233
  deaths from, 32, 106
  drug screenings, 104–105, 112, 121,
    125, 135, 177–178, 223–224
  impact of, 119–125
  interventions and, 149–196
  overdoses, ix–xi, 5–10, 23–24, 37, 40,
    44, 56–57, 67, 103–104, 121–127
  potency of, 111–118
  research on, 99–148
  taking as prescribed, 137–148
optogenetics, 202–206
overdoses
  amnesia and, ix–xi, 23–24, 37, 121–122,
    151–153, 177–178
  damage from, ix–xi, 5–10, 37, 40, 44,
    121–127, 177–178
  deaths from, 32, 106
  fentanyl overdose, xv–xvi
  heroin overdose, 23
  number of, 103–104
  opioid overdose, ix–xi, 5–10, 23–24, 37,
    40, 44, 56–57, 67, 103–104, 121–127
  seizures and, 178
oxygen deprivation, 42–43, 46, 72, 208

**P**

Pain Medicine Center, 217
pandemic, 223–225, 229
Parkinson's disease, 33–37, 86, 97, 102,
    175, 193, 211, 270
Pavlovian memories, 108, 268
PBS, ix, 232
Pedroia, Dustin, 185
Penfield, Wilder, 16–18, 199–202
PET scans, 92–93, 114, 208, 211, 214, 224
Pfizer, 175
Picower Institute for Learning and
    Memory, 190
Pink Floyd, 201
PKMzeta protein, 204–206
place cells, 79–80
plaques, 22, 88–93, 145–146, 189–193,
    209–213
postoperative cognitive dysfunction
    (POCD), 213, 231
post-traumatic stress disorder (PTSD),
    59, 72, 96, 172, 206
"Power of Observing," 12
priming process, 268
prion disease, 11–12, 23–26, 40–41, 64,
    209. See also Creutzfeldt-Jakob disease
procedural memory, 18–19, 27, 71, 121–
    123, 134–135, 187, 268
Project Inception, 203
Project Seahorse, 142, 154–157, 175, 183,
    216–218
Prusiner, Stanley, 12, 64, 209
PubMed, 37, 47, 65, 157, 183–184

**Q**

Quigley, Lawrence, 63

**R**

Raman, Vinod, 131–132
Ramirez, Steve, 202–206, 224
Ramón y Cajal, Santiago, 51–55, 94–96, 170
Ranck, James, Jr., 79
Real-world Assessment, Prediction, and
    Treatment Unit, 108
rebirth, 169–176

reconsolidation, 108–109, 203
retrograde amnesia, 19, 70, 75, 269. See
    also amnesia
reward memories, 108–109
reward-prediction error, 107
Rivers, Owen, xiii–xviii, 27–29, 101–102,
    151–157, 160–170, 174–176, 181–182,
    207–208, 211, 215–220, 224–227,
    232–234
Robins, Lee, 104–105
Rosen, Howie, 196

**S**

Sacktor, Todd, 204
Sahay, Amar, 87–88, 91–97, 170–173,
    195, 224
St. Elizabeth's Hospital, 141
Sam (patient), 84–85
Santa Clara Valley Medical Center, 34
schizophrenia, 16
Scoville, William, 16–18, 21–22
Scripps Mercy Hospital, xiv
Seeley, Bill, 64, 139
seizures
    auras and, x
    epileptic seizures, 15–16, 193
    grand mal seizure, 19, 70
    overdose and, 178
    pinpointing, 82–83
    seizure disorder, 57
semantic memory, 19, 172, 267, 268
Semon, Richard, 201–202, 205
senile dementia, 183–184
Shimamura, Arthur, 232–233
short-term memory, 269
Simonides, 79–80
Small, Juan, 10–11, 32–33, 37, 43–44, 65
Soldiers' Home, 62–63, 212, 214–216,
    224, 226
Somerville, Nick, 41, 45–46, 59–62,
    65–67, 129, 134, 158
spatial memory, 83, 122, 234. See also
    memory
spatial navigation, 187–191, 201–202. See
    also navigation process

speed cells, 79–80
Spencer, Diana, 69
Stanford University, 194, 195
STAT news, ix, 67
stroke, 8, 10–11, 32
substance use disorders (SUDs), 60–61,
    106, 128–131, 228
substantia nigra, 35, 36
Suthana, Nanthia, 82–86
synapses, 53, 89, 107, 204–205, 210
synthetic heroin, 34–35

**T**

Tanzi, Rudolph, 173–174, 195, 212
tau tangles, 22, 89, 92–93, 145–146, 157,
    190–192, 207–215, 220, 224, 227–228
Theta, 83, 85–86, 191
Tiffany (patient), 178
time, tracking, 80–81
Tonegawa, Susumu, 204–205
Tonight Show, The, 25, 37
transient global amnesia (TGA), 138–139
traumatic brain injury, 72, 91
Tsai, Li-Huei, 190–193, 195, 205, 224
Twitter, 159, 202

**U**

UMass Boston, 24
United Kingdom Dementia Research
    Institute, 210
University College London, 73, 77, 108, 122
University of California (UCLA), xiv, 27,
    38, 64, 82–83, 90–91, 101, 140, 144,
    161–165, 169, 193, 233
University of Connecticut School of
    Medicine, 12
University of Pennsylvania, 111, 114, 161,
    208, 217, 232
University of Vermont, 187
University of Wisconsin School of
    Medicine and Public Health, 145

**V**

VA Boston Healthcare System, 72, 75
vascular cognitive impairment, 26, 269, 270

Verfaellie, Mieke, 72–76
VICE documentary, 158, 217
Villeda, Saul, 193–196
viral encephalitis, 69–70
virtual reality (VR) games, 83–85
virtual reality (VR) tool, 175
VisCube, 207
Volkow, Nora, 97, 102, 110

**W**

Wait Wait . . . Don't Tell Me!, 88
Warren, Elizabeth, 185
Washington University Department of
    Psychiatry, 104
Wearing, Clive, 69–72, 233
Wearing, Deborah, 70–72
West Nile virus, 42, 228
West Virginia Hospital, 133
West Virginia University School of
    Medicine, 109, 120, 136
Wheliaw, Ann, 200
Wheliaw, Bessie, 200
"White Christmas," 199
Who's Who, 123
William A. Hinton State Laboratory, 40
Wilson, Barbara, 71, 72
Winchester Hospital, 4
Wish, Eric, 105
"Wish You Were Here," 201
working memory, 85, 164, 191, 269

**X**

Xie, Zhongcong, 212–216, 220, 224

**Y**

Yale University, 23, 229
Yassa, Michael, 91, 144–145

**Z**

Zabar, Yuval, 5–10, 24–25, 32–33, 46, 65,
    119–120, 124, 176

# MURDER, SHE WROTE: PRESCRIPTION FOR MURDER

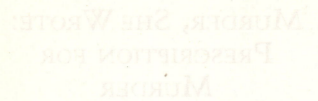

This Large Print Book carries the
Seal of Approval of N.A.V.H.

# MURDER, SHE WROTE: PRESCRIPTION FOR MURDER

## JESSICA FLETCHER & DONALD BAIN

### *Based on the Universal television series created by Peter S. Fischer, Richard Levinson & William Link*

**THORNDIKE PRESS**
*A part of Gale, Cengage Learning*

GALE
CENGAGE Learning·

Farmington Hills, Mich • San Francisco • New York • Waterville, Maine
Meriden, Conn • Mason, Ohio • Chicago

Copyright © 2013 by Universal City Studios Productions LLLP. *Murder, She Wrote* is a trademark and copyright of Universal Studios. Licensed by NBCUniversal Television Consumer Products Group 2013.
Thorndike Press, a part of Gale, Cengage Learning.

**LIBRARY OF CONGRESS CATALOGING-IN-PUBLICATION DATA**

Names: Fletcher, Jessica, author. | Bain, Donald, 1935– author.
Title: Murder, she wrote : prescription for murder / by Jessica Fletcher & Donald Bain.
Other titles: Prescription for murder
Description: Large print edition. | Waterville, Maine : Thorndike Press Large Print, 2016. | © 2013 | Series: A Murder, she wrote mystery | Series: Thorndike Press large print mystery | "Based on the Universal television series created by Peter S. Fischer, Richard Levinson & William Link."
Identifiers: LCCN 2016003885 | ISBN 9781410479242 (hardback) | ISBN 1410479242 (hardcover)
Subjects: LCSH: Fletcher, Jessica—Fiction. | Women novelists—Fiction. | Women detectives—Fiction. | Murder—Investigation—Fiction. | Large type books. | BISAC: FICTION / Mystery & Detective / Women Sleuths. | GSAFD: Mystery fiction.
Classification: LCC PS3552.A376 P74 2016 | DDC 813/.54—dc23
LC record available at http://lccn.loc.gov/2016003885

Published in 2016 by arrangement with New American Library, an imprint of Penguin Publishing Group, a division of Penguin Random House LLC

*For dear friends James and Jeannette Vann, with whom we share a love of all things jazz, and who took us by the hand and showed us Tampa, Florida, their adopted city. Jimmy Vann is a superb neocubist artist who was named Tampa Artist of the Year in 2011, and two of his works proudly adorn the walls of our home. His murals celebrating life in Tampa, and particularly the contribution made to the city by its first black patrolmen, are must-sees when visiting Cigar City. Thank you, Jimmy and Jeannette, and happy fifty-fourth anniversary. (You can enjoy Jimmy's art and fanciful art-inspired items at www.jamesvannart.com.)*

# ACKNOWLEDGMENTS

Many thanks, of course, to our dedicatees, Jeannette and Jimmy Vann, who introduced us not only to Tampa, Florida, but also to most of the people mentioned below. We are grateful to the following, all of whom shared their time and love of Tampa, as well as a good deal of information, not all of which made it onto these pages: Charlie Miranda from District 6, chairman of the Tampa City Council; Major Gerald Honeywell, District 3 commander, Tampa Police Department; Captain Diane Hobley-Burney, Sector E commander, Tampa Police Department; Ronna J. Metcalf, executive director of the Life Enrichment Center; and Kathy Steele of the *Tampa Tribune*.

Thanks, too, to those we encountered along the way and who generously let us take advantage of their knowledge: neurologist Dr. Greg Scott, cigar fanciers Marilyn and Ed Dunn of Thonotosassa, and David

Couzens of the Sandpearl Resort in Clearwater Beach.

We thank them all and remind readers that any errors are ours.

# CHAPTER ONE

Vaughan Buckley's voice was full of enthusiasm. "Our marketing folks have set up an ambitious publicity blitz in Florida for your new book," my publisher told me. "Naturally, you're the centerpiece of it. We've got TV and newspaper interviews arranged, and a series of speaking and signing appearances. It kicks off on December first."

"Oh, my," I said.

"A problem?"

"I just prefer not to be traveling with the holidays coming up. There's always so much to do, and —"

"I understand," Vaughan said, "but the pub date is November fifteenth, perfect timing for people buying holiday gifts, and it's vitally important that we take advantage of every possible publicity opportunity. The fact that you've set the novel in Florida means that we'll be focusing our first marketing efforts there. I hate to pull you away from the Christmas season in Cabot Cove — I know how much it means to you — but I'll see to it that you're only gone a week."

The novel to which he was referring featured a recurring character I'd introduced in a previous book — a police sergeant from Boston who'd settled into what he assumed would be a relaxing retirement in the Sunshine State but soon found himself knee-deep in murder at his retirement resort. As much as I love my publisher and appreciate everything he's done for my career, I was poised to decline. But because I consider myself a professional, I decided on the spot that I had an obligation to help sell my new book. As the writer, I was part of the publishing team, and that meant doing what was expected of me.

"Back in a week?" I said lightly. "That's a promise?"

"My hand is over my heart as I say it," Vaughan replied with a chuckle.

"Okay," I said. "You'll send me the itinerary?"

"I'll e-mail it first thing in the morning."

The itinerary arrived as promised. The promotional tour was scheduled to kick off in the Miami area, move up the coast to Fort Lauderdale and Palm Beach, and end in Tampa.

As I reviewed it, I made a call to my dear friend and Cabot Cove's favorite physician, Seth Hazlitt, whose recent travels had frequently found him in Florida. "Looks like I'm headed to Tampa first week in December," I said, and explained the nature of the trip and the schedule.

"Is that so?" he said.

"Just thought you'd want to know, since Tampa seems to have become your home away from home recently."

"I wouldn't go that far," Seth said, "but as it happens, your call is very timely."

"Why?"

"I just got off the phone with Susan Shevlin, booked a flight and hotel the second week of December in Tampa."

"That's wonderful," I said. "You really know your way around that city by now. Maybe if our paths cross there you can show me the sights."

"And introduce you to Al."

"Dr. Vasquez?"

"Ayuh. I've told him a lot about you already, Jessica. He'll be delighted to meet a real-live bestselling author."

"Even if she writes about murder?"

"*Especially* because she writes about murder. We'll coordinate our trips."

11

Dr. Alvaro Vasquez had been one of Cuba's leading medical researchers studying Alzheimer's disease. Seth had met him on a trip to the island nation, having traveled to Havana with a group of U.S. physicians under a new policy established by the current administration in Washington to open up travel to Cuba for select groups, including medical practitioners, artists, religious organizations, and journalists. I was surprised when he'd announced his decision to join the physician group and make the trip. He certainly was no fan of Castro's regime, and I'd heard him speak fondly of the day when the Cuban people would be free. Seth was also critical of what he considered the United States' ironfisted embargo against Cuba: "All this pigheaded embargo of Cuba accomplishes is to give Mr. Castro an excuse for why Cuba is in such desperate straits. He blames us and the embargo, and nobody challenges him, at least publicly."

But Cuba had proven a siren song for my old friend, and he'd visited there again and again, returning each time to wax poetic about his new friend Dr. Vasquez.

Then, about a year ago, newspapers and TV newscasts announced that Dr. Alvaro Vasquez, one of Cuba's leading medical researchers, had defected. His stature within its medical community had become world famous, and to have someone of his importance elect to leave was obviously a blow to Castro and his regime. According to the news reports, Dr. Vasquez and his wife had attended a medical conference in Bern, Switzerland. Rather than returning home, they'd flown to Washington and asked for asylum. Their application was under review, but no one doubted that someone of Vasquez's stature would be approved.

Naturally, Seth was excited about the news and was especially pleased when Vasquez called him a few weeks after arriving in the States. "Al is anxious to get together with me," he proudly told me after the call.

"You two have really bonded," I said.

"Ayuh, that we have. We get along because he appreciates straight talk from me. Lots of people he works with fawn all over him, tell him what he wants to hear. That's not me."

"No, it certainly isn't," I agreed lightly. "Have you made plans to visit him?"

"No, but I will. He and Ivelisse have decided to live in Tampa, Florida. I'll wait until they're settled in before setting up a visit." Seth had since made several visits south to see his friend, and now the op-

13

portunity had arrived when we were both going to be in the city at the same time.

Seth gave me his itinerary and said, "Mind a suggestion, Jessica?"

"Of course not."

"You'll be coming off your book tour and needin' some rest and relaxation, R and R, as they say in the military. I think you should plan to rest up in Tampa while I'm there, enjoy some downtime. The long-term forecast for early December here in Cabot Cove is for snow and more snow and sub-zero temperatures. Be nice to bask in some sunshine and enjoy fancy drinks with little umbrellas in 'em."

"Sounds appealing," I said. "I'll think about it."

"You do that. If you decide to stay in Tampa, call Susan. She'll book you into the same hotel I'm staying at. Should give the tongues of the rumormongers here something to wag about. Got to run. Let me know what you decide."

I thought about what he'd suggested for the next couple of days before making the decision to extend my Florida trip for a week. It meant cutting into my pre-Christmas activities, but the contemplation of some rest in a more favorable climate became overwhelmingly appealing. I'd also

14

become increasingly interested in Seth's friend and was eager to meet Dr. Alvaro Vasquez. I'd Googled him a bit and was impressed with his expertise and standing among his fellow physicians.

According to what I'd read, his quest for a cure for Alzheimer's disease was admired by his peers, although there seemed to be some sort of mystery surrounding his work, a sense that there were those in the medical research community who questioned his methods. He was known to be extremely secretive about his work, something that other Alzheimer's researchers found arrogant. One even went as far as to brand Vasquez a "grandstander" more eager for fame and fortune than in contributing to science.

Seth had dismissed the harsh views when I'd pointed them out to him. "Medical research is a highly competitive field, Jessica. I'd bet my last nickel this came from a jealous colleague."

"But, Seth, isn't it important for researchers to share important developments in their work so faster progress can be made toward effective treatments?"

He shrugged. "In a perfect world, perhaps. But medical research is every bit as big a business as any other," he said. "If Al gave

15

away his findings, his company might lose the edge they have. Then funding could be compromised and the research delayed, or worse. No, Jessica, I think Al has it just right. Once he's found the results he thinks will be effective, and has them patented or protected in some other way, that's the time to lay his cards on the table, but not yet."

The idea of a medical company patenting its results didn't sit well with me — although I understood that the efforts to come up with treatments or even a cure for any disease deserved to be rewarded — but perhaps I'm naïve when it comes to my expectations for medical research. I didn't argue with Seth, reserving my judgment until I'd met this paragon of Alzheimer's research he was so thrilled with.

I dragged out some of my warm-weather clothing and packed for the trip, although I read that portions of Florida were experiencing record-breaking cold temperatures. Jed Richardson, who owned a small fleet of Cessnas based in Cabot Cove's airfield, flew me to Hartford, Connecticut, where I caught a plane to Miami. Two members of Vaughan Buckley's publicity staff met my flight, and we set off on a whirlwind tour that occupied every minute of the day, and evenings, too.

Coral Gables was gloriously sunny, if a bit cool, and the line of book buyers snaked around the corner from the stop we made at Books & Books. Even though I love staying home surrounded by my friends and the familiar routines of my daily life in Cabot Cove — and leave reluctantly when travel is planned at particularly busy times — I always find myself energized when I'm on the road. Staying in new places or even ones I've visited before is stimulating. There are new sights to see, new foods to taste, and, best of all, new people to meet.

In Miami, I was delighted by the obvious contributions to the city made by its Latin American community and was able to steal some time from the book tour to spend an afternoon in a section called Little Havana, with its colorful art galleries, theaters, restaurants, and cigar stores. I even picked up a set of Cuban dominoes as a souvenir.

Everywhere we went we met crowds of enthusiastic readers, whether they bought their books in a store, borrowed them from the library, or downloaded e-books. I especially enjoy meeting readers. Hearing about their reactions to my novels, what they liked and, occasionally, what they didn't, is helpful to this writer, when so much of my time is spent alone at the computer. All writers need to get out in the world to see what's going on. After all, it's our experiences away from the keyboard that make up the foundation of our stories.

We ended up in Tampa, as scheduled, where the weather was indeed chilly, and I was glad that I'd included a couple of sweaters and a fleece jacket in my suitcase. As we'd seen at each of my stops, fans came out to meet me, and I gave a talk and signed books for several hundred buyers at a large Barnes & Noble. A reporter, Kathy Steele, interviewed me for the *Tampa Tribune,* resulting in a lengthy front-page article accompanied by photographs, and I did two TV interviews and a radio call-in program. My final appearance was at the Tampa Life Enrichment Center, where I spent an hour with men and women taking a class in creative writing, an enriching experience for me, and I hoped for them.

"You're a real trouper, Jessica," Vaughan Buckley's publicist said in the hotel lobby as he and his colleague prepared to head for the airport for a flight back to New York. "Vaughan said to thank you for accommodating his marketing plan."

"It was wonderful," I said, "but I admit I'm ready for some downtime here in Tampa."

"With that colorful character Dr. Hazlitt? We've heard about him."

My publisher had met Seth on a few occasions when he'd traveled to Cabot Cove to confer with me about my books, and apparently he'd shared his views of my good friend.

I laughed. "I never think of Seth as a 'colorful character,' but I suppose in his own way, he is. He's certainly been behaving in unexpected ways lately." I checked my watch. "Five o'clock," I said. "I'd better get to my room and freshen up. Seth and I are meeting at six."

"Vaughan said to send him his best."

"I'll do that," I said, wondering what new surprises my old friend might have in store.

# CHAPTER TWO

The hotel's dining room was small and intimate, with crisp white napery, heavy flatware, and a vase of pink roses on each of the fifteen tables. Seth, who had arrived before me, chatted with a waitress by a small bar meant more for fulfilling drink orders than for directly serving customers.

"She says our table will be ready in a few minutes," he told me in greeting. "You don't look any worse for wear after your book tour."

"It went well," I said. "No hitches, and the people I met were lovely. You look well yourself."

"There you go again, Jessica, trying to kid a kidder. The last week back home was busy. Seems that everybody in Cabot Cove decided to come down with the flu at the same time. Tough strain this year. I even had a few patients end up in the hospital. Had me running around like a headless chicken, but they're all on the mend now. Good thing, too. We got a real cold snap, lots of snow. There's something to be said for those snowbird folks who escape cold winters and head south."

"It's pretty chilly here in Florida, too," I said.

Which was true. Temperatures had dipped to near freezing every night since I'd arrived, threatening the citrus crop and causing Floridians to walk around bundled in heavy coats and sweaters.

"I'd call the weather here pleasant," was Seth's reply.

I decided to change the subject. "Have you seen your friend Dr. Vasquez?"

"Only briefly. He's been occupied with his research lab."

"A research lab must cost a lot of money," I said.

"Doesn't seem to be a problem. Al has funding from K-Dex, a pharmaceutical company in Tampa. He introduced me to K-Dex's founder and CEO, a nice chap named Peters, Bernard Peters, a real go-getter who has great faith that Al is making important strides in coming up with a definitive treatment of the disease."

"If he succeeds, it could be worth millions, even billions," I commented.

"It takes millions to find cures for diseases," was Seth's response.

It occurred to me that since Al had defected to Tampa, the Cuban government had not only lost a distinguished citizen, but stood to lose bragging rights for pioneering a medical breakthrough, to say nothing of the money that it could generate for the island and its people. I expressed my thoughts to Seth.

"Ayuh, Al is very aware of what it means to the Castro regime to have a breakthrough in Alzheimer's research happen here in the States after so much of the initial research was done in Cuba," Seth replied. "I had a long talk with him about that very subject. Al's philosophy is that it shouldn't matter where the breakthrough occurs, just as long as it happens."

"I imagine the Cuban government isn't happy that he's defected," I offered.

"No, I imagine they're not," Seth said.

"How do you feel about it?" I asked. "If Dr. Vasquez did most of his research in Cuba, shouldn't the Cuban people share in the rewards of his work?"

I didn't ask it to put my friend on the spot, but he reacted as though I did. He muttered something about it being premature to think about such things, then said, "Are you finished with your third degree, Mrs. Fletcher?"

I laughed. "I didn't realize I was interrogating you."

"Well, you were."

"In that case, what do you say, sir? Ready to take our table? I'm famished."

He motioned for the hostess, who came to us. "I reserved a table for four," Seth said. "Hazlitt."

I looked at him quizzically.

"I took the liberty of inviting two others to join us," he said, "to slowly introduce you to the local cast of characters. Hope that's all right with you."

"Of course it is," I said as we were led to a window table. "Who are they?"

"A couple of folks I've met on my previous trips to Tampa. Oona — that's Oona Mendez, a terrific young lady. She works for the Cuban American Freedom Foundation, some kind of organization that lobbies the State Department. They've just started flights directly from Tampa to Havana, and that was one of their projects. Bright as a newly minted penny, and pretty, too. She's got herself a boyfriend — I suppose that's what you'd call him — a fella named Karl Westerkoch. He's older than Oona, a bit of a stuffed shirt. Not sure what she sees in him, but that's not my business. He's got a German name but talks with a British accent. Hard fella to figure out."

"What does this Mr. Westerkoch do?" I asked as our waitress filled our water goblets.

"Never did find out," Seth said. "Looks like he might be some sort of diplomat, dresses that way, very high in the collar, if you get my drift, doesn't say much, and when he does talk, he mumbles."

"Sounds lovely."

"Not sure I'd use that word," said Seth. "But she's lovely. Let's check the menu. I'm starvin'."

We'd no sooner begun to peruse our menus when Seth's guests arrived. He had been right: Oona Mendez was indeed a pretty woman — and a tall one, skirting six feet with the high heels she wore. She'd donned a form-fitting green-and-yellow silk dress, the narrow waist of which made her substantial hips and bosom seem larger. Her hair was jet-black and worn long, her café au lait face skillfully, albeit heavily, made-up.

Mr. Westerkoch, her companion, was slightly taller than Ms. Mendez. He was a foppish sort, graying hair worn long on the sides and cascading down over the collar of his blue button-down shirt. His paisley bow tie drooped fashionably. His blue blazer was double-breasted and hung loosely on his pencil-thin frame. He didn't look like a diplomat to me, but I was beginning to learn that Seth and I saw more things differently than I'd realized before. What I noticed most about Mr. Westerkoch at that initial meeting was his posture and mouth. He perpetually leaned forward as though being propelled in that direction, and there was a curl to his lip that could be taken as a physical trait or an editorial comment. It would take a while for me to decide.

After we'd finished with introductions and they'd been seated, Oona said, "I've been looking forward to meeting you, Mrs. Fletcher. Dr. Hazlitt tells me that you're quite famous as an author."

"Thank you," I said. "I've been fortunate in my career."

"That's very self-effacing," she said.

"One attribute Jessica doesn't have is an inflated ego," Seth said.

"That's refreshing in this day and age," Westerkoch said through his crooked lip.

"You know who I am," I said, "but I'd like to know more about you, Ms. Mendez."

"Please, it's Oona."

"And I'm Jessica."

I glanced at Westerkoch, who didn't offer his first name.

Seth jumped into the conversation. "Oona is a close friend of Al's," he said.

"Dr. Vasquez," I said.

"A dear, sweet man," Oona said, "speaking of self-effacing. As a Cuban American, I'm very proud of Alvaro and the work he's doing in medical research." She paused and then continued. "He is extremely pleased that Dr. Hazlitt has taken such an interest in his work. You've become one of his most trusted friends, Seth."

"I do seem to enjoy Al's confidence. He's not only a fine gentleman; I believe that he's onto a very important medical breakthrough." Perhaps to stave off more questions from me, Seth abruptly changed the subject and suggested that we order a bottle of wine.

"And so secretive," Oona added, winking at me.

"Who?" I asked.

"Alvaro, of course," she replied. She wagged her red-tipped index finger at Seth. "I really believe that only you and his lab assistant know what's really going on behind that fortress of a lab that Alvaro has built."

Seth seemed flustered at the comment, turned, and waved our waitress to the table.

We gave her our dinner choices from the menu. "We'd like a bottle of wine," Seth said. "What do you recommend?"

"Is there Cuban wine?" I asked.

Westerkoch, who hadn't said anything, rolled his lip into an even more pronounced snarl and said in a low voice, "Cuban wine is dreadful."

"It is not!" Oona said.

"Did you have Cuban wine when you were in Havana?" I asked Seth.

"Ayuh. Didn't think much of it, but . . ."

"But what?" I said.

"Well," he said, "Al knows this physician from Cuba, a neurosurgeon, I believe, who left Cuba years ago, settled in Oregon, and has a winery that produces what he calls Cuban wine."

"How can it be Cuban wine if it's produced in Oregon?" I asked logically.

"Cuba's climate is bad for the grapes," Westerkoch said, waving a hand in front of his face as if to dissipate a bad odor. "It's too hot there. Grapes need cool winters with cold rain. It only rains in Cuba in the summer."

"I suppose because the doctor was Cuban, he can claim it's Cuban wine," Oona offered as an answer to my question.

Seth asked our waitress, "What wine

would you recommend to go with the food we've ordered?"

She smiled and said, "We have a Cuban pinot noir that some folks say goes nicely with the paella, but you can't prove it by me. It does have a pretty label, though."

Seth ordered a California wine.

The conversation during dinner spanned a variety of subjects, the Tampa professional sports teams taking center stage. Westerkoch seemed to come to life when discussing the relative prospects of baseball's Rays and the National Football League's Tampa Bay Buccaneers. He also held a heated debate with Oona about whether restaurants in Havana have improved; and, of course, Cuba itself was a hot topic, and Fidel Castro, who in 2008 had ceded control to his brother Raúl. I listened with great interest as Oona, who was into our second bottle of wine, railed against the Castro brothers and the destruction they'd brought to her beloved homeland.

"They took what was a thriving economy and turned it into a Communist slum," she said.

"It wasn't any paradise before Castro came," Westerkoch said in a tone that did not invite debate. "It was a sinner's paradise when Batista was in power, nothing but gambling dens and brothels run by the bloody U.S. mafia."

"He's such a hypocrite," Oona said, referring to Fidel Castro. "He complains about the food in Cuba, but when *paladares* are allowed to open he complains that the owners are being enriched."

"*Paladares?*" I said.

"Private restaurants in homes," Oona explained, "where the best food is served. They were illegal until 1994, but the government charges them so many fees, as much as six hundred U.S. dollars every month, the owners can barely survive."

"Cuban food is pretty good here in Tampa," Seth said.

This prompted Westerkoch to mutter, "They put salami on Cuban sandwiches here in Tampa. Real Cuban sandwiches don't have salami."

And so went the rest of the table talk. Westerkoch's frequent and barely stifled yawns heralded the end of the evening for our two dinner guests. As Seth examined the check to determine how much of a tip to include, Oona said to me, "I hope you're up to par tomorrow, Jessica."

"I expect to feel fine," I said.

"No, no," she said, laughing, "on the golf course. I understand that you have plans to play golf tomorrow with Alvaro, our Dr. Vasquez, and your charming friend Dr. Hazlitt."

I looked at Seth, who smiled and said, "Forgot to mention it to you, Jessica. Al is a big golfer, got serious about it after coming to the States. I thought a morning on the course was a perfect way for you to get to know each other."

"But I —"

"We'll discuss it later," Seth said as we walked with the others outside where Westerkoch's car was parked.

Oona gave me a big hug and said that she looked forward to spending time together. "At Alvaro's house tomorrow night," she added.

Another event that Seth had "forgotten" to tell me about.

After they left, Seth and I repaired to the lounge, where he ordered a mojito and a glass of club soda with lime for me.

"Golf?" I said. "You know I don't play golf, and I wasn't aware that you did either."

"Another thing you don't know about me, Jessica. True, I haven't played in years, at least not in Cabot Cove. But I did play when I was a teenager and young med student — was pretty decent, as a matter of fact. Al persuaded me when I met him in

30

Havana to play a round with him, and I did pretty darn good considering how many years it was since I last held a golf club. I've been playing with Al whenever I visit him here in Tampa."

"So much for your golfing history," I said. "The last time I played was in — let me see — it was at least fifteen years ago. I hated it."

Seth laughed, tasted his mojito, and complimented the bartender on the drink. "As good as they make in Cuba," he said. He turned to me. "You'll love it out on the course, Jessica, fresh air, sunshine, good conversation. All you have to do is keep your eye on the ball and swing the club smoothly. You'll be fine, a regular Babe Zaharias."

"What will I wear?"

"Slacks, a sweater, what you have on right now. The country club'll provide you with golf shoes. Trust me. It will be a nice informal get-together. The perfect way for you to meet Al."

"What about this event at Dr. Vasquez's house tomorrow night?" I asked.

"A party to honor *you,* Jessica."

"I thought I was staying an extra week in Tampa for R and R."

"Can't think of a better way to rest and relax than bask in the spotlight at a party welcoming you."

I sat back, cocked my head, and said, "Something has come over you, Dr. Hazlitt."

"Oh? What might that be?"

"Ever since you befriended Dr. Vasquez, you've — well, you've changed."

"Still my usual charming self," he said, smiling broadly.

I returned the smile. The fact was that Seth *had* changed. There was a twinkle in his eye that hadn't been there before, the sort of twinkle that you see in Santa Claus at Christmas. Whatever was behind it — and it certainly wasn't off-putting — my extended week in Tampa was shaping up to be unusual, to say the least.

# CHAPTER THREE

Golf!

The last time I'd been on a golf course was during a charity tournament to benefit Cabot Cove's fire department — and more than fifteen years ago, as I'd told Seth earlier. I'd actually enjoyed the experience because we did more talking than playing, and I even managed to hit that little white ball straight a few times, although not very far. As I recalled, Sam Watson, a local financial planner and accountant, won the match, which surprised no one. He and his wife were avid golfers who played even in winter when the course wasn't covered with snow. For some the game is that much of an obsession.

I had, however, played many rounds at our local miniature golf course and had always done quite nicely.

As I settled in a chair by my hotel room window to read a few chapters in a novel before going to bed, I had visions of swinging a golf club and completely missing the ball, or managing to make contact and sending it into someone's head, causing massive brain injury leading to death. Would I be liable for manslaughter? Then I wondered whether a skilled golfer who deliberately aimed at an intended victim would be guilty of murder, one of those "what if?" moments.

Those bizarre thoughts got in the way of the novel, and I closed the book's cover, sighed, and tried to conjure reasons for begging off the golf date. I didn't come up with anything plausible. Sprained ankle? Carpal tunnel syndrome from working too much at my computer? A terminal case of prickly heat? None of the above. I decided I'd be a good soldier and go along with the plans for the day, wishing that my initial meeting with Dr. Vasquez took place at his house party rather than on a golf course.

After showering and dressing the following morning, I spent a few minutes in front of a full-length mirror swinging an imaginary golf club at an imaginary golf ball. I tried to remember the few tips Sam Watson had given me during my earlier foray on the golf course but came up blank. Keep my left arm straight? Or was it my right arm?

Keep my eye on the ball, or focus on where I wanted it to go? It was a hopeless exercise, and I abandoned it to go down to the restaurant where Seth was to meet me for an early breakfast. As usual he was there before me.

"Al is sending his driver for us at eight," he told me as we perused the buffet.

"Your friend Al obviously has plenty of money," I said as I scooped scrambled eggs on my plate and used tongs to pick up a single strip of bacon.

"Let's just say that he lives comfortably," Seth said.

"You told me that this fellow who heads a pharmaceutical company is bankrolling him."

"Ayuh."

"He must have great faith in Dr. Vasquez and his research."

"It appears that he does. You'll get to meet him at dinner tonight. In the meantime, I assume that you've been practicing your golf swing." There was mirth in his voice.

"Yes," I said in the same light tone. "I played eighteen holes before breakfast. You do realize, Seth, that I'll be embarrassing you."

He looked at me quizzically.

"Playing golf for real," I explained.

"Nonsense, Jessica. Nobody expects you to write bestselling novels *and* be a pro golfer. Al loves the game, but to be honest he isn't all that good at it. He's new to it, of course, and he can't seem to get enough of it. He's that sort of personality, throws himself totally into everything he does. Anyway, you might even beat him."

I laughed away that suggestion and looked out the window. It was an overcast day with low-hanging gray clouds. A stiff breeze sent palm trees into motion and kicked up dust in the parking lot.

"Looks like rain," I said idly.

"Wishful thinking, Jessica," said Seth. "You'd like our golf date to be washed out, called on account of rain like a baseball game."

He'd read my mind.

We waited in front of the hotel until precisely eight, when a long black Mercedes with tinted windows pulled up and a man emerged from the rear.

"Hello, Al," Seth said as the man closed the gap between us, shook Seth's hand, and took one of my hands in both of his.

"And you, of course, are the famous Jessica Fletcher," he said through a dazzling smile.

I'd envisioned Dr. Vasquez to be a smaller,

older man than the person standing in front of me; the one photo I'd seen of him in the newspaper after he'd defected had been blurry, and he'd been in the background. Instead he was tall and movie-star handsome, his dusky complexion a perfect scrim against which very white teeth and deep brown eyes sparkled. Black hair streaked with silver lay close to his temples, not a strand out of place. A thin black mustache curved perfectly over his upper lip. He wore white slacks, a teal polo shirt, white sneakers, and a tan sweater casually draped over his shoulders, the sleeves tied on his chest. I could see him as a tennis pro with whom his female students fell madly in love, or a luxury cruise ship captain holding forth at the dinner table reserved for special passengers. He was, as my friend Mara back in Cabot Cove would say, "a hunk." The famous, and infamous, actor Errol Flynn came to mind.

"It's a pleasure," I said. "Seth speaks of you so often."

"Positively, I hope."

"Of course."

He turned to Seth and slapped him on the shoulder. "The good Dr. Hazlitt here keeps me honest."

Seth's laugh sounded a tad uncomfortable

to my ears. "But not on the golf course. Al shaves a stroke or two off his score now and then."

Vasquez adopted a shocked expression. "Nothing is sacred with your straight-talking doctor friend," he said. He looked up into the menacing sky. "Shall we? I've reserved an eight thirty tee time. From the looks of things, we'll be lucky to get in only nine holes."

I kept my smile to myself.

I noticed as we prepared to enter the car that there was a second man in the front seat next to the driver. Neither man paid us any attention, looking straight ahead as we got in. I sat between Dr. Vasquez and Seth. The air-conditioning was running full blast despite it being a chilly morning; it felt like entering an igloo. The driver pulled away from the hotel without instruction from his boss and we joined the flow of traffic.

"Where are we going?" I asked.

"My club," Vasquez replied. "Hunter's Green Country Club. Excellent course, one of the best in the area."

"You've played them all, I assume?" Seth teased.

"When I've had time, which never seems to be the case." Vasquez's words were tinged with his Cuban heritage. "Life was more leisurely in Cuba," he said somewhat wistfully. "Not as good as here in the United States, of course, but more — leisurely."

"Did you get to play much golf in Cuba?" I asked.

"Unfortunately, no. We have a splendid course, the Varadero, built on property once owned by your du Pont family, but I never had the chance to use it. Their mansion, Xanadu, is now the clubhouse."

"Is Fidel Castro a golfer?" I asked.

Vasquez's smile was wide. "Hardly. He considers it a wasteful pastime of the rich, which I suppose it is for some." He laughed gently. "For me it is a way to escape the laboratory and to try to conquer something that is easier than finding a cure for Alzheimer's."

He sounded melancholy, and I wondered if his transplanted research program was not proceeding as successfully in the U.S. as he had expected.

"I wouldn't say becoming good at golf is easier than *anything*," Seth said.

"Sometimes I think you are right, my friend," Vasquez said. "But I keep trying."

I decided to go on the record before we arrived that golf was not something that I knew anything about, and that I would likely slow down everyone's game.

Vasquez patted my hand, then squeezed it. "Nonsense, Mrs. Fletcher. I have a feeling that you are being too modest. Surely hitting a tiny white ball is considerably easier than writing a bestselling mystery novel. You must tell me how you do that. I'm afraid that if anyone asked me to write a book, I would be at a complete loss."

"Now who's being too modest?" Seth said as the driver pulled into the entrance to the golf club. "When you finally come up with a cure for Alzheimer's disease, publishers will be clamoring for a book from you."

Vasquez laughed. "If that is so, perhaps Mrs. Fletcher will collaborate with me."

"But only if you call me Jessica," I said.

"And I am Al. Very American, yes?"

A security guard examined the ID card the driver presented and waved us through. Now that we'd arrived, I felt my heart racing a little faster. It had never occurred to me when I agreed to spend a week in Tampa with Seth that it would involve playing golf. If he had suggested a fishing expedition, I would have enthusiastically agreed. I do a fair amount of trout and salmon fishing back home. I enjoy tying a fly to my line and wading into the myriad cold, crystal clear streams and rivers that are within minutes of downtown. Even if angling in

Tampa meant deep-sea fishing, it would be something with which I was familiar.

*But golf?*

As we entered the clubhouse, Vasquez stopped a young woman, introduced me and Seth to her, and said, "These fine people are my guests today. Mrs. Fletcher will need a proper pair of golf shoes. I'll take care of Dr. Hazlitt."

She escorted me to the women's locker room, where I was fitted for a pair of splendid-looking white shoes and assigned a locker. A few minutes later she took me to a covered area near the first tee. Vasquez and Seth were standing there waiting for me, along with a short, rotund, pink-cheeked man decked out in golf attire.

"Jessica," said Vasquez, "may I introduce you to my good friend Bernard Peters."

I recognized the name; Seth had told me Peters was the CEO of K-Dex, the pharmaceutical firm that was financially supporting Dr. Vasquez's research. I hadn't expected to meet him until that evening.

"A real pleasure," Peters said. "My wife's a big fan of your books."

"I'm delighted to hear that."

"I think she has every one of them. She has a standing order at our local bookstore for your books when they're released."

"What every writer needs," I said.

Vasquez motioned for a middle-aged man sitting in a golf cart to pull up to us.

"This is Harry, the best caddie at Hunter's Green," Vasquez said. "Let's get started. Jessica, you and I will team up against these two old duffers."

"Oh, no," I said. "Why don't you and Mr. Peters be partners? I'd rather drive Seth crazy with my ineptitude."

"I wouldn't think of it," said Vasquez. "I'm sure you'll do me proud."

I tried my best to justify his optimism. I completely missed the ball on my first swing, but hit it on my second, sending it in a fairly straight line that traveled about thirty yards. After many other swings — some successful, some not — we reached the first hole, and to my amazement my first putt went in, which brought forth a round of applause from the others. Buoyed with that success, I proudly moved on to the second hole and the third, my confidence waxing and waning depending on the accuracy — or lack thereof — of my shots. Bernard Peters, who certainly didn't appear to be athletic, proved to be an excellent golfer, as well as a good sport with the slow pace I set. He didn't say much but had a ready smile and encouraged me each time it was my turn to play.

As we progressed, a certain tension developed with Seth. I know him well enough to pick up on subtle clues when he's annoyed, and it happened on the second hole. Vasquez took it upon himself to give me a golf lesson as I prepared to putt. He came around behind me and placed his hands on mine as they clutched the shaft of my club. Having him press into me from behind was discomforting, and I glanced over at Seth, whose expression was disapproving.

"I think I've got it," I said, creating space between me and Vasquez.

"Yes, I agree," he said. "You certainly do have it, Jessica, in more ways than one."

He did the same at the third hole, but my body language and unwillingness to allow him to get that close sent a message that he obviously received. Seth's displeased expression was gone, and we continued with the game.

Despite my early better-than-expected performance, I kept my eye on the heavens and my fingers mentally crossed. I must admit that having the match canceled due to rain was a pleasant contemplation. The sky appeared to cooperate, turning increasingly dark, almost black at times, and the chilly wind seemed to swell with every step we took. I was about to putt on the fourth hole when jagged flashes of bright white lightning lit up the horizon, followed by a deafening clap of thunder. I looked around at my companions. Peters frowned up at the sky, but Vasquez waved off the weather. "It'll clear up," he said. "These things don't last very long. Go ahead, Jessica. Just don't let the noise throw you off."

I'd experienced plenty of thunderstorms before, but nothing like the heavenly show that was about to take place. The cloud-to-ground brilliant white streaks came in rapid succession, followed by thunder that shook the earth around us. The lightning illuminated the dark sky as though it were created by a mad theatrical producer pulling switches. All was black; then the next bolt came, and the next.

"Let's go," Harry, the caddie, said. "This looks like a bad one."

"We can wait it out," insisted Vasquez. "Look! It's clearing up over there."

"C'mon, Al," Peters said, shoving his club into his golf bag. "It's not like this is the last game you're ever going to play."

The caddie collected the remaining clubs, and no sooner had we squeezed into the golf cart than the rain came pouring down, set in motion by the increasing winds. The light plastic poncho that the caddie handed each of us offered minimum protection. The cart jounced along the track that skirted the course. Each rumble of thunder caused me to wince, and I gripped the sides of my seat to keep from sliding off when we hit a bump on the path. I was greatly relieved when we reached safety beneath the clubhouse's overhang, where other golfers and their caddies had also sought cover.

"My apologies for not providing better weather," Vasquez said as we shed our makeshift rain gear and went inside.

"Saved by the bell," I murmured to Seth, "or in this case, Mother Nature." Aloud to the group, I said, "I'm afraid that golf will never be my game."

"You did splendidly," Vasquez said, and Seth, bless him, reinforced the compliment.

"Quite a storm," Seth commented as we filed into the clubhouse restaurant. Vasquez insisted that we join him for breakfast even though it was a second one for both of us.

Peters begged off: "I have a meeting to get ready for," he said. "See you all tonight. A pleasure to meet you, Jessica."

"What a nice fellow," I said as the maître d' held my chair.

"Peters? Yes, he's a fine fellow," Vasquez agreed, but there was something in his expression and tone that butted heads with his words. I glanced at Seth, but he didn't seem to have picked up on the contradiction.

Our table afforded us a panoramic view of the golf course, and we watched the theatrical majesty of the storm play out as a waitress took our order. After the bouncy ride to the clubhouse, I didn't think my system would stomach anything stronger than an English muffin and a cup of tea, but my male companions seemed to be energized by the squall and ordered accordingly.

"I've never seen a storm quite this violent," I said. "Does it happen often in Tampa?"

"Tampa's the lightning capital of the world," Seth answered. "I've read that as many as fifty people are struck here every year."

"True," said Vasquez. "I've learned since coming to Tampa that when a lightning storm erupts, it's best to seek cover — immediately. Storms here are not to be taken lightly."

I thought that we'd moved rather slowly to leave the course once the storm hit, but didn't express it.

"Look," I said. "The sun is out."

46

Although it continued to rain, sunlight cut through, creating a pretty pattern on the ground outside.

"I told you they don't last long. Typical of the weather here," Vasquez said. "It can change by the hour. Time for my morning cigar. Join me?"

I looked to Seth, who seemed ambivalent.

"Your doctor friend still hasn't adopted the habit of a good cigar after a meal," Vasquez said, "but I'm working on him."

"Too early for me," Seth said, as though discussing an alcoholic drink.

We accompanied Vasquez outside, where he lit up a very long black cigar, using a lighter that was more of a small blowtorch to ignite the cigar's tip.

"A Hoyo de Monterrey Double Corona," Seth said.

"I believe I gave you one, Seth."

"Ayuh, that you did, Al. I'm saving it for a special occasion."

"Every occasion is a special one for a fine cigar," Vasquez said as he sat back in a webbed chair under the overhang, drew deeply on the cigar, slowly blew its blue smoke into the air, and sighed with contentment. He turned to me and said, "I suppose you find it strange that a medical doctor would indulge in such an unhealthy habit."

"It crossed my mind," I said.

He laughed. "Everything in moderation," he said. "Right, Seth?"

"That's my creed," Seth said.

Fortunately the breeze, which had died down along with the storm, blew in a direction that carried the smoke away from me.

After twenty minutes of watching Vasquez indulge his love affair with his cigar, we returned to the car, where the driver and his front-seat companion awaited our arrival. They were beefy men in black suits, white shirts, and skinny black ties, bodyguards right out of central casting.

"I enjoyed our little round of golf even though it was abbreviated," Vasquez said, displaying his wide smile, as the car pulled in front of our hotel.

"And I enjoyed it, although I was sure I wouldn't," I said, not adding that I'd been silently gleeful when the rain came.

"I have a special musical treat for my guests tonight," said Vasquez, helping me from the car. "And authentic Cuban food."

"Seth has been raving about the food ever since he came back from Cuba," I said.

"And for good reason. It will be a pleasure to introduce you to the delights of my native country. But before the party, I would like you, and Seth of course, to be my guests this afternoon at my laboratory. I will give you a personal tour."

"I'd like that very much," I said.

"Good. The car will pick you up here at two. There isn't that much to see, just the usual laboratory paraphernalia, but I think you might find it of interest."

"I know I did," Seth said. He nodded in my direction. "You know, Al doesn't allow many people into his inner sanctum, Jessica. This is quite an honor for you."

"Then I'm grateful to be among the chosen," I said.

"Splendid!" Vasquez said, taking my hand. "I look forward to seeing you. Until then . . ." He raised my hand to his lips, kissed my fingers, and got back into the car.

"Charming fellow, isn't he?" Seth said as the car pulled away and we walked into the hotel.

"Hmm," I said, not exactly agreeing.

*Charming* certainly would be an apt description of Dr. Alvaro Vasquez, although there was a certain slickness that accompanied that adjective for me, which I kept to myself. It represented an unfair judgment, I knew, based on such a preliminary meeting, but initial impressions are often the most

49

lasting ones.

I told Seth that I needed a little break to freshen up before we went to the lab. "You go rest for a bit," he said. "Meet you in the restaurant at twelve thirty for lunch."

*Lunch! Hadn't we just had two breakfasts?*

My friend of many years walked to the elevators and disappeared into one. It struck me that he'd fallen under the spell of Dr. Vasquez, which in itself was certainly understandable. Although Seth was a general practitioner back home — a family doctor who practiced medicine in the broadest of senses — he'd always been keenly interested in the more esoteric disciplines of medicine, which stood him in good stead when having to refer patients to specialists. Seth kept up on the latest in medical research, constantly reading and attending conferences when time permitted. That he was vitally interested in Alzheimer's disease didn't surprise me, not only from an intellectual standpoint but because he had patients who'd fallen victim to the progressive, debilitating illness.

But he seemed to be unusually deferential to Vasquez, a side of Seth that I'd seldom seen in our many years of friendship.

I pondered that as I made my way to the covered patio off the restaurant, took a

50

table, and ordered an iced tea. It had turned into a lovely morning, the sun bright, the gentle breeze bracing. The deck was on a canal in which a number of expensive yachts were secured to floating docks, and I enjoyed watching the passing of a variety of small crafts as I sipped my tea.

Reflections on our golf outing soon gave way to thoughts of Dr. Vasquez and Seth's involvement with him, and I found myself becoming a little uncomfortable. Don't ask me to explain my feelings, because I'd be unable to point to anything tangible. Maybe it was Vasquez's defection from his homeland that added an unusual dimension to the situation. I also wondered about the relationship between Vasquez and Bernard Peters, CEO of K-Dex, the company financing Vasquez's research. The amount of money being put up was obviously substantial. Of course medical research is expensive, and probably as speculative as the stock market for those investing in it. But clearly Vasquez was living the high life, presumably on Mr. Peters's dime, as the saying goes, and I'd detected a certain disdain on Vasquez's part when referring to his benefactor.

*Stop questioning everything, Jessica,* I chided myself. *You're in Tampa for a week of relaxation, a peaceful respite before returning to Cabot Cove and the hectic pace of the holiday season.*

*Relax!*

*Stop looking for intrigue.*

*Just sit back and enjoy this trouble-free week.*

# CHAPTER FOUR

The same two gentlemen in black suits picked us up in the limo to deliver us to Dr. Vasquez's laboratory. As in the morning, they said nothing, simply nodded as we approached the car, held open the door, closed it, and drove off. While they weren't what you would call discourteous, their demeanor was disconcerting, and I wondered what their role in Vasquez's life was. Were they simply from a service hired for the occasion of Seth's and my visit? Or were they employees of our host or of his benefactor, on permanent duty at the disposal of Dr. Vasquez? The few words I'd heard them speak were in Spanish. It would be interesting if their lineage was Cuban, since they worked for a Cuban American physician. But there were many Hispanic citizens in Tampa, from an array of countries. *And why would it be significant if they were Cuban?* I asked myself.

My musings were interrupted by our arrival at the laboratory, a nondescript pale blue one-story building not far from an airport (the sounds of planes taking off and landing confirmed that). While one of the limo drivers opened the car door for us, the other went to the building and punched a code into a box next to the heavy white metal door. A male voice was heard through a tiny speaker. The man in black responded in Spanish. Moments later the sound of some sort of security bars being disengaged could be heard before the door opened and a young man wearing a white lab coat over street clothes stood in the entryway.

"Hello," he said as we approached. He looked at Seth and said, "Welcome back, Dr. Hazlitt."

"Hello, Dr. Sardina," Seth said.

"And you are Mrs. Fletcher," the young man said.

"Yes."

"I'm Pedro Sardina, Dr. Vasquez's assistant. He's waiting for you." Pedro Sardina looked as if he'd spent all of his life inside the laboratory. Of medium height, with a pale complexion, he had prematurely thinning black hair and wore large, thick glasses. The pockets of his lab coat were bulging with items he'd stuffed there and probably forgotten. And a plastic protector in his breast pocket appeared to have been an afterthought, since faint ink stains were visible along the bottom seam.

Once we were inside, Dr. Sardina secured heavy metal bars across the door and reactivated the security system by coding in numbers on a panel.

I said the first thing that came to mind: "It certainly is well protected here."

Sardina smirked. "Security is important," he said as he indicated that we were to follow him down a short corridor. The hum of a motor grew louder as we approached the door at the end of the hallway, where he again entered a code.

"Pedro?" Vasquez's voice asked through the speaker.

"Yes."

The now familiar sound of a bar being raised was heard, and the door opened.

"Ah, I am so pleased you've arrived," Vasquez said, beaming. He, too, wore a white lab coat, over a blue shirt and tie and gray slacks. "Come in, come in."

I don't know what it was that I expected, but somehow the lab didn't match my vision. It was modern, certainly, but rather small. On one wall, white cabinetry was topped with a long green glass counter on which sat a few machines, their functions unknown to me. Above, a stainless-steel shelf held colorful bottles of fluids, which I assumed were chemicals; they were unlabeled. Also on the shelf were boxes of latex gloves, cotton face masks, and containers of antibacterial wipes. On the opposite wall, a bank of three computers, each on a rolling stand, sat next to another cart holding an elaborate maze of interconnected wires snaking out from what looked like an oversized microwave oven. A sanitary hood — at least that's what I think it's called — was tucked in a corner. The hum we'd heard in the hall, which must have been some kind of ventilation system, was close to a roar in the lab. Since there were no windows or skylights — no natural light at all — overhead fixtures gave the room an eerie glow. Could a cure for Alzheimer's disease actually come out of this modest facility? Evidently Vasquez and his backers at K-Dex thought it could.

Sardina closed the door and lowered the bar behind us, its clang jarring in the confined space.

"Like Fort Knox," Seth said with a chuckle.

"A necessary precaution," said Vasquez.

"Why is so much security necessary?" I asked.

"Jessica," Seth said in a disapproving tone. Clearly he didn't want me to challenge Dr. Vasquez in any way.

"No, no, my friend," Vasquez said to Seth. "She may ask whatever questions she likes, although" — he wagged a finger at me — "I may not always answer." He led us around a large table to where Dr. Sardina had settled on a high stool and was peering into a powerful microscope.

"Security is necessary because I have, unfortunately, made enemies in my defection from Cuba, enemies who would like nothing more than to discredit my research and to find a way to contaminate my results. By restricting access to the lab, we know who has been here and we can account for their time and actions. I am assuming my work is safe in your hands, señora, is it not?"

"Absolutely!" Seth inserted.

"Of course," I said, smiling. I cocked my head at his assistant. "What is Dr. Sardina looking at?" I asked, hoping I wasn't being unduly intrusive.

"The latest result of an experiment we've been conducting for the past month," Vasquez responded. "Tell me, Jessica, what crops do you immediately think of when I

57

mention Cuba?"

I hadn't expected to be quizzed, and I laughed. "Well," I said, "since Seth has visited Cuba, I think of cigars. Tobacco. And . . . well, sugar, I suppose."

*"Excelente,"* Vasquez said, reverting to his native language. "Yes, sugar, Cuba's most famous and lucrative crop. Sugar thrives in Cuba as in no other place in the world. Its soil is limestone and goes deep into the ground, as deep as seven meters, ideal for growing sugar; as much as eight million tons are produced each year. The sugar harvest, the *zafra,* from November through June, is a cause for celebration in Cuba. The *macheteros,* the harvesters with their machetes, work round the clock to feed the world's love affair with sweetness. Of course, tobacco, as you point out, is also important to my country's well-being. Enough tobacco to make eighty million cigars is harvested each year. Impressive, yes?"

"I suppose so," I said, "but what do sugar and tobacco have to do with your research on Alzheimer's disease?"

Vasquez looked to Seth and said, "Your lovely friend asks many good questions."

"Ayuh, that she does," Seth concurred happily.

"Sugar," Vasquez said as though pondering the meaning of it. "Glucose. I don't wish to bore you with a lot of big medical words, Jessica, but glucose — sugar — plays a vital part in not only our entire body, but is especially important to the brain. It is widely known that people with diabetes have a greater propensity to develop Alzheimer's. It is also known that the brain is the primary consumer of glucose — as much as two-thirds of all glucose circulating in the body goes to the brain. Glucose is the major fuel for our cells, including those in the brain. As our bodies ingest glucose from the foods we eat, insulin must be released from the pancreas to keep the glucose level in check. But many, including diabetics, become what is called 'insulin resistant,' which means, of course, that the glucose is allowed to build up. The result? Diabetes. And, I am convinced, Alzheimer's disease." He laughed. "Am I becoming too technical for you?"

"Not at all," I said, "although I'm sure your research delves into the subject at far greater depth."

"Yes, of course. There is much more to it. Let me just say that the role that glucose plays in how brain cells use sugar and produce energy might well be at the root of Alzheimer's, especially in how it enables the development of beta-amyloid peptides, and protein strands called 'tau,' or tangles that are common with Alzheimer's patients."

"Now you've lost me," I said.

Vasquez turned to Seth and tapped him on the shoulder. "My colleague here knows what I am talking about."

Seth nodded. "Course, I'm not nearly as knowledgeable as Al is, but I do understand the underlying theory of what he's trying to accomplish. If he can establish a direct link between glucose — simple sugar — and the development of plaque in the brain that's synonymous with Alzheimer's disease, and can come up with a way to counter insulin resistance, it could lead to a cure."

I asked the obvious next question. "How close are you to accomplishing those things?"

"That, my dear lady, must remain a secret known only to me until I am ready to announce it to the world."

I immediately thought of Bernard Peters of K-Dex, the man and the company investing in Vasquez's work. Did Vasquez keep from Peters the progress he'd made — or hadn't made?

As I debated asking that question, Vasquez said, "My dear friend Bernie Peters is always asking me for an update on how the research is going, but he knew from the first day that I would not provide him or anyone else with regular progress reports. He expresses his unhappiness with this arrangement, but . . . Well, let me just say that he must live with it."

I thought to myself that were I Peters, I, too, would be unhappy being deprived of regular progress reports. How much was Vasquez's research costing Peters and K-Dex? It had to be in the millions. It was also obvious that K-Dex's money was not only going to supporting the lab and the research; it was fueling a lavish lifestyle for the researcher.

Vasquez commented while laughing, "Of course, I give Bernie an update now and then just to keep him happy and to keep the funds coming."

Sardina looked up from the microscope and winced, an expression that clearly said that he did not share in Vasquez's humor.

Vasquez took us on a tour of two other rooms in the laboratory complex. In one, an empty animal cage sat on the counter.

"We are anticipating a delivery of SCID mice — mice specially bred for medical research," he said.

I said nothing, but privately I thought, *I'm glad they're not here now.* There was no way

I would get into a debate on the merits of animal research — I had neither the experience nor the expertise to take up the cudgels in that argument — but personally I hated to see any animal, specially bred or not, made to suffer even if humankind benefited as a whole. Seth would disagree, I knew, but I couldn't help my feelings.

"Are you the only tenant of this building?" I asked.

"Yes. Bernie leased it in its entirety for me. I insisted that be the case when I agreed to continue my research here in the States. It's a small building, as you can see, but it is adequate to the task. Once we are to the level that demands a larger, more sophisticated space, we'll deal with it then."

We ended our tour in a small office at the rear of the building permeated by the odor of cigar smoke.

"Join me?" Vasquez asked as he removed one of his large black cigars from an elaborately decorated case he drew from his shirt pocket. He cocked his head and said, "It is not unheard of for ladies to partake in cigars, you know."

Seth, bless him, sensed my unease and said, "I think we'd better be getting back, Al."

"As you wish. The evening is shaping up

nicely at the house. Ivelisse has things under control, although she has plenty of help."

Sardina showed us out, and we were driven back to the hotel, where a tea service was under way. "I'd love a cup," I said with a sigh as I sank into an upholstered wing chair near a window overlooking a tropical garden.

"Well, Jessica. You've been keeping me guessing. What did you think?" Seth asked as we were served tiny tea sandwiches, pastries, and a steaming pot of tea.

"About Dr. Vasquez?"

"Of course about Dr. Vasquez. Who have we been spending the last few hours with?"

"And most of the day," I murmured to myself, taking a sip of my tea.

"C'mon, woman. What are you dillying about?"

"Well," I said, "I find him to be charming, dynamic, and good-looking, but maybe a little too much of each."

Seth's expression mirrored his surprise at my comment.

"He's too good-looking? Never heard anyone complain about that."

"Don't get me wrong," I said. "He's a perfectly wonderful host, and I certainly admire the work he's doing. If he succeeds in helping conquer Alzheimer's disease, it will be a monumentally important gift to mankind, and womankind, too. He'll probably receive a Nobel Prize."

"My thoughts exactly," said Seth, popping a salmon sandwich into his mouth. "I feel privileged to have been taken into his confidence."

"It doesn't surprise me," I said. "You share his obvious love of medicine and medical research, and as he says, you level with him."

"Try to."

"So let me ask you, is he as far along in his research as he wants others to believe? I'm referring specifically to Mr. Peters at K-Dex."

"You sound suspicious, Jessica."

"Nothing of the sort. It's just that I have trouble getting my head around his extreme secrecy, especially with the man who is making it all possible by funding the research."

Seth shrugged, bit down on a tiny cucumber sandwich, and took a sip of tea before responding. "He told you why he needs the security. You know it must be very embarrassing to the Castro regime to have lost such a shining light in medical research to the U.S. It wouldn't surprise me in the least if they made efforts to break into the lab and steal his work," he said. "As for K-Dex,

64

I understand the company was struggling before hooking up with Al. It had a series of setbacks, promising drugs failing to pass FDA muster, the sort of situations that plague all pharmaceuticals. From what I've been able to glean, Al's work generated a fresh influx of investors and could make the difference between K-Dex surviving or going under."

"And so Mr. Peters is desperate enough to allow Dr. Vasquez to proceed on his own terms."

Seth hesitated before saying, "I suppose you could put it that way."

"From what you've told me, Dr. Vasquez's work in Cuba was pretty far along."

"Call him Al. He prefers it. I'd say that your assessment of his work in Cuba is accurate, at least based upon what he's shared with me."

I chose a salmon sandwich.

"I was surprised at how small the lab is and that Al has only one assistant," I said. "It seems to me that research on something as daunting as Alzheimer's would — well, would demand a much larger lab and staff."

"It's the size of the man's brain, Jessica, not the size of his laboratory that counts," Seth said rather sharply.

I fell silent for a time and concentrated on

what was left of the sandwiches and pastries.

"Something bothering you?" Seth asked.

"Oh, no," I said. "I guess I'm accustomed to you being more skeptical. It's one of your curmudgeonly traits I most admire. But here you seem to have what might be termed blind faith in what Al is doing, the way Mr. Peters must or seems to."

I checked his face for a sign of annoyance. Instead, he appeared to be hurt. I quickly added, "I don't mean to disparage your belief in him, Seth, but — well, it's not like you to accept on the surface whatever someone tells you."

Which was true. For as long as I've known Seth Hazlitt, he's been a man who seldom takes others at face value. Not that he's unduly suspicious or dismissive of what others say, but he's quick to cast a critical eye on claims without confirming backup.

He sat back and gazed out the window, as though peering into another world that only he could see. When he finally turned back to me he asked, "Have you ever wanted to be someone else, Jessica?"

I pondered the question. "I don't think so," I said. "There are plenty of people that I admire, but wanting to be them? No, I can't think of anyone at the moment."

"I admire that," he said.

"Why?"

"Because it means you're supremely contented with who you are and what you've done with your life."

"Yes," I said, "I believe I am." I paused before adding, "Aren't you?"

"I suppose I am for the most part, but when I was a young man coming out of medical school, I had visions of making a big breakthrough in science, dreamt of coming up with a cure for cancer or heart disease, doing something monumental to benefit society."

"And that's exactly what you do," I said. "How many people in Cabot Cove owe their lives to you? How many mothers have healthy children because Dr. Seth Hazlitt was there to deliver their babies and see them through illness?"

He held up his hand. "I get the point, Jessica, and I won't argue with you. But Al's work is historic. It could mean better lives for millions of men and women. That he's allowed me into his world, considers me a colleague, is . . . How can I put it without sounding shallow? It's flattering; that's what it is. To be close to a man who had to escape a brutal dictatorship in order to find the freedom to pursue his passion gives me a sense of . . . a sense of *importance.*"

I didn't know how to respond. I understood what he was saying, and tapped into the emotions behind it, but I was surprised that this dedicated physician who meant so much to so many in our town would feel the need to rub shoulders with someone else to achieve a sense of worth.

"Sounds silly, right?" he said as he finished off a tiny cream puff and what was left of his tea. "Forget we ever had this conversation. Just a foolish old man talking. This aging sawbones is feeling his age, needs a good nap before the festivities this evening. You'll excuse me?"

"Of course. But, Seth, I —"

My words trailed behind him as he walked slowly from the restaurant and disappeared into the lobby.

A wave of sadness swept over me. I'd rarely heard him talk that way about his life and unfulfilled aspirations. Yes, he was getting older, as we all were, and age can generate a tendency to look back at what might have been. But Seth was shortchanging himself. He'd had a wonderful, meaningful career — and still did.

As the sadness abated, a feeling of resentment toward Dr. Vasquez took its place. Why? I couldn't explain it and willed it from my thoughts. I was reading too much into what Seth had said and the hold that Vasquez seemed to have over him at that point in his life.

"More tea?" the waitress asked.

"What? Oh, no, thank you. Please put the charge on my room."

As I sat in a red-and-gold wing chair in my suite, I felt a chill that had nothing to do with the room's temperature. A sense of foreboding had settled in, and I found myself dreading the dinner party at Vasquez's home.

I wished I were back home in Cabot Cove.

# CHAPTER FIVE

Seth had been downbeat when he left the restaurant that afternoon, and I expected his mood not to have changed. But there he was in the lobby, nattily dressed in a blue blazer, white shirt, and red bow tie, and looking every inch like the Seth Hazlitt I knew and loved. He was absolutely ebullient as he greeted me, took my arm, and guided me outside, where we would be picked up. The weather didn't match his upbeat mood, however. Dark, low-flying clouds heralded the approach of another storm, and I hoped the party wasn't planned as an outdoor event.

"You look splendid," I said.

"And you, Jessica, will turn every man's eye at the party. I have one request."

"Which is?"

"That you ignore my maudlin conversation this afternoon. Don't know what got into me. You'd think I was one of those

70

morose drunks cryin' in his beer, but all I had was tea."

"As you wish, sir," I said as the black Mercedes pulled up, manned by our usual escorts.

"Did you get a good nap in?" I asked after we were settled in the back and on our way to the party.

"Never took one," he replied. "Perked up the minute I got in the room. Anxious for you to see Al's house. It's really nice, backs right onto the water on Davis Island. He's got his own dock and boat there, gave me a little tour of the bay last time I was in Tampa."

"Have you met others who'll be there tonight?"

"Ayuh, a few. Oona Mendez and her disagreeable companion will show, I suppose; probably Al's son, Xavier — nice young chap, doesn't say much, a bit of a brooder, if you get my drift, like so many young men these days — and Al's assistant, Dr. Sardina, I imagine. Oh, and Bernie Peters mentioned on the golf course that he and his wife were invited. I should let you know that Al's wife, Ivelisse, comes off a little strange now and then."

"Strange?"

"Lovely woman, very gracious, but some-

71

times she can be, well, 'scattered' might be the right term."

"Thanks for the warning."

Alzheimer's naturally crossed my mind, although I didn't ask. Becoming forgetful isn't necessarily caused by a disease. How many times lately had I forgotten why I went to another room, or drawn a blank on someone's name, someone I knew well? I believe it's called "aging." Still, my writer's mind almost always creates scenarios even when none exist. Dr. Vasquez was doing research on Alzheimer's disease. His wife, according to Seth, was forgetful. Did her husband suspect that his wife might be falling prey to the very illness he was trying to conquer?

*Stop it, Jessica! You're not creating a plot for a book now.*

Seth's contagious, upbeat mood continued throughout the drive.

"That's Al's home there," he said enthusiastically as we turned onto a street called Adalia. We approached an imposing two-story white house with a balcony that ran the length of the second level, and a set of large glass doors leading into the first. As we pulled into the circular driveway, the sound of music with a Latin beat could be heard over the crunch of our tires on the

72

gravel. The car's door was held open for us as we exited, and we went up a short set of concrete steps to where a large man, also dressed in a black suit, white shirt, and black tie, stood. He recognized Seth and greeted him.

"This is Mrs. Fletcher," Seth said.

"*Sí,*" the man said. "She is on the list. The guest of honor, in fact." He gestured at the doors. "Please go in."

My mind immediately conjured disturbing reasons for why a party would generate a formal guest list, or need someone to be standing guard at the front door, but I forced the thoughts away, and we went inside. Dr. Vasquez, dressed in a white guayabera trimmed with layers of lace, and a woman I assumed was his wife, Ivelisse, stood in a sizable foyer that was decorated with white-and-black tiles on the floor and a series of vividly colored abstract paintings on the walls. A massive modern chandelier constructed of a maze of burnished brass pipes and a hundred small bulbs was suspended from the ceiling by a thin metal tube. *I'd hate to be under it if it fell,* I thought.

Vasquez smiled and stretched out his arms. "Ah, my guest of honor," he said. "Welcome." He turned to his wife, a stunning woman with closely cropped white hair who was wearing a silver metallic sleeveless top over tight black pants. "My dear, this is Jessica Fletcher, our honored guest for the evening, and you know Dr. Hazlitt."

Mrs. Vasquez smiled sweetly and extended her hand. "It is such a pleasure to have you in our home, Mrs. Fletcher."

"I'm honored to be here," I said, "and please call me Jessica." I cocked my head in the direction of the music, which was louder now. "That music is wonderful."

Vasquez laughed and moved his feet to the rhythm. "Ah, they play a *guaracha*. Cuban musicians," he said, "very good ones. They did as I did — came to America to get away from Castro and his fascist regime, which frowned on the Creole forms of our popular music. Here, they may play what they like — and what we like. Come, the party has already started." He took my hand and led us from the foyer into a spacious living room, where the festivities were under way. I took note of large French doors leading out onto a long deck landscaped with flowering plants and potted palm trees, and ending in a dock, where a cabin cruiser could be seen bobbing in the water. Inside, most of the furniture in the living room had been pushed back or removed to accommodate the party, but oversized cubist oil paintings added color to the few spaces along the walls that weren't covered with floor-to-ceiling bookcases.

The source of the upbeat tunes was a group of five musicians dressed in colorful flowered shirts earnestly playing the composition I'd heard from the foyer. Two dozen other guests were in the room — stylishly dressed women, and men in what might be called business-casual attire. A couple danced, and others cheered them on, clapping hands and occasionally shouting "Olé" as though witnessing a bullfight. The atmosphere was captivating, and I felt myself swaying to the infectious rhythm. I don't consider myself much of a ballroom dancer, although I do enjoy a turn around a dance floor now and then. My late husband, Frank, was a better dancer than I am and gave me the confidence to try new steps, sometimes successfully, sometimes not, but always enjoyable.

Vasquez pulled me across the room to where an elderly gentleman sat at a small wooden desk, rolling tobacco into cigars. "This is Adelmo, one of the finest cigar makers in Tampa."

"How do you do?" I said.

The old man rapped his knife on the table but didn't reply.

"We bring him here both to entertain our guests and to provide a welcome gift." Vasquez plucked two of the completed cigars from a box on the desk with a smile. "You will excuse me, Jessica, will you not?"

"Of course," I said.

"Please take one for Seth," he said and went through the French doors to the deck, where he was immediately joined by his lab assistant, Dr. Sardina.

I picked up a cigar but then put it back. If Seth wanted to smoke a cigar, he could come get one himself. I would not encourage a new habit.

Adelmo looked up at me curiously. Not wanting to appear rude, I asked him, "Are you using Cuban tobacco?"

"No, señora. One cannot bring Cuban tobacco into this country. But this is very fine Dominican tobacco, and the wrappers are from Connecticut. Perhaps you know this state?"

My eyebrows rose. "Connecticut? I had no idea tobacco was grown in New England."

"Shade tobacco," he replied, "very much like the Cuban leaf."

"Ah, yes, shade tobacco," said a stout gentleman who joined us at the desk. "Grown in a shed to mimic the Cuban tropical climate," he said to me. "And how are you tonight, Adelmo?"

Their conversation allowed me to make a graceful exit, and I rejoined Seth, who was standing with Mrs. Vasquez, her arm linked in his.

"Do you dance, Dr. Hazlitt?" Ivelisse asked.

"Not one of my talents," he replied. "I heard a band like this when I was in Havana. Somebody said it was salsa, but I wouldn't know one sorta music from another. I thought salsa was something to eat."

"I think you are too modest," she said, and turned to me. "Don't you agree, señora?"

"That Seth is modest? Yes, I certainly agree with that."

As we stood in the group enjoying the dancers, Oona Mendez, whom I'd met at dinner with her companion Karl Westerkoch, joined us. "Wonderful to see you again," she said.

"Likewise. We're enjoying the music and the dancers."

"Cuban music," she said reverentially. "It is so full of spirit, so joyous, so — so sensual."

"Does what they're playing have a name?" I asked.

"*Son.* It means 'sound' in Spanish. It is the basis for all Cuban music, the *danzón*, the habanera, the mambo, and of course salsa."

"All I know is that it sounds wonderful," I said. "Is Mr. Westerkoch here this evening?"

She pointed across the room, where he stood leaning against a bookcase, a drink in hand, his dour expression testifying to his mood, which was decidedly not a party one.

A uniformed waiter approached carrying a tray of hors d'oeuvres — tiny crab and chicken croquettes accompanied by a silver bowl of hot sauce, sweet peppers filled with mushrooms and ham, and lobster meat on small pieces of flatbread drizzled with pepper aioli.

"I hope you like things spicy," Oona said, spearing a croquette with a toothpick.

"As a matter of fact, I do," I said, "as long as it isn't too hot." I tried a flatbread with lobster and assorted spices. "Hmmm," I said, "just right. Delicious."

Ivelisse Vasquez excused herself to greet Bernard Peters and his wife in the foyer. They say that people tend to marry those who look somewhat similar, and that theory held up when it came to Mr. and Mrs. Peters. They were both short and round, her ruddy cheeks matched his, and they were dressed in clothing of the same color. Ivelisse escorted them into the main room and I was introduced to Frances Peters. She greeted Seth with a kiss on the cheek; he'd certainly gotten around and met people during his previous stays in Tampa.

Mrs. Peters was a cheerful lady who punctuated everything she said with a laugh. Her husband had been good-humored, too, on the golf course, but that certainly wasn't his demeanor this evening. He gave me a cursory hello and headed straight for the bar, behind which a uniformed gentleman plied his craft.

"Are you ready for a drink Ju— ah, Jan— Señora Fletcher?" Ivelisse asked.

"Please call me Jessica," I said.

"Thank you, Jessica. We're serving authentic Cuban mojitos and" — she paused and screwed up her face in thought — "oh, yes, authentic daiquiris."

"Daiquiris were Ernest Hemingway's favorite cocktail. Maybe if I have one, it will help me write like him."

Alvaro Vasquez, who'd returned inside, caught the end of the conversation. "Another of your American authors, Norman Mailer, once scolded JFK after the Bay of Pigs disaster. He supposedly told your president that the mistake he'd made was invading a country without understanding its music."

"I hadn't heard that," I said.

"Or without understanding one of its favorite cocktails. Come. Enjoy a daiquiri."

With that he led me by the hand to the bar, where Westerkoch was getting a refill.

"Good evening," I said.

He mumbled something in return, took his glass from the bartender, and walked away.

*What a strange egg,* I thought as Vasquez ordered a daiquiri for me. "Make it especially good for our guest of honor," he instructed.

I tasted my drink. It was delicious, although I had the feeling that it would go down too easily and reminded myself to nurse it throughout the evening.

"I want you to meet someone, Jessica," Vasquez said, leading me to where a handsome young man stood talking to Pedro Sardina and a lady I would learn was his wife, Ofelia, an attractive, slender young woman with an oval face and sad brown eyes.

Vasquez didn't wait for a pause. He cut into their conversation and made the introductions. "And this is my son, Xavier," he said proudly.

80

Xavier, a relatively short man compared to his father and mother, frowned at the interruption but managed a smile for me. "My father has told me a lot about you," he said.

"I'm afraid our initial meeting wasn't an ideal place for me to make a good impression," I said. "I'm not much of a golfer."

"Nonsense," Dr. Vasquez said. "Give her a few days of practice and she'll win all the tournaments."

"I don't play golf at all," Xavier said. "It's a stupid game."

I wasn't sure that I agreed with his assessment, but I didn't question it. Nor did his father, who forced a smile.

"I understand that you live in Tampa," I said, changing the subject.

"That's right, when I'm not in Key West."

"Oh? You go there often?"

"Xavier has a lady friend in Key West," his father said, "conveniently far away from his parents. Fortunately he has a plane that he can fly to see her."

Xavier glared at his father.

"A plane?" I said. "You're a pilot?"

"Yes," he said, never lowering his gaze from his father.

"What sort of plane do you fly?" I asked, hoping to move the conversation into a

81

neutral area.

"A Cessna one-seventy-two, a later model, four-seater."

"I ask because I have a private pilot's license," I said a bit self-consciously, "although I'm afraid I don't get to fly very much." I smiled at the Sardinas, whom Vasquez had completely ignored.

"That's a shame," Xavier replied, his attention back on me. "You'll excuse me, please," he said, bowing slightly and nodding at the Sardinas. "It was a pleasure."

He walked away and I saw dismay on his father's face.

"What a handsome young man," I said to break the tension.

"He takes after his mother," Vasquez said.

An awkward silence ensued.

"I met your husband at the lab today," I said to Ofelia. "I'm impressed with the work he and Dr. Vasquez are doing to find a cure for Alzheimer's disease."

"Pedro doesn't talk much about his work," she said in a voice so soft that I had to lean closer to hear what she'd said.

Vasquez laughed. "The best kind of assistant," he said, nodding at Sardina. "Less talk and more research."

"The daiquiri is delicious," I said, taking a small sip.

"Don Casimiro classic silver rum," he said. "Only the best."

As the band launched into its next tune, Vasquez once again left my side and strode to where Bernard Peters was engaged in conversation with Oona Mendez.

"C'mon," Sardina said to his wife. "I want another drink."

"A pleasure meeting you," I said to Ofelia Sardina.

She smiled and they walked away.

Left to my own devices, I sipped my daiquiri and glanced around at the other guests. Despite the pulsating, joyous music, the couples dancing, and the tasty canapés, there was a palpable tension in the room, and it was more than the father-son contretemps. My gaze fell on Ivelisse Vasquez, who stood alone in an alcove and seemed to be in a world apart. Her son, Xavier, had disappeared altogether, and I wondered whether he'd decided that the party, like golf, was "stupid." I watched as our host escorted a frowning Bernard Peters onto the deck.

The hors d'oeuvres became more plentiful, enough so that it was necessary to balance plates and glasses while using utensils that were passed with the food. Westerkoch had commandeered a small, tall glass-topped table on which to rest his plate, and I placed my plate there, too.

"Lovely party," I said.

"Like other parties the doctor hosts. Frankly, I don't care for the music. It's too loud, and there's a tribal aspect to it, people jumping up and down like savages."

"I, ah . . . Are you involved in medical research, Mr. Westerkoch?" I asked.

"No."

"I just thought that —"

"I'm a consultant."

A friend once told me that when someone says he or she is a consultant, it means they're out of work, but of course I didn't verbalize that cynicism. Instead I asked, "What sort of consulting do you do?"

"Government basically. You ask a lot of questions, Mrs. Fletcher."

I didn't take his comment as a challenge and laughed. "I suppose that's because I'm a writer."

"How convenient. I understand that you and your doctor friend visited Vasquez's lab this afternoon."

"That's right. I was pleased to be invited."

"Your friend seems to have successfully invaded Vasquez's inner circle."

"Has he? Invaded? That's rather an odd way to put it. All I know is that he feels a camaraderie with Dr. Vasquez."

"I get the feeling that it's more than that. What does your friend tell you about Vasquez's research?"

"Very little. Dr. Vasquez explained it to me in the simplest of terms. You and he are obviously acquainted. What has he told *you* about his research?"

"As little as possible." He turned toward the French doors, through which threatening weather could be seen in the distance. "I need some fresh air before the sky opens up."

*I seem to be repelling all comers,* I thought as Westerkoch abandoned the table we shared. But the next half hour gave the lie to that thought. Guests to whom I hadn't been introduced invited me into their conversational circles, and I began to thoroughly enjoy myself. Seth, who'd spent much of the night with others, eventually wandered to where I was listening to a joke told by the gentleman I'd encountered at Adelmo's cigar-rolling desk. Carlos Cespedes owned a cigar shop and factory in Ybor City, the Cuban section of Tampa. ". . . and so Fidel Castro, he has trouble sleeping and goes to his doctor. 'I have insomnia,' he says. 'What should I do?' And the doctor says, 'Try reading some of your speeches.' "

We all laughed at this dig at Castro's famous, impossibly long speeches. Cespedes was about to launch into another tale when Seth guided me away from the group and toward the French doors, through which Vasquez and Bernard Peters could be seen on the deck. From their body language and the movement of their arms and hands, it seemed an argument was in progress.

"Is there a problem?" I asked Seth.

"Appears that way, doesn't it?"

We had only a few seconds to witness the confrontation; then Peters threw up his hands, came through the doors, and stomped across the room in the direction of his wife. On the deck, Vasquez pulled a cigar from his elaborately decorated case and held it up to his nose, inhaling the scent of the tobacco leaves.

"Let's keep him company," I said.

When Vasquez saw us, he substituted his gregarious side for the upset he'd just experienced and said, "Good, good, I need someone to share a good smoke with. Please," he said, offering Seth his cigar. "It was a gift from someone special. It's Cuban."

"Not for me, Al, but thanks."

Vasquez raised his eyebrows questioningly at me.

"No, thank you, sir," I said. "I just thought some fresh air would be nice."

"You can't blame my cigars for the stuffy air inside, Jessica. Ivelisse and I have an understanding. I can smoke to my heart's content, but only outside, or in my office at the other end of the house. I had a professional air cleaner installed in it, like ones in restaurants. I am an agreeable husband, yes?"

"It certainly sounds that way."

He fired up his lighter and went through the elaborate ritual of lighting his cigar, blowing a stream of smoke into the air with a satisfied smile. "Come. Let me show you my latest gadget on the boat," Vasquez said. "Jessica, you haven't seen my new toy."

"Uh-oh," I said as a drop of rain landed on my nose. I looked up into the black sky. "Better get back inside," I said.

"Nonsense," said Vasquez, "just a few raindrops. I can't waste this good cigar." He drew deeply on it and watched the blue smoke curl up into the air.

"Sorry," I said, "but I don't have a cigar to save." With that I made for the French doors and stepped inside. I looked back to where Seth and Vasquez continued to stand together, Vasquez smoking, Seth saying something that I couldn't hear. As I watched them, shadows emerged from behind a shed farther along on the deck. I squinted until I could make out Westerkoch and Oona Mendez. They approached Seth and Vasquez, stopped for a moment to say something, and came inside.

"A storm's brewing," Oona said.

"So I see," I said.

A sudden shaft of lightning illuminated where the two men stood, followed by a low rumble of thunder.

"They should come in," I said. "They told me that lightning here in Tampa is particularly dangerous."

Seth slapped Vasquez on his back and headed inside.

"Why is he staying outside?" I asked.

"He wants to finish that cigar," Seth replied with a laugh, brushing a few raindrops from the sleeve of his jacket.

More lightning bolts turned the deck brilliantly white, as if giant klieg lights had been turned on. The light show drew others to the French doors, Westerkoch, Oona, and Ivelisse Vasquez.

"Hello," Ivelisse said, smiling at me. "I'm Ivelisse Vasquez. Have we met?"

Her comment startled me, as much as a

clap of thunder that made me jump.

Outside, Vasquez looked up as though surveying the heavens. He took another deep drag on his cigar and raised it, seemingly offering it to the gods. As he did, the brightest and most menacing of lightning bolts carved a jagged path from the sky to where the deck met the water. In its harsh light Vasquez looked like a Shakespearean thespian portraying Hamlet, a spotlight establishing his stage presence. Then, as we watched in horror, Dr. Alvaro Vasquez doubled over and dropped to his knees, the cigar flying from his hand. He pitched forward and lay still as the sky opened up and the rain came down in sheets.

# Chapter Six

Oona Mendez shrieked.

Karl Westerkoch said, "Damn," and pushed the door open a crack, allowing the sound of the pelting rain to reach inside.

Ivelisse Vasquez stood motionless, her face blank.

Seth wrenched open the door and ran out into the downpour. I followed.

He knelt over Vasquez and placed his fingertips against his neck. "Get an ambulance," he shouted to no one in particular. "Call nine-one-one!" With that, he straddled Vasquez and began administering CPR.

I looked back in the hope that someone would bring an umbrella, but no one moved until Xavier appeared carrying a tan raincoat. Seth climbed off, and Xavier spread the coat over his father, including his face. Seth pulled it back and again tried to discern a pulse in the neck. He shook his head and continued pressing on Vasquez's chest with rhythmic thumps. "Come on, Al. Don't give up," he told his patient. "Where's the ambulance?" he called out.

"Did you call for an ambulance?" I asked Xavier.

He ran back inside the house.

One of the waitstaff who'd passed hors d'oeuvres came to where Seth labored over the still lifeless body. She popped open a large black umbrella, which provided some protection from the elements, and passed it to me.

"Is he dead?" I asked Seth.

"I can't get a pulse, but I'm going to keep trying." Seth's face was red from the exertion, but he refused to stop his lifesaving efforts, even when another guest offered to take over.

"Come on, Al, breathe," Seth exhorted. "You can't die. You have too much important work to do. The world needs you. Ivelisse needs you. Xavier needs you. Pedro Sardina can't do it alone. Breathe, man, breathe."

"Another umbrella," I shouted at those standing at the French doors.

One of the security men heeded the call and brought a second one; between the two of us, we managed to shield Seth and the still unresponsive Vasquez from becoming further drenched.

Seth looked up at me. "Get inside, Jessica. No point in you getting soaked, too. Nothing you can do here. Where is that ambulance?"

I glanced behind me. The party guests were grouped around the glass panels watching the drama on the deck. Then Ivelisse prodded one of the waitstaff, who pushed opened a door. The waitress took a tentative step on the wet deck and then darted forward to retrieve the cigar Vasquez had tossed aside when he'd been struck down. She placed the cigar butt on a plate and returned inside.

A moment later, the second security man emerged from inside. "The ambulance is on its way."

"Thank goodness," I said, handing him my umbrella. "Please keep them as dry as possible," I said.

Someone opened a door for me and I entered the house, sopping wet and shivering against the clammy feeling of my clothes on my skin. People moved back away from the doors and gathered in small knots, speaking in low tones. I looked around for Ivelisse but didn't see her.

"What happened, Mrs. Fletcher?" a guest asked.

"I'm not certain, but he may have been hit by lightning."

"Is he — ?"

"I'm afraid I don't know. Dr. Hazlitt is doing everything he can. Tell me, do you know which way the kitchen is?"

He pointed and I followed his direction down a short hallway until I reached an ultramodern kitchen with a wall of identical cabinets with invisible pulls. A large marble-topped island dominated the center of the room, and I spotted the plate with the cigar butt sitting next to a stack of dishes waiting to be washed. Several of the waitstaff — those who weren't still in the living room — huddled around the island, apparently not sure what they should be doing. One fellow, seeing my doused state, jumped forward with a roll of paper towels. "Can I get you something else, madam?"

"Not unless you have a spare uniform I can put on in place of these wet clothes," I said.

"I'm sure we can find something for you," he said. "Beatriz," he called to a waitress, who hurried to a large case left on the side of the room.

I tore off two of the paper towels, wiped my face, and when no one was looking, swiftly folded the sheets over the cigar, wrapping it up carefully.

Beatriz offered me a white jacket, apologizing profusely that they didn't have a complete uniform to provide, but I was grateful for anything that was going to allow me to shed my wet blouse. I changed swiftly in an adjacent bathroom, dabbing myself dry with paper towels. I deposited my shirt, the remaining towels, and the paper-wrapped cigar in a plastic bag Beatriz had provided.

I thanked the kitchen staff and returned to the party room, wandering among the guests, searching for a familiar face. No one paid any attention to me, and I realized I was now partially incognito in my uniform jacket. I folded the plastic bag and tucked it on a lower shelf of a bookcase.

Xavier had returned to the party from a different part of the house. He was accompanied by a middle-aged woman wearing an apron. Ivelisse had retreated to her alcove but by this time seemed to have regained a sense of the moment and had started to cry. The woman in the apron put her arms about Ivelisse and gently led her from the room.

"She's the housekeeper," I overheard Oona say to Westerkoch.

"Does she even know what's going on?"

94

he muttered.

"Who? The housekeeper?"

"No! His wife."

"Who can tell?"

"I heard that's why he stole the formula from Havana," Westerkoch said, "to speed up the process to find a treatment. It doesn't look like he succeeded."

"Shush! Someone will overhear you."

"Who cares?"

The band members had stopped playing and were packing up their instruments. Adelmo, the cigar roller, had left his desk. Another blinding lightning strike followed by a crash of thunder drew an audible gasp from the guests. It was succeeded by the sound of sirens coming from the front of the house. The security man who'd been at the door when we'd arrived opened it, and two uniformed EMTs rushed in.

"Where?" one of them asked.

I stepped forward and said, "Follow me." I led them out to the deck. Despite the recent celestial fireworks, the rain had stopped as suddenly as it had started. The EMTs knelt next to Seth and relieved him of his task. One of them used a stethoscope on Vasquez and attempted to find signs of life.

"You see what happened?" the other EMT

asked Seth, helping him to his feet.

"There was a bolt of lightning," Seth said, breathing heavily, "but I'm not sure if it hit him."

"Are you okay, mister? Do you need to sit down?"

"Just winded," Seth managed to get out, sinking into a chair someone pulled over for him. "I'm a doctor, been trying to revive him. Too late, I'm afraid."

"Looks that way, Doc. I'm sorry, but there's no pulse," the other EMT said.

"Are you all right, Seth?" I asked, kneeling at his side. I looked into his ravaged face. His jacket was soaked through and he was exhausted. I couldn't tell if the drops of water on his cheeks were from the rain or tears.

"I couldn't save him, Jessica," he said hoarsely, raising his trembling hands and wiping his eyes.

"If you couldn't, no one could," I said. "You were right there when it happened, Seth. You've been working on him all this time. There wasn't anything more you could do."

Seth shook his head sadly. "What a loss for humanity."

*And what a loss for you, my dear friend,* I thought.

The EMTs left, returning a few minutes later with a gurney. Seth and I watched them carefully lift Vasquez from the deck, place him on the gurney, cover him with a lightweight blue tarp, and roll him into the house, where everyone stepped back to give them a path to the front door. As one of the EMTs opened the door, two men wearing raincoats came through it. One of them put up his hand to stop the EMTs; the second man showed them a badge.

"What's going on here?" a guest demanded, his eyes on the gurney. "The man is dead. Let them through."

One of the newly arrived men, a heavyset fellow whose sparse hair had been plastered to his head by the rain, announced, "Police." He asked the crowd, "Who's in charge here?"

When no one stepped forward, Seth did. "The deceased is Dr. Alvaro Vasquez. This is his house. He was hosting a party, and —"

"We know who the deceased is," the portly detective said. "Are you a friend?"

"Yes, I am," Seth replied.

The second detective, a considerably younger man, asked everyone to sit. He turned to Seth and asked, "Has anyone left the party?"

Seth was obviously taken aback at being asked the question. He looked at me in bewilderment.

"We wouldn't know," I said. "We're from out of town. We don't know everyone who was invited, but a man at the door had a guest list."

His partner surveyed the others in the room. "Does anyone know whether any individual has left the party?"

Nervous looks were cast among the guests.

Seth spoke up. "I think we're all confused why the police have been summoned," he said. "I'm a physician. Mrs. Fletcher and I were —"

"Who's Mrs. Fletcher?"

"I am," I said. "Dr. Hazlitt and I saw what happened to Dr. Alvaro. He was —"

"I'd like you two to wait over there," the detective said, pointing to the bar area.

"I don't understand why —"

I cut Seth off and urged him to accompany me to where the detective had indicated.

We sat on the two barstools and watched and listened as the detectives obtained the guest list from the security guard and began asking questions of the others. They were interrupted by the arrival of an elderly man.

"Hi, Doc," one of the detectives greeted the new arrival, who ignored the detective and went to the side of the gurney, pulled back the tarp to reveal Vasquez's face, grunted, and covered him again. He waved his hand and the EMTs wheeled the body of Dr. Alvaro Vasquez outside.

The heavy detective guided the man called "Doc" to where Seth and I waited.

"I understand you're a physician," the older officer said to Seth.

"That's right. Seth Hazlitt, MD, of Cabot Cove, Maine. And this is Jessica Fletcher, the mystery writer. We — she and I — witnessed what happened to Dr. Vasquez. And I administered CPR, unsuccessfully, as you see, until the ambulance arrived. And you are?"

The police officer answered. "Detective Machado, Tampa PD. This is Dr. San Martín, Hillsborough County ME. We got the call that there was an emergency at Dr. Vasquez's home."

I suppose my puzzled expression asked the question, *Why would the police be called?*

Machado picked up on it and answered. "Dr. Vasquez is a well-known person in Tampa. We're always called in on cases like this."

Now I understood. The police aren't routinely called to the scene of what appears to be a death by natural causes or an act of nature. But when a leading citizen, particularly one who is newsworthy and perhaps controversial, is involved, the police naturally take an interest. So does the local medical examiner.

"We'll do an autopsy, of course," Dr. San Martín said. "Is the victim's wife here?"

"I believe that one of the household help took her away, probably to her bedroom," I said. "She was in shock, as you can imagine."

"His son, Xavier, is here," Seth added. He scanned the room. "But I don't see him at the moment. He's probably with his mother."

Dr. San Martín pulled a business card from his pocket and handed it to Seth. "I'm leaving with the body, but I would like to speak with you about what you saw. Will you be staying in Tampa?"

"Ayuh, at least for a few days."

"Maybe you'd be good enough to call me in the morning so that we can arrange a time to get together."

"I'll do that," Seth said.

The doctor started to leave, turned, and said to Seth, "One of the EMTs told me that you said the victim had been struck by lightning."

"Just an assumption on my part," Seth

said. "We saw the bolt of lightning. It appeared that he had been hit, but I'm sure your autopsy will confirm or deny it."

"Yes, I'm sure it will," San Martín said. "Another lightning victim," he muttered more to himself than to us. "Welcome to Florida."

"Yes, I'm sure it will," San Martin said.

# CHAPTER SEVEN

The two detectives asked everyone to provide their names and contact information before leaving. One guest protested. "This is an intrusion into our privacy," he proclaimed. "We're guests at a party where the host was unfortunately struck by lightning and died. You have no right to ask for personal information. You're treating us as though a crime has been committed."

Detective Machado politely explained that it was routine to collect information about the people who are present when an unusual death occurs. Although his demeanor was nonthreatening, his steely expression said something else. The man reluctantly gave his name, as well as his phone number, and left. The other guests followed suit.

While Machado's younger partner went outside to the deck and did a cursory examination of where Vasquez had fallen, Machado returned to where Seth and I still sat at the small bar.

"The ME says he'll be speaking with you tomorrow," he told Seth.

"That's right," Seth said.

"Maybe you can tell me what you witnessed. It's *Dr.* Hazlitt, right?"

"That's right. Mrs. Fletcher here and I are from Maine. I'm here in Tampa visiting Dr. Vasquez, and Mrs. Fletcher decided to join me for a week. She'd been on a tour promoting her latest book."

"You're a writer?" he asked, eyeing my white uniform jacket. He must have assumed I was one of the staff serving the party.

"Yes. I write murder mysteries."

That brought a smile to his face. "You write about murders and I investigate them."

He and Seth talked for a few more minutes, and I took the time to sum up the homicide detective. I judged him to be in his mid- to late forties. He had a dusky complexion — I guessed that he might have a Hispanic background — and bore the remnants, albeit faint, of boyhood acne. He wasn't someone that I would term outgoing, but there was an openness that was appealing.

He eventually turned his attention to me.

"Did you observe anything strange at the party, Mrs. Fletcher?"

"Strange? In what sense?"

"I'm not suggesting anything. However, when someone of Dr. Vasquez's stature dies suddenly, we need to cover all the bases."

"Of course," I said. "No, nothing strange happened at the party." I wondered whether the unsubstantiated tense feeling that I'd experienced was worth mentioning and decided it wasn't.

Seth looked past me and said, "Here's Dr. Vasquez's son, Xavier."

"How's your mother?" Seth asked when the young man reached us.

"Resting," he said.

"I'm so sorry for your loss," I said touching his arm.

He looked down at my hand, and I withdrew it immediately.

Detective Machado introduced himself and asked a few questions, which elicited nothing of interest as far as I could tell. When he asked Xavier the same question that he'd asked me — had he noticed anything strange that evening — the son replied, "It's always strange around here. If you don't have anything else to ask me, I'd like to get back to my mother."

"Sure," Machado said, and handed Xavier

his card. He also handed one to Seth and said, "I doubt if I'll have anything else to ask you, Doctor, but give me a call if you think of something."

"Ayuh, I'll do that."

The departure of the body of Alvaro Vasquez, the EMTs, the ME, and the two detectives created a vacuum of sorts in the large room, like the air had been sucked out of it. Most of the guests had decamped, but Seth and I remained, together with Oona Mendez, Karl Westerkoch, Bernard Peters, and his wife, Frances. The band had finished packing up its instruments and departed. Two waitresses scurried about picking up plates and glasses, tossing anxious looks at the remaining guests as they ferried serving pieces to the kitchen.

Peters sat alone in a red leather wing chair, staring straight ahead, his hands outstretched as though asking for wisdom from an unseen source. His wife, Frances, stood next to him, her hand to her mouth — seemingly stifling a scream or a moan.

I leaned close to Seth and said, "I think we should go."

He nodded and stood.

We made the rounds of the remaining guests. Oona Mendez and Karl Westerkoch sat together on a couch. She said she hoped to see us again; he said nothing, simply nod-ded. I approached Bernard and Frances Pe-ters. "This must be a dreadful shock to you," I said to him.

"Unbelievable," he said. "How could this have happened?"

I understood why he would be especially shaken by Vasquez's death. They'd not only been friends of sorts, but Peters and his company, K-Dex, had lost perhaps their only lifeline to solvency, based upon what Seth had told me of the company's shaky financial status.

"Will you be staying in Tampa?" Frances Peters asked.

"For a little while," I said.

Seth came up and offered his condolences. His arrival seemed to prompt Peters into a more animated state. He got up and said, "You were extremely close to Alvaro."

"I wouldn't say extremely close," Seth said, "but we did get along. I considered him a friend."

Peters led Seth a few feet away and said in a low voice, but not so low that I couldn't hear him, "I would like very much to talk with you."

"Of course, whenever it's convenient for you," Seth replied.

"Tomorrow? At Alvaro's laboratory?"

Remembering his promise to meet with the medical examiner, Seth said, "I suspect that will be all right, only I might have another appointment. How about you call me at the hotel, and —"

Peters interrupted Seth with, "Dr. Sardina. He's gone?"

Seth shrugged.

"Have you see Dr. Sardina?" Peters asked me.

"Earlier in the evening," I said, "with his wife."

Peters's expression turned grim.

Seth repeated his suggestion that Peters call him at the hotel, and we went to the foyer, where the two men who'd driven us sat on a bench.

"We'd like to go back to the hotel," Seth said.

They slowly got up, and one opened the front door.

"Wait just a second," I said.

"What are you doing, Jessica?" Seth asked.

"Just be a moment," I said as I went back through the living room, pausing only to grab the plastic bag I'd stashed on the bookshelf. I opened it, checking to see that my blouse and the paper-wrapped cigar were still inside, and retraced my steps in the direction of the foyer, where Seth had observed what I'd done. He looked at me quizzically but said nothing until we were back at the hotel, where after changing into dry clothes, we settled in the lounge. Seth had a beer and I indulged in a glass of sherry. I'd barely touched my daiquiri at the party.

The upset of having just witnessed Dr. Vasquez's sudden death had set in, and we said little for a while, each of us immersed in our private thoughts. I ached for Seth at that moment. I knew how important his recently forged friendship with Vasquez was to him, and I wondered whether he would give vent to his emotions. Not that I expected it. He has a hard shell that he uses to mask his feelings, although they sometimes manage to slip through the cracks.

"I'd like to visit Ivelisse again while we're still here in Tampa," he said.

"I'm sure she'd appreciate that."

"I can't begin to imagine what this means for his research."

"I suppose it depends upon how far along he was and to what extent his assistant can carry on."

He grunted his agreement, took a sip of his beer, and said, "Hate to be nosy, Jessica, but what did you have in that plastic bag?"

"My wet blouse, for one thing," I said.

"You didn't seem to have noticed that I was wearing a waitress's jacket. Tomorrow I should make arrangements to return it to the company."

"I noticed. Thought it was very clever of you to have found dry clothes. Anything else?"

I laid the paper-wrapped cigar on the table and unfolded it. "I know," I said, "it doesn't make sense, but I couldn't help myself."

"I've heard that from you before. But why? No need for you to go around pickin' up cigar butts. Happy to buy you a brand-new one."

"Oh, Seth, you know it's nothing like that. It's just that . . . well . . . it was strange that the waitress came outside to retrieve it at Ivelisse's direction. Plus, it didn't look like the sort of cigar I'd seen Dr. Vasquez smoke earlier."

Seth examined it more closely. "Al didn't get to smoke much of it before he got hit," he said. "Just a few puffs."

"Look at the wrapper," I said. "I remember that you told me he'd given you a Hoyo something or other."

"Hoyo de Monterrey Double Corona," Seth said.

"Where did he get Cuban cigars?" I asked. "I thought buying Cuban cigars was illegal here in the States."

"Ayuh, that it is, but Al told me he had a source."

"Like knowing a drug dealer."

"Hardly the same, Jessica, but people who smoke Cuban cigars are very fussy. Know what President Kennedy did during the Cuban missile crisis?"

"No."

"The president enjoyed a certain brand of Cuban cigar and told his press secretary, Pierre Salinger, to go out and buy up as many of 'em as he could find. Mr. Salinger came back the next day and reported that he'd bought twelve hundred cigars, so Kennedy went ahead and signed the trade embargo with Cuba. He wasn't about to do that without his favorite stogies. Al always had a big supply of Hoyo de Monterrey Double Coronas on hand."

"But this one isn't that type," I said. "Look at the wrapper."

Seth took an even closer look. "Macanudo," he read from the soggy label.

"And it's not as fat or as black as the ones he usually smoked. At least I don't think it is."

"You're right about that," he said, "but I don't see why it would interest you. Must be a cigar that somebody gave Al as a gift."

110

"You're probably right," I said. "I was curious, that's all."

"Curiosity killed the cat," he said. "I suppose you know that."

"It seems I've heard it before from a certain physician friend of mine."

"Let me ask you a question, Jessica."

"Go ahead."

"Did you pick up that cigar butt because you question whether Al was killed by lightning?"

"No, I don't think so. Well, maybe. It just seems to me that —"

"That it's unlikely that lightning killed him? The autopsy will determine that, and maybe lightning *did* kill him. Al was in great shape. He'd had a physical just a few weeks ago and was told his heart and everything else about him was A-one. So if lightning *didn't* kill him, and it wasn't likely that he had a coronary, then what *did* kill him?"

# CHAPTER EIGHT

I carried that morning's edition of the *Tampa Tribune* to the dining room, where I met Seth for breakfast.

"Already read it," he said when I handed it to him. "Not much of a story."

The article was only three paragraphs long and reported that EMTs had been called to the home of Dr. Alvaro Vasquez, a prominent Cuban American physician and medical researcher who'd defected from Cuba and who lived and worked in Tampa. It gave the cause of death as a possible lightning strike. It went on to say that the thunderstorm that likely killed Vasquez was one of the most violent in memory and that it was unusual for such a storm to develop in the winter months. The piece ended by saying that the Tampa medical examiner had also been called to the scene and that an autopsy would be performed.

"I called Dr. San Martín," Seth said.

"We're meeting him at ten."

"We?"

"Ayuh. I told him that we both had witnessed Al's death and that you had superior powers of observation."

"You didn't."

"Of course I did. As long as you're here with me in Tampa, you might as well get involved. If you want to, that is. The doctor has already done the autopsy — did it last night. Eat your eggs before they get cold."

"And?"

"And what?"

"Did he say what the autopsy revealed?"

"No. He didn't say much of anything. He seemed anxious to get together, though."

As we talked, a tall young woman entered the dining room and came directly to the table. "Dr. Hazlitt?" she said.

Seth nodded.

"Sorry to interrupt your breakfast, Doctor, but the concierge pointed out to me who you were. I'm Peggy Lohman, *Tampa Tribune.*"

"What can I do for you, Ms. Lohman?" Seth asked.

"Mind if I join you?" she said as she slipped onto a vacant chair.

"Seems that you already have."

She ignored his comment and said, "I'm

doing a story on the death yesterday of Dr. Alvaro Vasquez. We ran a small piece in today's paper. We didn't have much to go on so we had to keep it short, but we're putting together a much longer piece today. I tried to get hold of someone in the family, but no one returns my calls. I can understand that they're upset, but this is a big story considering Dr. Vasquez's stature in the community, and I need quotes from people. I called the police — they were summoned to the house, which seems strange to me considering he died of a lightning strike — but they had nothing to say, which isn't unusual for them, at least in my experience. Anyway, I know that you and Dr. Vasquez were close friends and professional colleagues and —"

She spoke rapidly, the words tumbling from her mouth.

"Ms. Lohman," Seth said, "I know that you're looking for a story, but —"

The reporter turned to me. "You must be Jessica Fletcher," she said.

"That's right."

"I knew that you were traveling with Dr. Hazlitt."

"We're friends," I said.

"Of course. You were at the house when Dr. Vasquez was hit by lightning."

114

I started to respond, but she forged ahead.

"I think it's fascinating that you write murder mystery novels, and I know that you came to Florida to promote your latest book. I've always wanted to write a novel based on some of the cases I've covered, really juicy ones. I started a novel a few years ago but never seem to get back to it. Maybe you have some tips for me."

"I'm afraid I don't have any tips," I said, "but do you have a question for Dr. Hazlitt?"

If she was offended by my answer, she didn't show it. She said to Seth, "Dr. Vasquez was working on medical research here in Tampa, something to do with Alzheimer's disease. Right?"

"Ayuh."

"Pardon?"

"That's Maine talk for 'yes,' " I said.

"You are correct," Seth said. "Al — Dr. Vasquez was doing research on the disease."

"Hadn't he been doing the same research in Cuba before he defected?"

"That's my understanding."

"How did you meet him?"

To my surprise, Seth settled into a comfortable conversation with her. I'd seen him handle the press before. He's never been a fan of reporters, although he's quick to point out that despite the media's excesses, it's the only true check and balance on government that we have in our society. What most upsets him is when reporters badger people after someone has died in the hope of coming up with a bit of sensationalism. Ordinarily he would have politely, but firmly, dissuaded the reporter from pressing him with questions. But here he was answering her queries, good-naturedly and even enthusiastically. He seemed to bask in having become close to Dr. Vasquez and wanted the reporter to know that he was. I kept silent during the interview.

When she stood to leave, she said to me, "I really enjoyed meeting you, Mrs. Fletcher. Maybe while you're here in Tampa we could get together and, you know, talk shop."

"Talk shop? Oh, about your novel. I'm not sure I'll have the time — I plan to stay only a few days — but I do wish you all the best."

"Nice gal," Seth said after she was gone.

"You were certainly accommodating," I said.

"Just helping her do her job."

"I noticed that you couched your answers when she asked about Dr. Vasquez's research."

"Wouldn't be my place to talk about that.

116

I'd better call Bernie Peters. He said he wanted to speak with us today."

"He wanted to speak with *you,* Seth."

"Seems it's up to me whether I bring you along. Course, if you really don't want to —"

"It's just that I was thinking of making plans to head back home."

"I'd really appreciate it if you'd stay awhile, Jessica. We were together when Al died, and I figure that we should stay together until his death gets sorted out."

The truth was that I wanted very much to hear what the medical examiner had to say, and to follow up on what Vasquez's death meant to Bernard Peters and his company. Call me inherently curious. I don't mind; I've been called worse many times over the years.

"I'll be happy to stay," I said.

Seth gave me an "I knew you would" smile. "Good. I'll call Bernie. Meet you in the lobby in a half hour."

My emotions were decidedly mixed at that moment. I had no official reason for staying in Tampa. All I'd done was witness someone's death by lightning. Seth's question about whether Vasquez had, indeed, died from a lightning strike seemed to me nothing more than idle speculation. Still . . .

117

As I considered this, I realized that my friendship with Seth had taken a new and interesting turn. I'd had the misfortune of becoming involved in a number of real-life murders over the years — I hate to acknowledge how many — and was usually the one who smelled a rat, as they say, when everyone else was pointing to natural causes in someone's passing. And it was always Seth who chided me about being overly inquisitive and suspicious.

But here he was, eager to meet with people involved in his friend's death, and even chatting with a newspaper reporter. I wasn't sure what to make of it, any more than I knew what had intrigued me about the fact that Vasquez had been smoking a cigar that wasn't his usual brand that night. But now that I'd determined to go along with Seth, at least in the short term, I would give it my best.

The office of the Hillsborough County medical examiner was located on North Forty-sixth Street. Dr. San Martín's secretary told us that he was in a meeting but would be free shortly. We read magazines for fifteen minutes, until he came through the door, apologized for keeping us waiting, and ushered us into his large, messy office. There were file folders, magazines, and large

envelopes containing X-rays on every surface. Two piles of books leaned precariously
in a corner. In another corner, a six-foot-
tall classic wooden cigar store Indian cast
its angry look over the room.

"I appreciate you finding time for me this
morning," San Martín said as Seth and I
settled in chairs across the desk from him.

"Hope I can be of help," said Seth.

San Martín's expression questioned my
presence.

"I'm just along for the ride," I said. "Seth
didn't think you'd mind."

"I don't, of course, but it is a little disconcerting to have a writer in our midst. I trust
you aren't making notes for one of your
novels."

"I assure you that I'm not," I said.

"Good, because some of what I say this
morning isn't for public consumption."

He said it in a way that demanded a
response.

"Count on it," Seth said, and I agreed.

"As I told you on the phone, Dr. Hazlitt, I
did the autopsy last night. To be more accurate, I participated in the autopsy with a
colleague of mine."

"Come to any conclusions?" Seth asked.

San Martín paused before replying. "Yes,
I did, and my colleague concurs. Based

upon a gross examination of the deceased, I do not believe that he was struck by lightning."

"That doesn't surprise me," said Seth.

"Oh?"

"I don't know a heckuva lot about lightning and what it does when it hits somebody — I've only had two patients who were hit by lightning."

"That's probably more than most doctors up north see," San Martín said.

"True. And in my cases, both survived, but one was left in pretty bad shape, had neuropsychiatric, vision, and hearing problems."

"You know more than you think," said San Martín.

"Appreciate that," Seth said. "The thing is, I got close to Dr. Vasquez right after he fell, tried CPR on him. I got a good look at his face and neck. From what little I do know, when someone gets hit by lightning, there're usually burns on the head and neck. There weren't any burns on Al — Dr. Vasquez — nothing on his clothing or on any exposed skin. I also remember reading that only twenty percent of folks hit by lightning die on the spot."

San Martín smiled. "Everything you say is correct, Dr. Hazlitt. The keraunopathologists would be impressed."

I tried to pronounce what he'd said and failed.

"Keraunopathologists," he repeated. "Specialists in the pathology of lightning. Not many of them. At any rate, your observations are correct. Usually when someone is struck by lightning, burn marks are visibly evident, especially at the entry and exit points. Most people don't realize that a lightning strike has about ten times the kilovolts as your typical industrial electrical shock. That sort of power burns a victim pretty bad. It immediately turns the victims' sweat into steam." He paused for effect. "There were no burn marks on Dr. Vasquez."

"So the fact that there was a lightning strike at the moment he died was a coincidence," I offered.

"I'd say that's a fair assumption," San Martín said.

"Your autopsy ruled out lightning as the cause of death," Seth said, "but did it give you any clue as to *why* he died? He'd told me that he'd had a physical exam a few weeks ago and everything was fine."

"Do you know who his doctor was?" San Martín asked.

"Can't say that I do."

121

"Easy enough to find out. I'd like his input. To answer your question, his heart looked fine. But there was a marked change in the muscles supplied by his cranial nerves, specifically his breathing muscles. It looks to me as though he died from sudden and total respiratory failure."

"What could cause that?" Seth asked, his expression skeptical. "Are we back to thinking it *was* lightning?"

"No," said San Martín. "I can't say I'm an expert with cases of lightning strikes —" He smiled at me. "Keraunopathy. But I have autopsied my fair share of lightning victims. This, after all, is Tampa. Florida has twice as many fatal lightning strikes as any other state. Even the name 'Tampa' is said to stem from a Native American word for 'sticks,' which many believe refers to lightning. Nationally, death by lightning is the third leading cause of weather-related deaths."

"But you don't believe Dr. Vasquez was struck by lightning," I said.

"No, I don't. Lightning can cause severe injury to the cardiopulmonary system, but that isn't the case with Dr. Vasquez. Something else affected his cranial nerves and respiratory system, and did it with incredible speed. His death was instantaneous. The toxicology report might give us some answers. I've put a rush on it. Did either of you notice anything unusual about his behavior that day? I understand he was hosting a party when he died."

"That's true," Seth said. He looked to me. "Did you see anything unusual, Jessica?"

"Since I barely knew the man, I wouldn't have picked up on changes in his behavior. He seemed healthy and happy, full of life and spirit. I did, however, wonder why he stayed on the deck after the storm hit. I remember him saying when we played golf that when a storm approached, you'd better get to cover fast."

Seth laughed. "He didn't want to waste that cigar he was enjoying. Al did love his cigars."

"Not unusual here in Tampa," San Martín said. "I personally can't stand them, but to each his own."

Seth asked me, "Do you still have that cigar that Al was smoking?"

"Yes, I do."

"When Al fell, the cigar went flying. Mrs. Fletcher retrieved it after the cleaning staff picked it up," Seth explained.

"Silly of me, I know," I said. "He'd mentioned that the cigar was a gift from a friend. Apparently it was different from what he was accustomed to smoking."

I dug in my purse and pulled out the cigar, which I'd placed in a small plastic bag I'd gotten from the hotel. I handed it to Dr. San Martín, who turned the bag so that he could see the cigar through both sides.

"It's a little squished," I said. "It was in a puddle when the waitress picked it up."

"Tampa used to be the cigar capital of the world," San Martín said as he dropped the bag on his desk. "That's how Ybor City came to be. Cigars! There used to be a hundred and fifty cigar manufacturers in that section of Tampa alone."

"I'm looking forward to visiting Ybor City while I'm here," I said. "I understand it's . . . well, that it's very colorful."

"That it is," San Martín said. He stood and stretched. "It was a late night and these old bones are feeling it. Thanks for stopping by and sharing what you know. There's more to doing an autopsy than examining the body. Everything surrounding a death has to be taken into account."

"I'd appreciate knowing if you come up with any other conclusions," Seth said as they shook hands. "I'm not only a doctor; I was Al Vasquez's friend."

"You'll hear from me," San Martín assured him. "Thank you for coming in with Dr. Hazlitt, Mrs. Fletcher. May I suggest that you take the trolley when you visit Ybor City? It's part of the experience."

Seth and I decided to take advantage of the warmer weather that day by walking back to the hotel rather than hailing a taxi. He'd arranged to meet with K-Dex's Bernard Peters at noon at Vasquez's lab, and for the three of us to have lunch following the appointment.

"Shakespeare wrote about lightning," I said as we walked slowly.

"Say again?"

"Shakespeare," I said. "I remember when I taught Shakespeare back when I was an English teacher. Let me see if I remember it. 'To stand against the deep dread-bolted thunder? In the most terrible and nimble stroke of quick, cross lightning?' It's from *King Lear.*"

"He knew a lot, didn't he?"

"Shakespeare? He certainly did. Unless you meant Dr. San Martín. He's a lovely man."

Seth nodded his agreement. "Well," he said, "at least we know that Al didn't die of a lightning strike. I was pretty certain of that."

"A sudden respiratory attack, enough to kill him instantly," I said. "Have you ever seen that in a patient?"

125

"No, can't say that I have. It's not possible, as far as I'm concerned. There has to be some other explanation."

I stopped Seth and placed a hand on his shoulder as I raised a foot to shake out a pebble that had gotten into my shoe. As I did, I looked back from where we'd come and noticed a small silver vehicle driving very slowly. The driver had stopped when we did.

"Do you see that car?" I asked Seth.

"Which one?"

"The small silver one."

"What about it?"

"It was behind us when we took the taxi from the hotel."

"So?"

"Here it is again, driving slowly, as though trying to stay behind us."

"You feeling a little paranoid this morning, Jessica?"

I squinted at the car, trying to see if I recognized the driver through the tinted windshield. The car suddenly sped up, turned a corner, and was gone.

"Sorry," I said as we continued our walk.

"About what?"

"The car. Nothing unusual about a car being where we are twice in a day."

We'd almost reached the hotel when Seth

said, "Did you see the driver of the car?"

"Not clearly. It was a man. He was alone, I think. Why?"

"No reason. Just asking."

As we entered the lobby, we were stopped by a desk clerk. "Someone came by and left you this, Dr. Hazlitt." He handed Seth an envelope on which his name was handwritten.

Seth opened it, frowned as he read the note that was inside, and handed it to me.

Dear Dr. Hazlitt,

It is important that I speak with you. Please call me at my cell number as soon as possible.

— Dr. Pedro Sardina

He included the number.

"Sounds important," I said, handing back the note.

"It does, doesn't it? I'll call from my room. Meet you back here in a half hour."

The message light was flashing on my room phone. It was a call from Oona Mendez.

*I don't know what your schedule is like, today, Jessica, but I would like very much to meet with you at your convenience.*

She, too, left a phone number.

I told Seth of Oona's message in the taxi on our way to the laboratory.

"Seems we're popular folks these days."

"It appears that way."

"I reached Dr. Sardina. He'll be at the lab when we're there with Bernie Peters, but he said he didn't want to talk with him around."

"I wonder why."

"I suspect there's going to be some tense times between Peters and Sardina," Seth said. "Sardina knows how far Al got with his research, and Bernie obviously wants to know, too."

"I can't fathom why a smart businessman like Bernard Peters would allow Dr. Vasquez to keep his research results under such close wraps. After all, Mr. Peters's company is paying for it."

"I don't understand it either, Jessica, but I intend to find out."

The conviction with which he said it startled me. I had no idea that he'd decided to seek answers to that question, or any question, for that matter. I knew that Vasquez's sudden death had had a tremendous impact on Seth. He'd kept his emotions in check, but it was obvious to me that he was struggling with them. Despite the little time they'd spent together, Alvaro Vasquez had become a treasured friend, something that few people I knew could claim. Yes, Seth Hazlitt had a world of friends back in Cabot Cove, but few were truly allowed entry into his inner circle, and I thankfully counted myself among them.

I decided to push him.

"Care to elaborate?" I asked.

"About what?"

"About wanting to find out the situation between Dr. Vasquez and Bernard Peters?"

"You sound as though I shouldn't."

"Not at all, Seth, but I didn't realize that you had issues to resolve aside from naturally grieving over your friend's death."

He thought before responding. "The way I see it," he said, "Al devoted his life to finding a cure for Alzheimer's. He pursued the cause despite interference from Castro's totalitarian regime, and he showed guts when he and Ivelisse left Cuba and came here to continue his work. He was one hell of a fine man, and I want to make sure that his work gets the credit it deserves."

I started to say something, but he continued.

"There's more to it, though, Jessica. There's a real foul smell, the way he died. It wasn't lightning like everyone assumed. Sudden and complete respiratory collapse?

Never heard of such a thing. Doesn't make sense."

It was my turn to think before speaking. When I did, I asked, "Are you suggesting there might have been foul play?"

"I'm not suggesting anything, Jessica. All I know is that something's rotten in Denmark, only it's here in Tampa, Florida, and I want to know what it is. I owe it to Al."

# CHAPTER NINE

The guard who'd been at the door when we'd first visited the laboratory was on hand when we arrived to meet with Peters and Sardina. But this time the door to the building was open, and we weren't questioned as we approached. We entered and followed the narrow corridor back to where the lab itself was located. Peters and Sardina were there, and neither man looked happy.

"Good to see you," Peters said, shaking Seth's hand as we entered the lab. "Thanks for coming."

"No trouble at all," Seth said. He looked over at Vasquez's assistant. "Hello, Dr. Sardina," he said.

Sardina muttered what passed for a response and busied himself at one of the computers.

Peters indicated with a flip of his head that we should follow him outdoors, where he led us far enough away from the guard to ensure privacy.

"Hate to get personal," Seth said, "but I get the impression that you and Pedro Sardina were not havin' a pleasant chat about the weather."

Peters's tight lips and angry eyes confirmed that supposition.

"Anything I can do to help?" Seth asked.

"It's missing," Peters said flatly.

"What's missing?" Seth asked.

"Al's laptop computer, the one he used to chart the progress of his research."

"It can't just be missing," Seth said. "There's got to be a simple answer."

"You know the computer I'm referring to," Peters said. "I understand that Al shared some of the material on it with you."

"Ayuh, he did. I got to read some of the entries."

"That's more than he did for me," Peters said.

"Mr. Peters," I said, "I obviously have no knowledge of what transpired between you and Dr. Vasquez, but I have to ask a question that's been on my mind ever since I got here. Dr. Vasquez joked once that he kept progress reports from you. I can't help but wonder why you, as the source of Dr. Vasquez's funding, would be kept so much in the dark about his progress — and, I suppose, why you would put up with it."

Peters's smile was rueful. "Want a straight answer, Mrs. Fletcher?"

"Whatever answer you wish to give."

"I let Al get away with it because, frankly, I had no choice. His research was vitally important to me and to K-Dex. I'd known for years about his research in Cuba into the impact of sugar on the brain, and the role it might play in promoting the growth of beta-amyloids, a chief component of the plaques that are a definite hallmark of brain abnormalities in Alzheimer's patients. The same holds true of how glucose, and insulin resistance, could influence the unusual growth of tau proteins, another provable aspect of the disease. To be honest, I was taken in by Al's faith in his research. But who wouldn't have been? Every report that leaked out of Cuba said he was on the brink of a truly major medical breakthrough."

"And you believed those leaks?"

"I did. You might also have noticed that Alvaro Vasquez was a charming, manipulative man."

"Charming? Yes," I said. "Manipulative? I wouldn't know about that."

"Take my word for it," Peters said angrily. He made a fist and rammed it into the palm of his other hand. "I trusted him," he growled. "I had to. So much depended upon his research providing a leap forward. If he'd found a definite link between how glucose influenced brain cells and Alzheimer's, and had come up with a way to reverse it, it could have led to a cure, with K-Dex leading the way. Think about what that would mean to millions of people, Mrs. Fletcher. I never *stopped* thinking about it."

I thought for a moment that Peters might break into tears.

"Let's get back to his laptop," Seth said. "Surely it wasn't the only documentation of his research and the progress he'd made."

"I've been led to believe that it was," Peters said ruefully.

"What about Dr. Sardina?" I asked. "Would he know where it is?"

Peters's sad expression turned angry again. "I trust Dr. Sardina as far as I can throw him, the little weasel."

His harsh statement lingered in the air, and neither Seth nor I responded.

"I was questioning Sardina when you arrived. He's an arrogant young man, that's for sure. He claims that Vasquez kept him uninformed about how his work contributed to the big picture and that Al kept the overall progress reports to himself. Sardina would work on a specific project, give the results to Al, and that's the last he'd hear about it. When I asked him about the laptop, he told me that Al kept it under lock and key and took it home with him every night."

"Then that's probably where it is," I said.

"I can only hope, but I'm not sure I believe him. Of course, it might all be a moot point, depending upon how far along Al was. If he hadn't achieved the sort of results he was always promising, his progress reports won't be worth diddly."

There were other questions on my mind at that moment, all of them pertaining to why a businessman like Bernard Peters would enter into such a loose and problematic business arrangement with Vasquez. Of course, there undoubtedly were legal documents cementing Peters's interest in Vasquez's work. At least I hoped there were, for his sake.

"Have you gone to the house to see if the laptop is there?" I asked.

"I called and spoke to Al's daughter, Maritza."

"She's here?" Seth said.

"She just arrived from Cuba."

It had been in the back of my mind that the Vasquezes' daughter had not accompanied her parents to Tampa. I remembered a conversation Seth and I had had shortly after we'd learned that Al had asked the United States for asylum.

"The newspaper said that both he and his wife defected," I'd said. "Do they have any children?"

"Oh, they do," Seth had replied, "a son and a daughter. I met them when I was in Cuba."

"They didn't defect?"

Seth had hesitated before answering, and I'd wondered why.

"It's a bone of contention with Al and his wife," he'd finally said. "Really none of my business. His son came to the States more than a year ago, which didn't sit well with his folks. The daughter is in medical school and refused to leave Havana. You know how families can be. Kids have minds of their own."

"Where does his son live?" I'd asked.

"In Tampa. He'd gone to Miami from Cuba, according to Al, but moved to Tampa not long after he arrived in the States. I imagine that played a role in Al's decision to settle there."

"So the parents and son are reunited," I'd said.

"Seems so," Seth had said. "I'm sure that pleases Al and his wife."

136

"I would imagine it does," I had replied at the time.

But the prickly relationship between father and son that I had witnessed at the party made me wonder whether Al had regretted moving to live near his offspring. Perhaps the decision had been made because Ivelisse was close to her son, but choosing to live near one child came at a cost. Her daughter had remained in Cuba. The Vasquezes had never returned to their homeland. How long had it been since they'd seen Maritza?

"What did you say to Maritza?" I asked Peters.

"I expressed my condolences, of course, and I asked whether I could come to examine some of Al's belongings but didn't get anywhere. She said that her mother was in no condition to have visitors and that I should call back in a day or two."

Peters was obviously distraught, and I wasn't sure we should go through with plans to have lunch with him, but he settled it when he said, "Look, I have to cancel our lunch plans. I'm meeting with my attorneys to see if they can come up with a way to untangle this mess. If we can't, the company stands to go under. We'll do it another time."

"Of course," Seth said.

Peters went to his car and drove off, leaving us to decide what to do next.

"I suppose we should go back inside and talk to Dr. Sardina," I suggested.

Sardina was still at the computer when we walked in.

"Hope we're not disturbing anything important," Seth said.

Sardina looked up and shook his head.

"Mr. Peters has left," I said.

"Good," was Sardina's reply.

"We were talking about Dr. Vasquez's laptop computer, the one he used to keep track of progress," I said. "Did you help him input lab results?" I asked, already knowing the answer.

A rare laugh came from him. "Me? I think he would have chopped off my arm if he'd seen me go near that laptop."

"He let me take a look a few times," Seth offered.

"I know," Sardina said. "He evidently trusted you more than he trusted me."

His bitterness was palpable.

"I'm sure he trusted you," I said. "After all, you worked side by side with him every day."

"Need to know," Sardina said. "That was his favorite saying, need to know. He told me just enough to keep me interested. I

138

should have left ages ago."

His anger permeated the lab.

"Apparently, he didn't share any more information with Mr. Peters," I said. "He said if he can't find the laptop, the company will be ruined."

"Don't you worry about him," Sardina said sourly. "Bernie Peters ain't goin' to be missing any meals anytime soon," he said, putting on a southern accent.

"We understood the company invested millions of dollars in Dr. Vasquez's work and this laboratory," I said. "That's a lot to lose."

"And don't forget his home on Davis Island, and his boat, and all the other perks the great doctor received."

It was clear to me that Dr. Sardina had not been on the receiving end of any extra benefits and was resentful. "Don't you think those losses will affect K-Dex and Bernie Peters?"

"They would if they weren't well insured."

"What do you mean?" Seth put in.

"Peters had key-man insurance on Vasquez. Anything happens to him, the company recoups all its investments and Peters himself walks away with a tidy sum."

Seth pursed his lips and whistled. "Did Al know about this?"

"If I know, he knew," Sardina replied.

"Is this what you wanted to talk to me about?" Seth asked. "You called and said you wanted to get together."

Sardina pressed his lips together and stared at the computer screen. He then looked at me.

"Anything you want to say to me can be said with Mrs. Fletcher present," Seth said.

Sardina looked directly at Seth and said, "How much do you know about Alvaro's research?"

"Some," Seth said, "but from what he told me, he was about to reach a major advance, one that could lead to new pathways for drug trials."

Sardina's smile was small but said volumes.

"Tell you what, Dr. Hazlitt. You buy me a nice lunch and I'll tell you things about Dr. Alvaro Vasquez that I'm sure he never told you about himself."

# Chapter Ten

Sardina drove us to a strip mall on the outskirts of Tampa and pulled in front of a restaurant whose sign promised an Asian buffet. "Hope you like Chinese food," he said as we entered the large, busy place and found an isolated table away from others. "I come here a lot. Good food, reasonable prices."

We took turns going through the multitude of hot and cold buffet lines, one of us staying behind to secure the table while the others filled their plates. Once we were all seated with our food, Sardina said, "I'll tell you right off the bat that I was no fan of Dr. Vasquez."

"I sorta gathered that," Seth said.

"Don't misunderstand," Sardina said. "I'm sorry that he's dead."

"Yes, I'm sure you are," I said. "How long had you worked for him?"

"A little over a year. I can't believe I stayed

as long as I did."

"You indicated back at the lab that you wished you'd left a long time ago," Seth said.

"That's right." He tasted a few items on his plate before continuing. "Dr. Vasquez — he told everybody to call him Al, but not me; with me it was always Dr. Vasquez, very formal." He said the name again, this time with disgust. "Yes, I should have left long ago. No, I never should have gone to work for him in the first place."

"How did you meet him?" I asked.

"I knew him in Cuba, Mrs. Fletcher. We didn't work together there. He was into his research, and I — well, I'm not a medical doctor. I have a PhD in infectious diseases. We ran into each other now and then. Dr. Vasquez —" He grinned. "Now that he's gone, maybe I can call him Al like everyone else. Al was in favor with the Castro regime, got plenty of perks because of it. Ofelia and I were invited to a couple of parties at his house. Nice place — not what he has here, but a lot better than where we lived."

"I visited Al's home in Cuba, too," Seth said.

"He told me that you did." Sardina looked at Seth quizzically. "You and he really struck up a friendship, didn't you?"

"Pleased and honored to say that we did."

"He thought a lot of you."

Seth nodded, struggling to keep his emotions in check.

"How did both of you end up in Tampa?" I interjected, giving Seth a chance to compose himself.

Sardina turned his attention to me. "Ofelia and I left a few months before Al and Ivelisse defected. We attended a conference in London and came here instead of returning home. That was before the government put a tourniquet on foreign travel. We were lucky to get out."

"If the government further tightened restrictions on travel after you left, how was Dr. Vasquez able to make his escape?"

"Al had connections," he said, and stopped.

I had the feeling that he wanted to say more but was editing himself. I asked a different question. "Why did you decide to come to work for Dr. Vasquez?"

"Necessity. I thought once we got to the States, I wouldn't have a problem finding work in my field. Well, I was wrong. As much as Cubans have assimilated into U.S. society, it doesn't mean we're welcomed with open arms. My degrees weren't recognized here, and all I could manage to find was a low-level job in a lab at a university. It didn't pay much, and I had a run-in with my supervisor, who knew less than half of what I know and refused to listen to my suggestions. It was around that time that Al called and asked if I wanted to work with him on his Alzheimer's research. I jumped at the chance. He was paying a lot more than the job I had. Besides, working on finding a cure for a major disease was really appealing. The reality turned out to be less so."

He seemed to be collecting his thoughts, and we ate in silence until he spoke again.

"Al —" He chuckled. "I can't get used to calling him that. Dr. Vasquez was — how can I put it? — he was not an honest man."

"In what way?" I asked.

"In every aspect of his life."

I could feel Seth, who was sitting next to me, stiffen. I put my hand on his arm to keep him from blowing up. "That's quite a condemnation," I said.

"And a truthful one, Mrs. Fletcher. Alvaro Vasquez was a smooth con man. I'm sure you saw that the few times you were with him. He lied to everybody — me, his wife, his kids, and especially Mr. Peters."

"Did he lie about how his research was going?" I asked, glancing at Seth to gauge

144

his reaction to what Sardina was saying. Seth had had nothing but praise for Vasquez, personally and professionally, and I knew it must have hurt to hear his friend disparaged like this.

"I'm afraid so," was Sardina's reply.

"Now, hold on a second," Seth said, dropping his fork noisily onto his almost empty plate. "I'd like to know what you base that on."

Sardina, sensing Seth's pique, held up his hands in mock defense of himself. "Please don't misunderstand, Dr. Hazlitt," he said. "I know that you and he were friends."

"I'm not talking about our friendship," Seth said. "I'm talking about his research. Are you claiming that he wasn't honest about his research, that he lied about it?"

"Yes, sir, I am."

"But how could you know?" Seth paused and then continued. "You told us back at the lab that you weren't privy to how the research was progressing, that you only knew bits and pieces on, as you put it, a need-to-know basis."

"That's true," Sardina said, "but that doesn't mean that I was completely ignorant about the bigger picture. I hated the way Al strung Mr. Peters along, always asking for more money for a new phase of the research even when there wasn't a new phase. I was with him plenty of times when he did it. He'd get more money from Mr. Peters, and when he left, Al would laugh about it."

"I'm shocked to hear this," Seth said, and his face reflected his anguish. I wondered whether he was thinking the same thing I was, that what Sardina was saying didn't necessarily represent the truth. After all, Peters had said that he didn't trust the young researcher. We were hearing one side, and I've always believed in waiting to hear both sides before coming to a conclusion. Of course, the "other side" of the story was Dr. Vasquez, and he wasn't in any position to refute Sardina's claims.

I wondered whether there was more to Sardina's negative view of Vasquez, perhaps a personal motive. I decided to ask.

"What about Dr. Vasquez's personal life?" The question came to mind because of what Dr. San Martín had told us about the circumstances of Vasquez's death. It certainly wasn't a new thought for me. It had been rattling around in my brain since our meeting with the medical examiner. If Vasquez hadn't died of a lightning strike, and since the autopsy had revealed what Seth considered an almost impossible circumstance — a sudden and total collapse of Vasquez's respiratory system — there was the possibility of foul play. I hated to even consider that option, but it couldn't be ruled out.

"What do you mean?" Sardina asked.

"He seemed to be a pleasant, well-liked man," I said. "Did he make enemies?"

"According to him, he had enemies from Cuba threatening to scuttle our work. That's why we were locked up tighter than a drum. Frankly, I think he just didn't want anyone else to discover what he was really about." Sardina motioned for a waitress to bring the check.

"Did he have personal enemies, as well? People without a nationalistic motive?"

Sardina snorted. "Let me just say that there wasn't a woman who was safe from his advances. I often think that his infatuation with Ofelia was why he hired me in the first place. He didn't make any bones about being attracted to her, and she's had to fend him off more than once. I imagine there were a lot of men who took a dislike to Alvaro Vasquez."

I thought back to the way Vasquez had tutored me on the golf course, pressing in close as he instructed me.

I grabbed the check when it was delivered by a pretty young Asian waitress, waving off

Sardina's and Seth's offers to pay. "Let me," I said.

When we were in Sardina's car, he asked where he could drive us.

"Our hotel, if you don't mind," Seth said.

"Yes. That would be helpful," I said to Sardina. "I have a call to make," I reminded Seth, referring to the message Oona Mendez had left on the answering machine of the phone in my hotel room.

"I have some calls to make, too," Seth replied.

Sardina dropped us off in front of the hotel, but before he left, Seth leaned into the car through the open front window. "Mind a bit of advice?"

"Go ahead," Sardina said.

"I suggest that you keep your negative comments about Al to yourself. The man is dead and can't defend himself. He deserves your respect."

If Seth's harsh words impacted Sardina, he hid his reaction well. He simply said, "The truth is always hurtful, Dr. Hazlitt. I'm sorry if I've upset you."

# CHAPTER ELEVEN

I called Oona Mendez when I got to my room and arranged to meet her at King Corona, a café in Ybor City on East Seventh Avenue. Before leaving, I called Seth and told him where I was going.

"Did she say what she wants?" he asked.

"I'll know soon enough. Have you had a chance to digest what we heard at lunch today?"

"If you mean the spareribs and fried rice, yes. As for what the young Dr. Sardina had to say, I'm still getting over it."

"He certainly had a litany of negative things to say about Dr. Vasquez. I wonder to what extent his claim that Vasquez made inappropriate advances to his wife colors his view."

"I wonder the same thing, Jessica."

"Care to come with me to meet Oona?"

"No, I think I'll catch me a nap."

"Sounds like a good idea," I said. "I'll

check back in with you when I return."

I left plenty of time between leaving the hotel and meeting Ms. Mendez so I could take the historic Tampa streetcar to Ybor City as Dr. San Martín had suggested. Up to now, my plan for a week of R and R in Tampa following my hectic book tour had involved neither rest nor relaxation, and I was determined to change that. It was a lovely, crystal clear day in the city, and it felt good to be on my own, breathing in the fresh air and feeling the sun's warmth on my face. I picked up a map from the concierge and figured out where the closest streetcar stop was, only a few blocks from the hotel. I waited with a group of tourists until the next car came along, its bell clanging, the sound of its wheels on the rails and the brake the motorman applied reminding me of San Francisco's famed cable cars. I took one of the hardwood seats — whoever designed them did not have comfort in mind — and we lurched forward, passing the imposing Tampa Convention Center and the Tampa Bay History Center building, grinding to a halt at the Florida Aquarium, and then up to Ybor City, the Ybor Channel on the right, until reaching Eighth Avenue in the heart of this unique section of Tampa.

A brochure I took from the trolley told me that Ybor City was settled in 1886 by cigar makers Vicente Ybor and Ignacio Haya, who'd moved their thriving cigar-manufacturing business to Tampa from Key West. With a railroad, a port, and a climate that functioned as a natural humidor, cigar manufacturing flourished, turning Tampa and Ybor City into the cigar capital of the world. That lasted until the 1960s, when embargos against Cuban tobacco and declining cigar consumption sent the cigar-manufacturing industry into a steep decline.

Despite the hard benches of the streetcar, I thoroughly enjoyed the ride along the red-brick streets, taking in the large old-fashioned globe streetlamps and the period buildings with their wrought-iron balconies. I got off at a stop near the Don Vicente de Ybor Historic Inn and browsed this former real estate office that was built by Vicente Ybor in 1895. It became a health clinic until a businessman converted it into an inn in 1998. It was like stepping into an earlier era, and I could almost hear the voices of guests speaking Spanish and detect the scent of their cigars.

After that pleasant break, I walked a few blocks to the King Corona, where Oona was already waiting at an outdoor table. She was smoking a cigarette and had a large cup in front of her.

"Hope I'm not late," I said.

"Right on time," she said. "Tea? Coffee? A cold drink?"

"What are you having?"

"Tea, creamy vanilla rooibos tea, red tea, from Africa, a specialty here."

A waitress appeared, and I told her I'd have the same.

"King Corona's not fancy," Oona said, "but it's good, serves the real thing when it comes to simple Cuban food. Hungry? The Cuban cheese toast is always good."

"Oh, no, thank you, I just came from lunch."

"A good one?"

"Lunch?"

"Yes."

"Very good. Asian."

"You and Dr. Hazlitt?"

"Yes, and Dr. Sardina."

Uplifted eyebrows accompanied "Oh?"

"He seems like a nice young man," I said. "Naturally he's upset at Dr. Vasquez's sudden death. I think he must also be uncertain of his future."

"As we all are. Upset at Alvaro's death, that is. Your friend forged quite a friendship with Alvaro, didn't he?"

152

"Seth? Yes, he did. Dr. Vasquez's death has shaken him, as you can imagine."

"I find it interesting that Alvaro shared so much of his research with Dr. Hazlitt."

"Why? Seth is a medical doctor. It seems natural to me that they would be able to discuss complicated scientific investigations easily."

"It wasn't like Alvaro to be open about his work with anyone."

I smiled. "Seth has a way of inspiring trust in people. He's a wonderful physician and a fine gentleman. He's held in very high regard back home."

"Maine."

"Yes. Cabot Cove, Maine. Have you ever been up north to New England?"

"I can't say that I have. Jessica, you do know what my job is here in Tampa?"

"Only what you told me at dinner. Something to do with —"

"The Cuban American Freedom Foundation. Because the U.S. doesn't have formal diplomatic relations with Cuba, we represent Cubans in America and work to foster better Cuban American relations. Our main office is in Miami; there are more than nine hundred thousand Cubans living there. Tampa has the second-largest community. Our organization works closely with all branches of the U.S. government, including the Treasury."

"Sounds like an exciting job."

"Boring most of the time," she said, and laughed.

But her good humor faded quickly and her expression turned serious. "I'm still grappling with Alvaro's death. All I keep thinking is how ironic it is that he was killed by lightning. It's almost as though his charismatic personality acted like a target, inviting the lightning to strike him."

I wasn't sure that I agreed with her dramatic explanation of her friend's death but said nothing.

"He was a marvelous human being, Jessica."

She blinked back tears, and I thought of what Sardina had said, that no woman was safe from Vasquez's advances. Had Oona fallen for his obvious charms? Had he been her lover?

"I'm so sorry you've lost your friend," I said. "A sudden death is always difficult to comprehend. You have my sympathies."

"Thank you," she said, clearing her throat. She was composed when she added, "Alvaro's death brings with it certain complications."

I nodded, listening.

"It wouldn't surprise you, I'm sure, to know that the Cuban government would very much like his research returned to Cuba. His defection wasn't taken lightly by Castro and his cronies."

"So I've heard."

"Almost no one has been allowed to leave the country since then, certainly not any doctors or other medical personnel. The loss of Alvaro's research and the glory it would have brought the Cuban government was a terrible blow."

Did she know what Sardina knew, that Vasquez's laptop on which he was thought to have kept track of his research's progress was missing? Or was it?

I was about to ask when an old man, bent and limping, approached carrying a fistful of cigars. "Cigar?" he asked in a weak, singsong voice. "Best cigars. Robustos, Don Diegos. Cheap, too."

Oona waved him away.

It was my turn to smile. "Ybor City might not be the cigar capital of the world any longer," I said, "but they seem to be offered everywhere I look."

She ignored my observation and said, "I know that Dr. Hazlitt — what an absolutely charming man — was taken into Alvaro's inner circle, so to speak, and was privy to the status of his research."

I thought back to what Karl Westerkoch had asked me at the party about how much Seth knew. I had a feeling Oona was probing for the same information, and I was sorry Seth had decided not to accompany me.

"I really don't know the extent to which Seth was taken into Dr. Vasquez's confidence, Oona. I suppose you'd best ask him."

"Yes, of course, I should do that. Did Dr. Sardina have anything to say at lunch about Alvaro's research and how far he'd progressed in finding a cure?"

I shifted in my chair and finished what was left in my cup. What had begun as a pleasant conversation about an unpleasant subject, Vasquez's death, was turning into a bit of an interrogation.

I fudged my answer. "He spoke about it, of course, but didn't say anything specific."

"What about Bernard Peters at K-Dex? Have you been in contact with him?"

I'm uncomfortable lying, always have been, and hate being put in a position where it might be necessary. Oona's questions were best answered by the people involved, Dr. Sardina and Bernard Peters among them.

"Seth and I had a brief chat with him this

156

morning," I said and left it at that.

"He must be beside himself," Oona said. "It's my understanding that his company, K-Dex, has sunk millions into Alvaro's research."

"I really wouldn't know about that."

"But your friend Seth must be aware of it, considering how close he became with Alvaro."

I said nothing.

She must have sensed my growing unease with the questions, because she shifted subjects. She leaned closer to me and said, "There's more riding on Alvaro's research than money."

"Well, of course," I said. "If his research was successful, it would have a major impact on the lives of people with Alzheimer's and their families."

"That's not what I mean," she continued in the same conspiratorial tone. "The disposition of his research could have serious ramifications with regard to the tenuous relationship we have with Castro's Cuba."

"I hadn't thought much about that," I said, which was true.

"I'm sure I'm not breaching any secrets," she said, "to tell you that the Castro regime has stepped up its efforts here in Tampa and

Miami to sow discontent among Cuban Americans."

"I wasn't aware of that happening."

"Oh, yes. There are Cuban Americans in both cities whose sentiments are still with Castro. Well, that isn't strictly true. Some of them don't pledge an allegiance to anyone. They do it for the money. The bottom line is that the Cuban regime will pay almost anything to get its hands on the research. We can't let that happen."

"We?"

"Our government."

"Are steps being taken to ensure that Dr. Vasquez's research stays here and doesn't fall into Cuban hands?"

"Let me just say that the key is to find Alvaro's notes."

Did she know about the allegedly missing laptop? It seemed to me that she did.

"Is there a problem finding his reports?" I asked, this time doing the probing myself. Had Peters told her what he'd told Seth and me?

She paused before asking, "Do you know of a problem with that, Jessica?"

"How would I?"

"I just thought that your Dr. Hazlitt might have shared something with you."

The lame street peddler returned offering

158

cigars and lighters. Oona again told him to leave, but I reached in my purse, withdrew a five-dollar bill, and handed it to him. He opened his almost toothless mouth into a smile and allowed me to take a lighter, a red one, from his hand.

*"Gracias,"* he said.

*"De nada,"* I replied.

I examined the lighter. It was similar to the one I'd seen Vasquez use to light his cigars, more like a blowtorch than any lighter I was accustomed to seeing.

"For cigars," Oona said. "It shoots out a flame."

I tried it and saw that it certainly did.

"Thinking of taking up cigar smoking?" she asked playfully.

"You never know," I said, dropping the lighter into my purse and getting up from my chair. "I really should be going."

"I'm glad you found time for me," she said, rising and shaking my hand.

"I'm sure we'll see each other again before I return home," I said.

She handed me a card on which her office contact information was listed. "Please call me if you hear anything that bears on what I've said."

"I can't imagine what that might be, but I certainly will stay in touch. By the way, I meant to ask you something about your friend Mr. Westerkoch."

"Yes?"

"He told me he was a consultant. What organization does he consult for?"

"Various agencies," she said with a small smile. "Thanks again for coming."

I watched her walk away and disappear around the corner.

Our conversation had raised more questions than it had answered.

That Dr. Vasquez's research would have political overtones had come as a surprise, although I suppose it shouldn't have. I could understand that laying claim to his research would be of considerable interest to the Cuban government, but it seemed to me that the ones with the most to lose were Bernard Peters and K-Dex, unless what Dr. Sardina had said about key-man insurance was true.

A chill in the air reminded me that it was time to get back to the hotel. I was eager to find out how Seth's afternoon had gone, and if other friends and acquaintances of Alvaro Vasquez were quizzing him. I left King Corona and retraced my steps to where I'd gotten off the streetcar. I hadn't realized how long Oona and I had talked, and it was starting to get dark. I was alone at the streetcar stop in front of the historic inn, except for a man wearing what the young people call a "hoodie." He leaned against a building a dozen feet from where I stood, trying, in my opinion, to appear casual. When I looked in his direction, he turned away from me, dropped a cigarette he was smoking, crushed it with his sneaker, and walked away. I watched as he crossed Eighth Avenue and got into a car — a small silver sedan that looked like the one I'd noticed earlier in the day. The car quickly pulled away and sped past me, the driver and young man looking straight ahead.

The streetcar arrived and I slipped onto a bench, gripping the back of the seat in front of me. I was on edge. Ever since arriving in Tampa to meet up with Seth, I'd experienced this sort of unease, nothing tangible, no single incident to which I could point. Of course, witnessing Alvaro Vasquez's sudden death had played a part, but as upsetting as that was, it couldn't explain the tension I'd felt before that awful event.

I was relieved when I reached the hotel. The first thing I did upon entering my room was to call Seth. There was no answer, so I left a message. Strange, I thought, that he would have gone out without leaving word for me. I waited fifteen minutes and tried his room again. Still no answer. He'd said he was going to take a nap. Was he still asleep? If he was, he would have awakened to the sound of the ringing phone. Seth was used to being called at odd hours by patients or the hospital and was a light sleeper. I made one more attempt before going downstairs. I poked my head in the bar and restaurant looking for him, before I approached the front desk.

"Did Dr. Hazlitt leave any messages for me?" I asked the young man.

"No, Mrs. Fletcher. Nothing here."

"I'm concerned about him," I said. "I've called his room three times with no response."

The young clerk smiled. "He probably stepped out for a while."

I shook my head. "He wouldn't do that without leaving me a message. Could we possibly go to his room and see if he's all right?"

A few minutes later the clerk arrived with an assistant manager, who accompanied me to Seth's room. We knocked, louder each time. No response. The manager looked at me for approval to use his master key to enter. I nodded.

He opened the door and stepped back to allow me to enter. I did so with trepidation.

I'd conjured the dreadful scenario of walking into the room to find my dear friend dead of a heart attack or stroke and drew a deep breath of relief once I saw that he wasn't there.

"Looks like he decided to take a walk," said the manager.

"Yes," I said, "it does look that way. I'm sorry to have bothered you."

"Not a problem, Mrs. Fletcher. Better safe than sorry."

He stood at the door, and I realized he expected me to leave with him, which I understood. It wasn't my room; to have allowed me to stay would have been a breach of hotel security. I rode down the elevator with him, thanked him again, and went to the lounge. Although I'd peeked in there earlier, I was hoping Seth might have gone in while I was upstairs. I'd just come from the lounge when the concierge, who'd been absent from his post when I'd returned from Ybor City, greeted me. "Good evening, Mrs. Fletcher."

"Good evening. You haven't seen Dr. Hazlitt lately, have you?"

"As a matter of fact, I have. He left about a half hour ago."

*Thank goodness,* I thought with a sigh. "Was he alone?" I asked.

"No. He was with someone."

"Do you know the person he was with?"

He shook his head. "Afraid not."

"A gentleman?"

"Yes."

"Could you possibly describe him for me?"

"Regular-looking fellow, wore a suit. He was — I suppose you could say chubby."

"Thank you," I said, and headed for my room.

It had taken me time to get over the fear I'd felt when entering Seth's hotel room. I sat by the window in my suite and tried to imagine whom Seth might have left with, and why he didn't leave any message for me. The concierge's vague description had been no help. While I don't usually track down where Seth is — after all, he's entitled to his privacy and to make his own arrangements — I decided to call him on his cell phone. I retrieved my phone from my purse and discovered that it was turned off. When I activated it, the tinkle of little bells told me that I had a message. It was from Seth.

"You ought to get in the habit of leaving your phone on, Jessica," he said. "No sense havin' one if you don't keep it on. Anyway, I'm on my way out of the hotel, having a drink with Bernie Peters. He called me and said he had something important to talk about. Would have invited you, too, if you had your phone on. Speak with you later."

I called him back only to be connected to his message center. "I'm back in the hotel," I said. A minute later my phone rang.

"Hello, Seth."

"Jessica. You called?"

"Yes. I'm sorry my phone was off earlier. You're with Bernie Peters? I'm disappointed I missed you."

"Just about to leave. Should be back at the hotel in fifteen, twenty minutes."

"I'll be here," I said; then, as an afterthought, I added, "Seth, on your way back, please pay attention to your surroundings."

"Why do you say that?"

"Because I think we're being followed."

# CHAPTER TWELVE

"What's this about us being followed?" Seth asked after we'd been seated at a table in Bern's Steak House. We'd decided to treat ourselves to a leisurely early dinner, and the concierge had recommended Bern's, although he did caution that we'd have to dress up, which we did.

"That same silver sedan I noticed when we left the medical examiner's office was parked near where I was this afternoon in Ybor City," I explained. "There was a man in the front seat, and another lounging on the street, who I'm sure was observing me. When I looked his way, he got into the car and they drove off."

Seth looked at me over the top of his menu and raised his eyebrows. Actually, the menu was called the *programme du jour,* all eighteen pages of it.

"I know," I said, "you think I'm being paranoid."

"Not at all, Jessica," he said. "You know the old saying."

"Just because I'm paranoid doesn't mean I'm not being followed."

He smiled and returned to reading the bill of fare.

"Seth," I said, "I have this feeling of unease that I can't shake."

He lowered the menu to the table and placed a hand on mine. "We were witness to a shocking event, Jessica."

"It isn't Vasquez's sudden death that has me upset," I said, "although I know it must be a terrible time for you, grappling with the loss of a good friend."

"Thank you for understanding," he said. "I've been working hard to get on with life in a normal fashion. As a doctor, I should be accustomed to death — I've certainly been exposed to it enough times — but when it's someone you felt close to" — he paused for a moment to rein in his emotions — "well, then it's another story." Seth's eyes were moist. He coughed to clear his throat and took a sip of water before asking, "So if it's not Al's death upsetting you, what is it?"

"It's the people who are alive that concern me. Today, Oona Mendez kept pressing me about the research and how much *you* knew about it."

"What did you say?"

"I said she should ask you directly."

"Good for you. Not that I would necessarily share any information with her. Why would it be her business?"

"She said she's looking at it from a government standpoint and told me that there are members of the Cuban American community here who are agents for the Castro regime."

"Wouldn't surprise me. Al felt the same way. That's why security was so tight at the lab. I'm a little annoyed that Peters and Sardina haven't kept that up. Careless of them. Probably why the laptop went missing. If Castro comes up with the research results, I blame those two."

"Even before Dr. Vasquez's death, Karl Westerkoch questioned me at the party, asking me how much you knew about the research. Bernard Peters, who was kept in the dark despite his and his company's support of Vasquez and the laboratory, is frantic to find the laptop to ensure the value of his investment. Pedro Sardina has nothing but bad things to say about Vasquez. And now we know that Al wasn't hit by a lightning strike. You say the way he died is highly unusual, almost medically impossible. I may be making too much of this, but it seems to me that all these things add up to good reason for me to be ill at ease."

"And there's that mysterious silver car you say is following you."

"Yes, that, too."

"Well," he said, "it's not as though I haven't been havin' some of the same feelings. I had an interesting time with Bernie Peters."

"That's right. I forgot to ask about it. I've been so busy with my own thoughts that everything else gets lost."

"It seems that Bernie and his lawyers are going to bring a suit against Al and the lab."

"Can they do that? He's dead. And what do they base the suit on?"

"They're demanding an accounting of the research and taking possession of it. Bernie leveled with me. He says he's invested every cent he has in Al's research, has almost bankrupted K-Dex, and has even mortgaged his home to keep the money flowing. He's a desperate man, Jessica."

"Then Sardina must be wrong about Peters covering his bets with insurance."

"Bernie says Sardina is not to be trusted. I'm of the same mind. I don't believe he told us the truth at lunch. I think he lied to us about Al to cover mistakes he made himself or to get back at a boss he didn't like. Bad-mouthing a man who's dead is pretty low."

"In that case, I can't imagine K-Dex would have any problem prevailing with a suit. Peters's company paid for the research. It belongs to them."

"Depends on how the agreement between Bernie and Al reads. Of course, chances are that the laptop is at the Vasquez house. Bernie says he's being stonewalled by Al's son, Xavier."

"Have you tried to arrange a visit? I have a feeling that because you were close to Alvaro, they might treat you differently."

"No, not yet, but I think it's time that I did. I'll call first thing in the morning."

"Any luck?" I asked when Seth joined me at breakfast.

"I got hold of Al's daughter, Maritza. She says her mother still isn't up to seeing visitors but that it was okay for you and me to come by."

"Both of us?"

"Ayuh. I said we'd be there in an hour. I think it's time we rented a car. It's being delivered here at the hotel." He looked at the menu. "Too bad they don't have blueberry pancakes. I could go for a stack about now."

"They'd never be as good as Mara's," I said, referring to our favorite luncheonette in Cabot Cove, where the blueberry pancakes were a specialty.

"True. So I guess I'll go for an omelet. I'm hungry — always am after I have a big meal the night before. Let's eat and get moving."

Before getting into our rental car, I surveyed the parking lot in search of the silver sedan. Seth noticed what I was doing and asked, "Any sign of our tail? That's what they call it, don't they, a tail?"

"No sign of it, and yes, they call it a tail. Maybe I've been imagining things."

"Mebbe," he muttered as he got behind the wheel of our bright yellow Toyota and fumbled while looking for the ignition. "But you're usually pretty observant."

"If someone *is* following us," I said, "he'll never lose *this* car. Couldn't you find one a little less colorful?"

"Took what they gave me," he replied as he found where the key went and started the engine, and we headed off for the Vasquez house on Davis Island.

"I feel awkward coming to the house," I said.

"Why?"

"It's so soon after his death. You're sure his daughter said it was all right?"

"Not only said it was all right — she seemed to welcome a visit from us."

I sat back and took in the sights as we crossed the Hillsborough River on Kennedy Boulevard and eventually came to Davis Boulevard, which took us back across the water, Hillsborough Bay this time, and onto the island. As we pulled into the circular gravel driveway, I saw that one of the black-suited security men stood at the front door, arms folded across his large chest, a formidable gatekeeper. He grunted as we walked past, but he didn't stop us and looked on placidly as Seth rang the bell. We heard movement inside the house, and a few moments later a pretty young woman dressed in jeans and a pink T-shirt with sequins at the neckline opened the door.

"Dr. Hazlitt," she said without smiling. "It's good to see you again."

"Hello, Maritza," he said. "I'm sorry we get to meet again under such circumstances."

"Yes," she said solemnly and stepped back inside to allow us to enter. Although it was another unusually chilly day in Tampa, the air-conditioning was going full blast, the way it had been in the limo the other day. I had the macabre thought that a body could be preserved in the house for a very long time, but Maritza seemed comfortable in her T-shirt. I was glad that I'd worn a sweater.

We followed her into the living room, where I was surprised to see Ivelisse Vasquez sitting by a window, a red blanket wrapped around her, her attention directed at the outdoors.

"Mami, it's Dr. Hazlitt and his friend," Maritza said.

"Jessica Fletcher," I provided.

"And Jessica Fletcher," Maritza said.

Mrs. Vasquez turned slowly and appeared to be trying to focus on us. We approached, and Seth extended his hand, which she took.

"Hello, Mrs. Vasquez," I said, also taking her hand, which felt like holding a delicate bird.

"Thank you for coming," Ivelisse said, her voice weak. "Did you know my husband?"

Seth and I glanced at each other before Seth said, "Yes, we met in Cuba, and I've spent time with him here in Tampa."

"Oh, yes, of course," she said, returning her gaze out the window.

Seth gestured to a footstool next to Ivelisse, but I indicated that he should sit instead. I stood behind him, acutely aware that I was not a close family friend.

"He's gone, isn't he?" Ivelisse said absently.

"Al?" Seth said. "Yes, he's gone. Jessica and I are sorry for your loss."

Maritza, who'd remained at a distance, approached and said, "Time for some rest, Mami. You need to rest." She gently helped her mother stand. Seth stood, too, and secured the blanket, which had started to fall from Ivelisse's shoulders. Maritza led her mother from the room. "I'll be back in a minute," she said.

"I wonder if Xavier is here, too," Seth said, looking around.

"I don't suppose he's making funeral arrangements yet," I said. "The medical examiner will want to hold the body until the lab results come back."

Seth muttered something in response and pursed his lips.

"Do you think that the ME's finding that he wasn't killed by lightning has been reported to the family?" I asked.

He shrugged. "Under ordinary circumstances, it would be, but since the ME doesn't have a definitive cause of death yet, he might hold up informing the family until he does."

Maritza reappeared.

"I don't think that my mother has accepted the reality of my father's death," she said.

"Sometimes it takes a very long time for that sort of reality to set in," I said. "I know it did when my husband, Frank, died."

"Has Mr. Peters come by?" Seth asked.

She made a sour face. "No, he hasn't," she said sharply.

"Is your brother here?" Seth asked.

"No. Xavier flew to Key West this morning. He had business to attend to there. He said he'll be back tomorrow."

"Would it be okay with you, Maritza, if I spent some time in your father's office?" Seth asked.

I wondered what her reaction would be to his directness. Would she find it an untoward request?

"Sure," she said. "I'll show you where it is."

"You don't have to bother," Seth said. "Your dad and I spent some time there together."

"No, I'll go with you," she said as she led us down a hallway past bedrooms to the rear of the house. "Papi thought so highly of you, Dr. Hazlitt."

"I'm flattered that he did. And he was very

proud that you are following in his footsteps. Are your studies going well?"

Maritza stiffened. "I'm not following in his footsteps," she said. "My studies are in a completely different area." As we entered the office she asked, "Is there anything in particular you are looking for, Dr. Hazlitt?"

"No," Seth said, "I just wanted to think about your dad in this setting. He always seemed especially comfortable here."

"He always needed a retreat, as he called it, a place to escape, to get away from everything and everyone."

"Aside from your mother's shock at his death," Seth asked, "has she been all right otherwise?"

Maritza's raised eyebrows, and the stream of air that came from her lips, answered the question.

"Lately, she's seemed to be, well, a little forgetful," Seth said.

Maritza shook her head and straightened her shoulders, "Mami is fine, just fine," she said. "She has what you call 'senior moments,' that's all." She forced a laugh. "I guess it comes with getting older."

I knew what Seth was thinking — that in Ivelisse's case, age in itself didn't account for her slippages in memory, and that as a medical student Maritza would likely be attuned to her mother's symptoms. However, it was clear that she didn't want to believe her mother was failing, or in any case talk about it. "Was it difficult for you to get permission to leave Cuba?" I asked her.

"No. Despite what too many people think, the Cuban government respects when a family member dies, even when it's in another country. No, I got my card right away; I didn't have any trouble at all."

"I'm pleased to hear that," I said, a little surprised that it had been so easy for her to obtain a white card, allowing her to leave the country. From what I'd read, permission was routinely denied most people, no matter how valid their reasons for traveling out of Cuba.

"I have some things to do," Maritza said. "Make yourself at home. I'll be back in a moment."

"I didn't think she'd be so defensive about the Cuban government," I said after she was gone.

"I suppose she's right about what other people assume about Cuba," Seth said as he went behind Vasquez's massive desk, its edges inlaid with small, colorful tiles, and opened a drawer.

"Seth," I said, "I don't think you should be —"

He put his finger to his lips. "Just stand by the door and wave if you see her coming back."

It ran through my mind that I was aiding and abetting something, if not illegal, then certainly questionable. But I did as Seth suggested while he quickly opened and closed drawers in the desk. The last one he opened was the middle drawer. "What's this?" he said.

"What's what?"

"This envelope. It's addressed to me."

I went to him and looked at the envelope. The handwriting said, "Dr. Seth Hazlitt."

"It's Al's handwriting," he said as he slipped it into the inside pocket of his tweed sport jacket.

"Seth, I don't think you should do that."

"Why not? It's addressed to me, in Al's own hand. He obviously meant it for me, so I'll take it."

Seth got up from behind the desk and opened a closet, scanned what was in it, shook his head, closed it, and said, "Nothing." He returned to the desk and slumped in the large leather swivel chair behind it.

"What were you looking for?" I asked.

"Al's laptop, of course," he snapped.

"It's obviously not here," I said, "at least not in this office, but I'm not surprised. It's silly to think we could just come in here and find it waiting for us. We don't even know if he brought it here the day he died. Dr. Sardina said he took it home with him every night, but maybe he's exaggerating. Besides — and I don't mean to be critical, Seth — even if we do find the laptop, it doesn't belong to us. It would be the property of his estate, or belong to Bernie Peters."

Seth heaved a big sigh. "I know, I know. You're right, Jessica, and I wasn't intending to keep it. But if Al achieved some sorta breakthrough with his research, it has to be put in the right hands, people who can carry it further and put an end to Alzheimer's. Bernie Peters should have the results in his hands. He financed it. What bothers me is that if Al *did* bring the laptop here every night, then where is it?"

"One of his family members may have taken it," I offered.

"Makes sense," he said, "but what do *they* intend to do with it?"

Our conversation was interrupted when Maritza returned.

"I have a question, Maritza," Seth said.

"If I can answer it, I will."

"Your father used to bring his laptop home with him from the lab."

"Yes?"

"He showed me entries he'd made the last time I was here in Tampa."

"I don't understand the question," Maritza said.

"Well," Seth continued, "I was wondering whether you know where it is."

"Where what is?"

"His laptop computer."

"How would I know? I live in Cuba. I have no idea what my father did or how he ran his research. A laptop? I never heard anything about it."

Her denial of knowing anything about the laptop didn't ring true to me.

"Maybe Xavier would know where it is," Seth said.

She shrugged.

I was about to suggest that we leave when the doorbell sounded. Maritza jumped up to answer it. In her absence, I walked to a small table in a corner of the office where something had caught my attention. Idly, I picked up a thick brochure and unfolded it.

"What are you looking at?" Seth asked.

"This is an aeronautical chart," I said, "a sectional chart, actually."

Seth looked over my shoulder at the paper I held. "Pilots use these?" he asked.

"Yes. They give you everything you need

180

to know about navigation, airports, radio frequencies to use, the height of obstacles like radio towers and mountains, all the essentials."

"Looks like a lot of gobbledygook to me," he said.

"They're not as confusing as they look once you become familiar with them."

"Maritza said that her brother was flying to Key West. Wouldn't he need them?"

I checked the date on the chart. "This one is out-of-date," I said. "I'm sure he has new ones."

I refolded the chart and laid it on the table. "Seth, I think it's time we left, don't you?"

"I do," he agreed, and we turned toward the office door.

"Wait a minute," I said, and returned to the table. "I just realized something."

"What?"

"This sectional is for Cuba."

"And?"

"I just wonder why he has a sectional chart for Cuba."

"You'll have to ask him when he gets back," Seth said, and led us down the hallway to the foyer, where Maritza stood at the front door talking to Dr. Sardina's wife, Ofelia.

"I just thought I'd stop in and see how Ivelisse is doing," Ofelia said.

"As well as can be expected," Maritza said. "She's resting now."

"I can wait." Ofelia walked past Maritza and settled in the living room.

"Ofelia," Maritza called after her, "is something wrong?"

"No, nothing's wrong. Did they — ?"

Maritza nodded.

"I miss him already," Ofelia said.

"Has your husband gone away?" I asked.

"Can I get you a cold drink, Ofelia?" Maritza said quickly, and I had the feeling that she wanted to divert the conversation to a different topic.

"That would be nice," Ofelia replied.

When Maritza left the room to get Ofelia her drink, I said to her, "We had a pleasant lunch with your husband. He took us to an Asian buffet restaurant and —"

"He told me," Ofelia said.

I wanted to ask about her husband's negative view of Vasquez, which I assumed he'd shared with her, but wasn't sure how to approach the subject. Instead, I said, "There seems to be some concern about Dr. Vasquez's research and how far he'd gotten with it. I imagine your husband will be

182

asked about that many times in the coming days."

"Yes, I suppose he will," she said. She turned to Seth. "Did Pedro tell you much about the research, Dr. Hazlitt?"

"Can't say that he did," said Seth.

"But Dr. Vasquez did."

Seth thought before answering. "Yes, Al — Dr. Vasquez shared some of his results with me. Your husband was — how shall I say it? — your husband seemed to be disappointed in how little was shared with him."

"Oh, that's not true," Ofelia said, straightening as though to enhance her denial. "Alvaro always took Pedro into his confidence."

"I was mistaken, then," Seth said.

"Your husband's away?" I asked again, this time without Maritza interjecting herself.

"Just for a day. He'll be home soon."

"Business?"

"Ah, yes. He's on business."

"Having to do with the research?" I managed to ask before Maritza reappeared carrying a tall glass of what looked like lemonade.

"Oh, thank you so much," Ofelia said, taking the glass and sipping the drink. "This tastes like fresh lemons. I forget how good it can be with fresh lemons."

"I always use fresh lemons," Maritza said, taking a seat next to the other woman.

"We should be going," Seth said, taking my arm. "Please contact us at the hotel if there's anything we can do."

We got in the yellow rental car and pulled away.

"What do you think?" Seth asked.

"It's obvious that Maritza didn't want us asking questions about where Dr. Sardina has gone. It struck me that Xavier Vasquez and Sardina are both gone for the day. Do you think that — ?"

"That Sardina went to Key West with Xavier? Certainly a possibility."

"Where are we going?"

"Back to the hotel, where I can read what's in that envelope Al left me."

We retraced our trip over the bay and onto the mainland. As we approached a stoplight, Seth swerved suddenly into the adjoining lane to avoid a pothole in the road, and stopped. A vehicle behind us also swerved but in the opposite direction, and ended up stopped at the light in the lane we'd previously occupied. I glanced over at the occupants of the dark blue car. I couldn't see the driver clearly, but the person in the front passenger seat was a young man wearing the same sort of gray hoodie that I'd seen on the man at the streetcar stop in Ybor City. He glanced at me for a second before turning away quickly so that his face was obscured.

"Seth," I said, "that's the same young fellow . . ."

The light changed and the blue car pulled away sharply and sped ahead of us.

"What say, Jessica?" Seth asked as we slowly moved forward.

"The passenger in the car next to us was the same young person I'd seen in Ybor City, the one who I was certain was eyeing me."

"Is that so?" Seth said, and accelerated.

I put my hand on his arm. "Please don't try to follow him, Seth. He's going much too fast."

"Not giving me a chance to be a real detective," Seth said, but he took my advice and slowed down.

We parked the car when we reached the hotel and settled on a sofa in the far end of the lobby. Seth pulled the envelope from his pocket, opened it, and unfolded the single sheet of paper it contained. His face was set in a hard scowl as he read, and even though we were in a public place, I wished he'd read aloud for my benefit. When he finished, he handed the paper to me. It was a typewritten letter to him from Dr. Vasquez.

My dear friend and brother-in-arms Seth:

It is my hope that you shall never read this letter, for if you do, it will mean that I am no longer alive. You see, my new-found best friend, I am surrounded by enemies, people on all sides of me, people who will not shed a tear should I meet with an unfortunate accident, or die from other than God's will. I can trust no one, not colleagues in the medical profession, not businesspeople, not even my own family.

As you know, my life has been devoted to finding a cure for Alzheimer's disease. It wasn't always easy in Cuba. Funds were scarce, and I had to beg, borrow, and sometimes even steal to keep the research going. When I had a chance to leave Cuba and continue my work in the United States, I jumped at it, especially since the work I'd done in Cuba had enticed the pharmaceutical company K-Dex to commit to generous funding for me. Bernard Peters has been my lifeline, although I must say that his motives as well as his larcenous character have disappointed me greatly of late.

I have been meticulous in documenting my research in the hope that one day what is contained in my documentation will convince others to carry on when I am gone. But because of my distrust of others, I am placing a burden on you, Seth, as a man of character, and in whom I see a totally honest and straightforward human being. You would be excused for wondering why I would place such trust in a man with whom I have spent so little time. Call it instinct. Call it intuition. Call it what you will, my friend, but whatever it is, I trust it.

I have written this letter after our aborted golf outing with your beautiful and charming friend Jessica Fletcher, and intend to give it to you this evening at the end of what I trust will be an enjoyable evening of good music, good drinks, and good conversation. I will, of course, instruct you to not open it while I am alive. But should I die, I ask you to retrieve the copies I have made of what has been entered on my laptop regarding the progress in the laboratory. I have regularly copied my notes to these little devices — thumb drives — and have carefully stored them. They are located in Tampa Mini-Storage on S. MacDill Avenue. The code for my combination lock on unit number 61 is 7-2-9.

I know that you'll do the right thing, Seth. I am counting on it. Of course, I won't be around to thank you for taking on this weighty task, so I do it now, profusely.

Your friend,
Al

I finished reading and looked at Seth, who rubbed his eyes.

"This is remarkable," I said.

He nodded.

"To have generated such trust and respect in such a short time is — well, it is nothing short of — remarkable."

"I'm at a loss for words," he said.

"I understand," I said.

Neither of us said anything for a few moments. Finally, I said, "There's something to this beyond his obvious trust in you."

Seth cocked his head.

"Based upon this letter, I'd say that there's a strong chance that your friend Al was . . ." I hesitated.

"Yes?" Seth said.

I took a deep breath. "That he was murdered."

# CHAPTER THIRTEEN

We made some wrong turns as we headed for the self-storage facility — by his own admission, Seth can be directionally challenged at times. Eventually we found South MacDill Avenue and drove along until arriving at a modern commercial building with a large red sign, TAMPA MINI-STORAGE.

"Do we have to check in with someone?" I asked as Seth found a parking space.

"Probably not. I've rented space in the storage facility outside Cabot Cove, and unless you want to use one of their carts, you just go to your unit. Should be the same here. Besides, it's not like we're breaking any laws. We have the combination to the lock. Just act like we belong here. They must have plenty of customers and can't remember what everybody looks like."

Buoyed by his confidence, we crossed an impressive, up-to-date lobby, gave a wave to two attractive uniformed people behind the desk, and headed straight for the area where the rental units were located. We walked down a long hallway with an immaculately clean floor. On either side of us, the walls were lined with shiny corrugated metal doors marking the storage lockers.

"Chilly in here," Seth said with a shiver as we searched the numbered doors until we reached unit sixty-one. We looked up and down the hall. We were alone. Seth took a slip of paper from his pocket on which he'd written the combination. He rotated the lock's dial, first all the way around, then to the right to the numeral seven, then left to two, and finally to the right again to nine. One end of the hasp pulled free of the lock. He slipped the lock off and opened the door. Lights had automatically come on when we entered the hallway, and an overhead light in the unit was also on.

The unit was large; the contents of a studio or small one-bedroom apartment would easily have fit. But the only thing in it was a small round table on which a smoked plastic box rested, like a featured piece of art in a museum. That was it. Nothing else was in the pristine space.

We heard footsteps in the corridor and I softly closed the door. Seth approached the table and ran his fingers over the top of the box. He glanced at me before lifting the lid and peering at what was inside. I stood next to him and shared his view. There were three black devices about an inch and a half long and three-quarters of an inch wide.

"What do you call these things?" Seth asked, picking one up.

"Flash drives," I said. "Or thumb drives. I use them all the time to back up my files."

"Thought you did that on those floppy disks."

"No one uses floppy disks anymore," I said. "Thumb drives are a lot easier to use and hold more data. Dr. Vasquez must have taken what was on his laptop and transferred the information to these."

"Can't believe those little things would hold all that much."

Seth's inherent interest in and understanding of the latest medical research doesn't translate to computers. He's always complaining about his computer and his inability to master even its most rudimentary processes. Fortunately, he has a local computer technician in Cabot Cove who's ready to rush to Seth's aid at a moment's notice, and a nurse in his office whose skills are better than his.

"What do you intend to do with them?" I asked.

"Take 'em with me, I guess."

"Maybe it's better to leave them here," I said.

He pondered my suggestion before saying, "I want to know what's on them."

"You'll need a computer for that."

"You didn't bring yours?"

"Of course I did. I always travel with my laptop. You can plug the thumb drives into my computer and store what's on them on my hard drive."

He took the three thumb drives and put them in the side pocket of his jacket.

"Why not take the box?" I asked.

"Don't see any use for it," he replied. "Let's go."

We peered out the door to be certain the corridor was empty, then relocked the door and walked confidently through the lobby and to the car.

"Seth," I said.

"What?"

"Dr. Vasquez's letter to you. The police should know about it. He obviously feared for his life to have taken the time to write the letter and secure the thumb drives the way he did. We should tell Detective Machado."

"I don't know, Jessica. I wouldn't want anyone else to know that I have his research notes. Seems to me it's nobody's business except mine, until I decide what to do with them."

"I understand what you're saying, Seth, but you can't keep from the authorities that Vasquez was afraid that something would happen to him. It could be crucial to their investigation." What I didn't say was that if someone killed Alvaro Vasquez for the research notes, our lives could be in danger, too.

Seth drove faster than usual in the direction of the hotel.

"You know how to transfer information from one of these little thumb drives to another one?" he asked.

"Yes, it's easy. You plug one into my laptop, take the information that's on it and save it on my hard drive, then plug in a blank thumb drive and transfer the information back onto it."

"Tell you what," he said. "Let's stop and buy three new thumb drives, swing past the hotel, do what you say you can do, make me a second set, and then I won't mind sharing the letter with the police as long as I have a copy of Al's research notes."

Which is what we did. We purchased the drives from a Staples store we passed, went to my room at the hotel, transferred the material, and in less than an hour were back in the car with the originals in Seth's jacket pocket, the copies in his room's small safe.

We had as much trouble finding police headquarters as we did locating the self-storage facility but finally pulled into its parking lot and entered the building. A heavyset uniformed black man sat behind a Plexiglas shield. An open area beneath the shield allowed small items to be passed back and forth; it reminded me of a bank.

"Is Detective Machado in?" I asked.

"What's it about?" the officer said through a speaker in the Plexiglas.

"Dr. Alvaro Vasquez's death. I'm Jessica Fletcher, and this is Dr. Seth Hazlitt."

He picked up the phone and informed the detective who we were and that we wanted to see him. "Take a seat over there," he said, pointing to a wooden bench. "He'll be out in a minute."

Too nervous to sit, Seth and I perused various notices hanging on the walls. One sign proclaimed that the Tampa Police Department was able to translate English into thirty-eight other languages, including Yiddish, Polish, Haitian, Creole, Hindi, Serbian, Tagalog, Urdu, Dutch, Czech, Chamorro, Arabic, and Farsi.

"What in the world is Chamorro?" Seth

mused aloud.

"I think it's a language spoken in the Pacific on places like Guam."

"You would know that," he said.

"Just don't ask me to speak it," I said.

Detective Machado appeared as we were admiring a large, colorful graphic of a speeding train cutting through a street map for District One, with the headline: CUTTING THROUGH CRIME. HAVE YOU GOT YOUR TICKET? Accompanying the detective was a man whose face was familiar, although I couldn't quite put a name to it.

"Mrs. Fletcher?" the man said.

"Yes. I know you from somewhere, but —"

"Carlos Cespedes," he said. "We met at Alvaro Vasquez's party."

"Yes, of course. You own a cigar factory. How are you?"

"My cigar factory isn't so big anymore. People smoke fewer cigars these days, more cigarettes. But this is a minor complaint." His already elongated face became more so. "How tragic," he said. "To think that our friend is no longer with us is — it is so unfortunate, so *triste,* sad."

"It certainly is," Seth said.

Machado said impatiently to Seth and me, "Is there something you want to talk to me about?" He looked fatigued.

Mr. Cespedes pulled a business card from his shirt pocket and handed it to me. "Please come visit me at my shop and factory," he said. He lowered his voice and added, "I must speak with you about Alvaro's death."

Cespedes went through the door, looking back nervously.

"You know him, huh?" Machado said.

"As he said, we met at Dr. Vasquez's party," I replied, wondering why he was at police headquarters conferring with the detective, and what he wanted to discuss with us about Vasquez's death.

"Come on in," Machado said. "Your timing is good. I intended to call you anyway."

He led us to an empty room that I assumed was used to question suspects. It had the requisite large window, which I was sure was a two-way mirror. The table was scarred and stained from countless wet cups, and probably cigarette burns from a time before smoking was banned indoors. The four wooden chairs were spindly. Seth and I sat on one side of the table; Machado took a chair opposite us.

"Tell you why I was going to call," he said. "I'm sorry to have to tell you that we're treating Dr. Vasquez's death as a possible homicide."

"Oh?"

"Looks like the victim wasn't struck by lightning after all, according to the ME."

Neither Seth nor I said anything. I knew that Seth was thinking what I was — that it might not go over well for the lead detective on the case to know that the medical examiner had told us that Vasquez wasn't hit by lightning before he'd been informed.

"Then what *did* kill him?" Seth finally asked.

Machado shrugged his large shoulders and rolled his fingertips on the table. "They're doing the toxicological studies as we speak. Of course, it could end up that he had a heart attack or a stroke, only the ME says he doesn't think that's the case. He told me that he had a conversation with you, Dr. Hazlitt."

"Ayuh, that's right, we did have a chat."

"Sometimes the ME and us here in law enforcement don't always connect the way we should. What'd he tell you?"

Seth was on the spot, and I wondered how he'd handle it.

"He didn't have too much to say," was Seth's reply. "We sorta had a medical conversation, you know, doctor to doctor." He smiled.

Machado did not. "I thought that because you were close to the victim, witnessed when he died, and are a physician, that Dr. San Martín might have told you things that he hasn't passed along to us yet."

I said, "Excuse me for injecting myself, Detective, but why would your medical examiner keep things from the police? That strikes me as highly unusual."

Machado, who wasn't a man for whom smiling came easily, managed one and said, "Doc San Martín sometimes gets his back up when it comes to passing on information. Doesn't like to be rushed. Plus, like any PD, we occasionally have somebody leak something to the press that shouldn't be leaked. The doc has raised holy hell about it more than once. For me, I think he's wrong to take it on himself to decide what to give us when and what to keep under his hat, but there's not a lot I can do about it." Seth started to respond, but Machado added, "And I'll tell you another thing. The PD isn't the only source of leaks. The ME's office has had its share, too." Machado's voice mirrored the fatigue on his round face. He sounded very much like a man under siege, someone in need of a vacation. I knew how he felt.

"This is all very interesting," Seth said,

"but you said you intended to contact us about it. Why?"

"A couple of reasons. First off, Dr. Hazlitt, I'm told that you and the deceased were pretty chummy."

"Friends? Yes, we were," said Seth.

"Good enough that you made multiple trips to spend time with him."

"I don't know that I would call four trips over several years 'multiple.' I was simply visitin' a friend."

"Pretty expensive flying back and forth — to visit a friend."

"Not terribly expensive," Seth said. "Wouldn't you spend some money to enjoy time with a good friend?"

"Oh, sure, and don't get me wrong. I'm not accusing you of anything. It's just that because you and the victim were close, he might have told you things that would help me with my investigation."

It was the perfect time to bring up the letter that Vasquez had left for Seth, or at least to mention Vasquez's fear for his life. Seth picked up on the cue.

"Matter of fact, there is something you might find helpful."

Machado came forward in his chair and picked up a pen from his desk. "Shoot," he said.

"Al — Dr. Vasquez feared for his life."

"Whoa," Machado said. "When did he tell you that?"

"He, ah — well, to be honest with you, Detective, he wrote it to me in a letter."

I smiled at Seth. I was relieved that he'd decided to be completely honest and not to try to slant the story.

"When did he write you this letter?"

"A few days ago."

"And you knew about it the day he died and didn't come forward with it?"

"He wrote it a few days ago, but I didn't receive it until this morning."

Machado wrote something on a lined yellow pad before asking, "He mailed it to you here in Tampa? Where? Your hotel?"

"No, I —"

Obviously Seth didn't want to admit that he'd taken it from Vasquez's home office without permission, but he answered the question. "He left it for me at his house. Mrs. Fletcher and I were there this morning."

"Wait a minute," Machado said. "He dies a couple of days ago but leaves a letter for you at the house?"

"That's correct. He said that he intended to give it to me the night he died, after the party Mrs. Fletcher and I attended. Unfortunately he didn't live long enough to follow through."

"Okay," said the detective, "I think I understand. Where is the letter?"

"I have it right here," Seth said as he fished it from his inside jacket pocket and handed it across the desk.

Machado opened the envelope, placed a pair of half-glasses on the bridge of his nose, and read. When he was finished, he said, "He doesn't say who he thought might kill him. Did he give you any names?"

"No," Seth said. "This was the first I learned of his concern."

Machado said to me, "Where do you fit into all of this, Mrs. Fletcher?"

"I really don't fit in at all," I said, "aside from being Dr. Hazlitt's friend."

"You're more than that," Machado said.

"Meaning?"

"I Googled you. From what I read, you not only write big bestselling murder mysteries; you've been involved in more than a few real ones yourself — at least according to newspaper reports I got off the Internet, lots of stories about some of those cases."

"It's nothing I brag about," I said.

"Hey," he said, "I'm impressed. If you ever quit writing for a living, maybe you could become a cop."

"I think it's a little late for a change in career, Detective Machado."

"No offense," he said. "Look. If the ME is correct, that Dr. Vasquez might have been done in by someone else, I've got a real hot potato of a case on my hands. Vasquez was well-known in Tampa and controversial, too. My boss, Major Stacks, is already getting pressure from local Cuban American groups."

"Why?" I asked. "As far as the public knows, Dr. Vasquez was killed by a lightning strike."

"Tampa's no different from anyplace else," Machado said. "We've got our share of conspiracy buffs. Word gets around that some people think that maybe Vasquez wasn't killed by lightning and their paranoia shifts into high gear. Our liaison officers who work with the Cuban community tell us that some people are convinced that Vasquez was killed by our government to get his research and keep it out of Castro's hands. We've also got pressure from the feds."

"The FBI?"

"Right. So my boss tells me in no uncertain terms to wrap this up."

"Why would the FBI be interested in a local death?" I asked, hoping I wasn't sounding too naïve.

"Because of who the victim was, Mrs. Fletcher. Vasquez defected from Cuba. It made all the papers. He supposedly was sitting on some medical breakthrough that the Cubans would want back, and the feds obviously want to make sure that that doesn't happen."

"I certainly understand," Seth said. "That's why we stopped by, to let you know about the letter."

Machado glanced at the sheet of paper again. "So Vasquez tells you that he left you copies of his research. Did you go to this self-storage place?"

"Yes." Seth pulled the three original thumb drives from his jacket and handed them to the detective.

"It's all on these?" Machado asked.

Seth nodded. "As far as I know."

"You haven't looked at them?"

Seth shook his head, and I knew he was relieved to be telling the truth.

"What else is in that storage place?"

"Nothing," I answered, "just a small table and the box in which these were housed."

"You take the box and table from there?"

"No," Seth said. "We left them."

Machado exhaled a long, poignant stream of air and stood. "Let's go," he said.

"Where?"

"The storage place."

"Is it really necessary for us to go with you?" I asked.

"Can't make you, but I'd appreciate it."

A half hour later we were walking down the hallway toward space number sixty-one, whose door was wide-open.

"You locked it when you left?" Machado asked.

"Ayuh, I'm sure I did," Seth said.

We stepped inside. The smoked plastic box was gone; only the table remained.

Machado abruptly turned and left the room, with us following. We went to the lobby, where Machado flashed his TPD badge to a young woman at the desk. "I need to speak with the manager," the detective said. A few minutes later, a man appeared and introduced himself, checked Machado's badge, and said, "Not again."

"What do you mean by that?" Machado asked.

"You're the second cop — law enforcement officer — who's been here this afternoon."

"Who was the other one?"

"FBI."

"FBI? You're sure?"

"He had his credentials. He wanted us to

open one of the lockers."

"Number sixty-one," I said.

"That's right. He said that the renter was deceased — Dr. Vasquez. I'd read that the doctor had died, hit by lightning, so I opened the locker and this FBI agent took what was in there, not much, just a table and a box on it. He took the box."

The manager accompanied us back to the open storage space, and Machado looked around. There was nothing to see aside from the table, which he dismissed with a cursory running of his hand over it. He thanked the manager and we retreated to Machado's unmarked car.

"Looks like you were right about the FBI being interested," Seth said as we drove back to police headquarters.

"The agent must have been after what Dr. Vasquez wanted Seth to have, the thumb drives he gave you," I said.

Machado grunted and drove the last few blocks in silence.

"Appreciate you coming in," he said to us as we prepared to get into our yellow rental car.

"What will you do with those thumb drives we gave you?" I asked. "They contain what could be very valuable information about Dr. Vasquez's medical research."

"We have experts to check them. I want to know whether he names names or says anything about fearing for his life. There's probably nothing, but then again . . . Anyway, thanks again for your cooperation. You can still be reached at the hotel?"

"Ayuh," Seth responded.

"How long do you plan on staying in Tampa?"

"Hard to say," Seth said.

"I'll stay in touch," Machado said, and walked away.

When we'd gotten in the car I said, "Either someone other than Dr. Vasquez knew that he'd rented that storage space, or —"

"Or what, Jessica?"

"Or someone followed us there, and that someone is involved with the government."

We drove a little farther before I asked, "Just how long *do* you plan on staying, Seth?"

He ran his tongue over his lips before answering. "As long as it takes to find out what really *did* happen to Al."

# CHAPTER FOURTEEN

Seth was eager to get back to the hotel to read what was on the thumb drives, but I suggested that we first stop at Carlos Cespedes's cigar shop and factory. "He said that he wanted to discuss something about Dr. Vasquez's death," I said. "As long as you're committed to getting to the bottom of it, we should follow up on every possible source of information."

"Is that what you'd have people do if you were writing this as a novel?" he asked.

"Yes, I suppose I would," I said.

"Makes sense," he said.

I retrieved Cespedes's business card and read off the address to him, then consulted the street map that came with the rental car and gave directions.

Cespedes Fine Cigars was located in downtown Ybor City. The owner's business card billed it as a cigar factory and shop, although from the looks of it the use of the term "factory" was a misnomer. In reality, it was no more than a storefront with two small tables and a few chairs on the sidewalk in front. A sign next to the door offered coffee, cold drinks, and "Authentic Cuban Pastries." An older woman and a younger man occupied one of the tables. Both had large white mugs in front of them, and both were drawing on cigars.

We found a parking space in a lot across the street. As we got out and waited for traffic to clear before crossing, a family of four, mother, father, and two youngsters, emerged from Cespedes's shop, each carrying a small plastic shopping bag.

"You don't figure they bought cigars for the kids, do you?" Seth mused.

"Only chocolate ones, I hope," I said as we took advantage of a break in traffic and walked across.

The door to the shop was open, and we entered. A long counter to the left held a cash register and clear plastic boxes containing an assortment of cigars. Stacks of colorful cigar boxes filled shelves mounted on the wall. A young woman sat on a stool browsing through a magazine.

*"Buenos días,"* she said.

*"Buenos días,"* I replied. "Is Mr. Cespedes in?"

She pointed to the rear of the shop and went back to reading.

Our view of the back of the shop was obscured by a row of large barrels in which tropical plants bloomed. Once we reached them, we could see beyond to where two older men sat at wooden tables rolling cigars. They looked up for a moment before returning to their tasks, and I recognized one as Adelmo, who had been making cigars at Alvaro Vasquez's party. I was about to ask for Mr. Cespedes when he appeared through a slit in a red curtain.

"Ah," he said, smiling and coming to us, his hand outstretched, "you came, you came. I am so pleased."

"Thank you for the invitation," Seth said. "So this is your cigar shop."

I judged Cespedes to be in his late sixties or early seventies. A short, balding man with a sizable paunch, he wore a red-and-white checkered shirt and tan slacks and had a malleable face on which he adopted a hangdog look. "It used to be much more, I am afraid. I once owned a whole building here in Ybor City, a real factory. I even had lectors."

My puzzled expression prompted him to explain.

"Lectors," he said. "Readers. While my *ta-*

*baqueros* and *tabaqueras,* the cigar rollers, do their work, the readers sit high above them and read aloud from the newspapers, or short stories. It is a very Cuban thing, very educational, yes?"

"I saw and heard the lectors when I was in Havana," Seth said.

"Ah, Havana," Cespedes sighed. "I miss it."

"You never go back?" I asked.

"Once — no, twice — many years ago. It is very different now that the *imbécil* Castro is there. I would go shoot him myself if I could."

My attention drifted to what the men were doing at their tables.

"Ah," Cespedes said, "too much from me about Castro, huh? You're interested in how my cigars are made."

One of the rollers, or *tabaqueros,* as I now knew, picked up a small rounded knife and banged it on his table. Adelmo did likewise.

Cespedes laughed. "They welcome you with their banging, a Cuban custom."

I remembered that Adelmo had rapped his knife on the table when I'd said hello at the party. At the time, I hadn't realized he was responding to my greeting.

"That is the *chaveta,*" Cespedes continued, "the knife used to smooth and cut the tobacco leaves, smooth the filler tobacco, cut the tips. They say they circumcise the tips, like a baby. See? He rolls the tobacco into a tube that goes into the wooden mold. Then he takes the solid cylinder — we call it the 'bunch' — and lays it on the wrapper and uses a tiny bit of vegetable glue to secure the second wrapper. Then he glues the cap into place and trims any excess tobacco."

"How many can he roll in a day?" I asked.

"For those medium-sized cigars, maybe one hundred, maybe a little more. For the bigger, fatter cigars, not so many." He gave out a plaintive sigh. "Everything is so different now in Ybor City. Once there were a hundred and fifty factories rolling a quarter of a million cigars every year. Now, for me, there is only this." He took in his shop with a wave of his hand. "It is a shame that you weren't here last month for the cigar festival. A team rolled the world's longest cigar, a hundred feet long. You can check it in that Guinness book."

I laughed as I envisioned a hundred-foot-long cigar.

Seth broke in with, "You said that you wanted to talk to us about Al Vasquez's death, Mr. Cespedes."

"Yes, I do."

A family of tourists came through the front door.

"Please, come with me," Cespedes said as he parted the red curtains. "We can talk better in here."

Behind the curtains was a small office. Large posters of famous cigar labels of the past dominated the walls. A calculator surrounded by piles of papers sat on a desk. Family photographs in silver frames were lined up on a table behind the desk.

"You were Alvaro's good friend," he said as Seth and I sat in director's chairs with floral-patterned canvas seats and backs, while he perched on a stool.

"That's right," Seth said. "I assume that you were a close friend, too."

" 'Close friend'? We were friends, acquaintances, and I suppose you could say business partners."

"What sort of business were you and Al in?" Seth asked.

"He never told you? It is hard to explain. As you can see from my shop, what I once had is no longer. Now I make ends meet. But it wasn't long ago that there was plenty of money from when I sold my factory building. The company that bought it turned it into a handsome social club. You can see it on Avenida Republica de Cuba. Once my father had a hundred *tabaqueros* and *taba-queras* rolling cigars in our factory, and the best lectors reading the latest news, and short stories, too, by your Papa Hemingway and Agatha Christie and other great writers. We treated everyone well. Our cigars were among the best in the world. Now members of the club drink and eat and dance and hold their meetings in what was our building. It is a good place, the club, but I am sad every time I drive by."

It was obvious to Seth and me that Cespedes would get around to addressing Vasquez's death only after he was finished lamenting what had happened to Ybor City and his family's cigar factory. We listened patiently until Seth again asked, "What sort of business were you and Dr. Vasquez in?"

Cespedes didn't hesitate. "His research, of course," he said.

Seth and I looked at each other before Seth said, "The research? How were you and Dr. Vasquez involved together in his research?"

"I invested in it, five hundred thousand dollars, what I received from the sale of my building."

Seth shook his head as though to clear it. "Let me get this straight," he said. "You invested in the research that Dr. Vasquez was doing to find a cure for Alzheimer's disease?"

"Yes."

"But what about — ?" I started to say.

Seth finished the sentence for me. "Bernard Peters."

"Oh," said Cespedes, "Mr. Peters is also an investor."

"Does he know about *your* investment in the research?" I asked.

"No, no," Cespedes said. "It was very important to Alvaro that my investment be kept a secret from everyone, anyone. I have never even told my wife."

"Do you know if there were other independent investors like you?" I asked.

Cespedes shrugged. "Probably. Alvaro, he needed the money. He liked the high life, you know? The boats, the women, the good food. He liked to entertain his friends. Peters only gave him so much, and he expected it to go toward the laboratory."

"What were you to receive for the half million dollars you invested?" I asked.

"Ten percent. I wasn't sure whether to make the investment. Those savings were all I had. But to be on the ground floor, as you say, of a cure for that terrible disease, was a privilege I could not pass up. Alvaro told me that he was only allowing very few people to invest in his work."

"I assume you had a good written agreement with him," Seth said.

"Yes. Alvaro gave me a letter saying that when he found a cure, I would receive ten percent of all the money it would make. He told me never to show the letter to anyone."

"Just a letter?" I said, unable to keep incredulity from my voice. "And you trusted him that much?"

"Yes, of course," Cespedes replied. "He was my countryman — he was a man of great reputation, a man whose character was above reproach. He was — he was Cuban, my friend."

"This is what you wanted to talk to us about?" Seth said.

"Yes. You are a medical doctor and Alvaro's close friend. I asked Dr. Sardina about the research and whether Alvaro had found the cure."

"And what did he tell you?" I asked.

"He said that it would be some time before the results could be evaluated. Yes, that is what he told me. I thought that maybe you could tell me more."

Seth grappled for an answer, finally saying, "I don't have anything to tell you at the moment, but I might have information later. I'll get in touch with you when I do."

*"Gracias, gracias,"* Cespedes said. "Alvaro told me that you are a fine and upstanding man and medical doctor."

"And I appreciate his kind words about me. We really should be going."

"Of course, of course. Before you do, may I ask you a question?"

"Go right ahead."

"There is the rumor in Ybor City, with some of my Cuban friends, that Alvaro might have been — how shall I say it? — that he might have been killed by an *asesinato,* not by the lightning."

*"Asesinato?"* Seth said. "Assassination?"

"Yes, yes."

"Who would assassinate him?" I asked.

Cespedes sighed deeply and shrugged. "The DI," he said, "Castro's intelligence agency. They have agents in Florida, many here in Tampa. They want to destroy our CAFA."

"Which is?" I asked.

"Our Cuban American Freedom Alliance."

"Al told me about that," Seth said to me. "It's sort of the Cuban government in exile here in Florida, groups that want to topple Castro and return to Cuba."

"Do you have any proof?" I asked Cespedes.

"Proof?" His laugh was cynical. "No, no proof, but there are the rumors, many rumors."

"Well," Seth said, "there are always rumors. Like I said, Mr. Cespedes, I'll let you know if I find out anything about Alvaro's research. Thanks for letting us drop in. It was — interesting."

Seth was silent as we got in the car and headed back to the hotel.

"What did you think?" I asked as he joined the flow of traffic.

"Gorry, I don't know what to think. I think I'd like to know who *else* invested in Al's research. I'm beginnin' to think that my good friend might not have been as much on the up-and-up as I thought."

# CHAPTER FIFTEEN

Once settled in Seth's hotel room, I plugged in my laptop and inserted the first of three thumb drives, and we settled back to read the words on the screen. One of the first files we opened simply contained a short list of names. Carlos Cespedes was among them.

"Do you suppose these are the others who invested in Dr. Vasquez's research without Bernie Peters's knowledge?"

"Mebbe." Seth's expression was worried. He wrote down the names on a lined legal pad.

There were numerous separate documents, most of them containing a lot of long medical and scientific terms that I didn't understand but knew that Seth did. We read in silence; Seth made an occasional note on his pad.

After a half hour he said, "Stop it there, Jessica."

He rubbed his eyes and paced the room.

"Have you learned anything so far?" I asked.

"Not much, except that one of Al's earlier experiments didn't pan out the way he'd expected."

"It must be frustrating doing medical research," I said. "I imagine there are lots of dead ends."

"That's for sure. Hungry?"

"As a matter of fact, I am."

"Let's order up," he said.

I consulted the room service menu and read off items to Seth, who opted for onion soup and a Crab Louie salad. I was in a mood where making choices was difficult and simply ordered the same.

"I can't make sense out of this," Seth said as he sat on a small couch and massaged his temples. "It doesn't compute for me that Al would sell a stake in his research to someone like Cespedes. How could he do that? Bernie Peters is the one who bankrolled the research once Al came to the States. I'd understand it if Bernie approved having Cespedes provide additional funding, but to do it in secret?"

I had been thinking a great deal about what Seth had said in the car, that maybe his friend wasn't as honest and straightforward as he'd thought — as he'd *hoped.* I didn't want to rub it in by reinforcing that possibility, but Seth spared my having to mention it.

"How could Al do such a thing? Sounds to me like he was selling Bernie out from under." Seth brought his fist down hard on the couch's armrest. "There's got to be a reasonable answer to this, Jessica, because I will not accept that Al was conning people."

"Mr. Cespedes willingly and blindly entered into a business deal with Vasquez that wiped out what savings he'd accumulated from the sale of his building," I said. "He wouldn't be the first person to have been enticed to invest in something fraudulent."

"Well," Seth said, "Al was . . ." His words trailed off.

I knew what he was about to say.

Alvaro Vasquez had defected from Cuba amid much fanfare. His reputation in Cuba was that of a pioneering medical researcher who was on the verge of conquering a particularly devastating disease. He'd settled in Tampa, Florida, and opened a laboratory in which he could continue his much ballyhooed work. It didn't surprise me that Cespedes, and others like him, would have succumbed not only to Vasquez's reputation, but to the Cuban physician's personal

220

charisma as well.

One of many questions I had was whether Vasquez had approached Cespedes, or Cespedes had approached him seeking to invest in his research. And would it have made any difference? The great showman P. T. Barnum once said, "There's a sucker born every minute and one to take him." The movie *The Producers* also came to mind. Zero Mostel plays a shady theatrical producer who sells fifty percent shares in a play to multiple wealthy widows, convinced that the play would flop and none of the duped investors would have to be accounted to.

Had Vasquez played that same con game, selling pieces of his research in the hope that . . . what? What could he have hoped for?

I told Seth what had occurred to me.

His silence said that his thinking was along the same lines.

Our dinners arrived and we watched the news as we ate. It was toward the end of the newscast that a local anchor reported:

Breaking news in the death of Dr. Alvaro Vasquez. A credible but anonymous source in Tampa law enforcement has told this station that the police are now treating the esteemed physician's death as a possible homicide, based upon a report they received from the Tampa medical examiner. Dr. Vasquez, who was originally thought to have been struck by lightning, was a prominent medical researcher who'd defected from his native Cuba and settled in Tampa. The investigation is ongoing.

"Detective Machado was certainly right about there being leaks from the ME's office and his own department," I commented after we'd turned off the television.

"Seems that way," said Seth. "Let's get back to seeing what was on the laptop."

We spent hours more reading Vasquez's entries, and I had to fight to maintain interest. The words became a blur at times, and I excused myself now and then to go splash cold water on my face. When the final entry had been read, and Seth had finished making notes, he sat back, rubbed his eyes, and said flatly and wearily, "He failed."

"What do you mean?"

"Al failed, Jessica. According to what he's written — and he's written plenty, as you can see — every effort he made to find a breakthrough in definitively linking sugar and insulin resistance to Alzheimer's was unsuccessful. He outlines several more research pathways to follow up on, but to date, he's got nothing."

I didn't know what I had expected to learn from reading Vasquez's notes, but it certainly wasn't what Seth had just announced.

"You're sure?" I asked. I was well aware that Seth was highly knowledgeable when it came to medicine, but maybe — just maybe — there were aspects of Vasquez's research that were beyond his expertise.

He nodded glumly. "Yes, I'm sure," he said, "unless there were trials he ran that aren't represented on these thumb drives. But I don't think so. His last entry was the day before he died."

"He knew that his research had failed when he hosted the party at his house and wrote you that letter. But it sounded from what he wrote that he still felt that what he had achieved might be worthwhile in the hands of others."

"Ayuh, he did indicate that, Jessica. And it's not unreasonable. Knowing what blind alleys a researcher has gone down can save other researchers a lot of pointless planning, not to mention time and money. But that's not what's on my mind at the moment."

"What is?"

"Whoever killed him — and I'm assuming that the ME and the police are right in considering his death a homicide — did that person know that Al was engaged in these financial shenanigans, and even more important, did that person know that Al's research didn't pan out?"

"You mean, if someone like Cespedes suspected the money he gave Al was squandered, and that he wasn't going to share in the profits from the lab's research, that would give him a strong motive for murder?"

"It's something to think about," Seth said.

"Well, Cespedes may have had an inkling something was wrong, but evidently he didn't know for sure," I said. "He was hoping to learn more information from you. We should look up the other people on that list we found in the file and find out if any of them will admit to giving Al money on the side."

"I have their names," Seth said, drawing a circle around them on his legal pad. "But there's another scenario to consider."

"What's that?"

"We may be the *only* ones who know that Al was unsuccessful. Maybe the person who killed him thought the key to a cure was in here." He bounced one of the thumb drives up and down on his palm. "That person could have wanted Al out of the way, in order to benefit from the research."

"Bernie Peters?" I blurted out.

"Maybe. All I know is that I'm not leaving Tampa until I find out the answer."

I looked at my friend of many years. He looked weary. His color was gray, and his voice had lost some energy. But I also saw in his eyes a steely determination.

I didn't know what was in store for me over the coming days in Tampa, but I was sure that it would be nothing like the idyllic life I usually led back in Cabot Cove. And I also knew that I'd be there for Seth no matter where his investigations led us.

That's what friends are for.

# CHAPTER SIXTEEN

"Sleep well?" I asked Seth the following morning when we met for breakfast.

"You always did have a sense of humor, Jessica," he said in a voice deepened by a lack of sleep.

"I withdraw the question," I said. "I didn't sleep well either."

"No reason you should if I didn't. Seen this?"

He handed me that morning's edition of the *Tampa Tribune,* opened to an inside page. I read the headline: "Intrigue on Davis Island: Dr. Alvaro Vasquez a Murder Victim? Police Think So."

The article took up the full page and was accompanied by a photograph of Vasquez.

It was written by Peggy Lohman, the reporter who'd come to our table at the hotel a few days earlier.

"No need to read it," Seth said. "Nothing new in it, plenty of background on Al's career and his defection from Cuba, lots of quotes from anonymous sources in the police department. She did interview Sardina."

"So he's back," I said. "What did he have to say?"

"Not much. The reporter asked him how Al's research was going, and Sardina said that it was going fine but that he wasn't in a position to discuss it."

"Spoken like a politician," I said. "Did he say anything else?"

"Only that Al was a wonderful man and mentor and that he missed him and expressed his condolences to the family."

"Not exactly what he told us."

"We weren't the press."

"I'm concerned about these leaks from the police. Have you heard more from Dr. San Martín?"

"As a matter of fact, I have. Had a strange call from him first thing this morning. He wants to meet with us today."

"Did he say why?"

"No, but he doesn't want us to come to his office. We're meeting him at some restaurant outside of downtown called the West Tampa Sandwich Shop. One o'clock. I have the directions."

"That *is* strange," I said.

"Everything about Al's death is strange. I'm concerned about the set of thumb drives that I gave to Detective Machado. Seems like the police department is a sieve. Nothing's secure. Maybe I shouldn't have left them with him."

"You didn't have any choice once he read about them in the letter."

Seth grunted his agreement.

"So," I said, "what's on our schedule today?"

"I called Al's house. We're going there at ten."

"Why so soon again?"

"I want to talk with Ivelisse, see if she knows anything helpful."

"Whom did you speak with?"

"Al's son, Xavier. He just returned from Key West."

"How did he sound?"

"Fine. I thought he might balk at having us come by. He actually sounded pleased. Hard to read him."

I glanced at my watch; eight o'clock.

"Feel up to a walk after breakfast?" I asked. "I need a little exercise."

" 'Fraid not, Jessica. Between my aching back and bad knee, I'd be lucky to make it a block, especially with the pace you like to set. Zach Shippee tells me I'm his annuity."

Shippee was Seth's chiropractor in Cabot Cove.

I couldn't help laughing.

"Nothing to laugh at, Jessica. You'll get there one day, too."

"I wasn't laughing at you, Seth. It's just that what Zach said was —"

"Might be funny to you, but not to me. Eat your breakfast and take your walk. We'll meet up in the lobby at nine thirty."

I didn't make any further mention of Seth's aches and pains. He was clearly not in one of his better moods, and I'd learned years ago to back off when that was the case. We finished breakfast in relative silence. He headed for his room, and I winced as I saw him walk in obvious pain across the restaurant.

I ventured outside and took some deep breaths. It was a lovely beginning to the day in Tampa, sunny, a clear blue sky, and a refreshing breeze setting the palm trees in motion. I wondered what the weather was like back in Cabot Cove and reminded myself to place a few calls to catch up with friends.

I had forty-five minutes before I'd need to meet Seth and chose to take a short stroll through a wooded wetland adjacent to the hotel. I found a narrow dirt path leading into the undeveloped land and hesitated. While the wooded area was appealing, I wasn't sure whether it was wise to venture into unfamiliar territory. Were the paths all marked? How large was this plot of land anyway? Naturally, I thought of alligators — you can't be in Florida and *not* think about them. What was the possibility of coming across one? Highly unlikely, I decided. Alligators liked to be around bodies of water, and from what I could see from my vantage point, there didn't seem to be any ponds or lakes in this densely packed patch of land.

I progressed slowly, stopping every now and then to admire the variety of palm trees that lined my path, and clumps of flowers that would suddenly appear from behind a tree, vivid splashes of color in what was otherwise a monochromatic landscape. I moved through an area of trees on which someone had affixed a handwritten sign — *gumbo limbo* — which I assumed was to identify the species of those particular trees, their smooth bark peeling off in broad sheets to reveal red trunks beneath. Beyond was a swath of tall grass that looked like hay. It intruded on both sides of the path, narrowing my passage, and as I walked past it, my bare leg brushed against some of the strands. "Ouch," I said as I looked down to see a thin red line on my calf. I examined the grass more closely and saw that the strands had sawlike teeth, sharp enough to have broken the skin.

My leg stung, and a single, tiny drop of blood appeared at the end of the scratch. I considered returning to the hotel but noticed up ahead a clearing in which a wooden bench was situated next to a small pond. *Perfect,* I thought as I approached the clearing. I sat on the bench and used a tissue from my purse to blot the drop of blood.

A shaft of sunlight through the palm fronds reached where I sat, and I closed my eyes and tilted my face up to catch the warm rays. I was enjoying this moment of peace when I heard what sounded like muffled footsteps. I opened my eyes and looked back at the path I'd just taken. The sounds stopped. I glanced at the pond, hoping I hadn't awakened a sleeping alligator. I was wondering if climbing on top of the bench would provide any protection when I heard the sounds again. This time, I stood, straightened my skirt, and tensed. A moment later the source of the footsteps appeared — Karl Westerkoch. I was both surprised to see him and relieved that it was a familiar face.

"Good morning," I said.

"Good morning, Mrs. Fletcher. Communing with nature this morning?" he said in his pinched, British-tinged voice.

"I guess I am. The vegetation in Florida is so different from what we have in Maine."

"A very different climate."

"Yes. I'm surprised to see you here."

"Oh? Did you consider this your private domain?"

The man was seemingly incapable of being pleasant.

"I wasn't suggesting that it was," I said.

He came to the bench, sat, crossed his long legs, and looked up into the sun coming through the trees. "In the interest of full disclosure," he said, "I came because I saw you walk in here."

"You followed me?"

"You might say that, although it does sound terribly cloak-and-dagger, doesn't it?"

"To you, perhaps."

"Come, sit," he said, patting the bench beside him. "I'm really quite harmless."

"Mr. Westerkoch," I said, "since you've admitted to following me into this lovely grove of trees, I'd like to know why."

He ran his fingers over the crease in his slacks, moved them to his neck to adjust a tie that wasn't there, and said, "Please, sit down. I have a crick in my neck this morning and it's painful to have to look up at you while we talk."

My first inclination was to bid him a good day and walk away — he was that disagreeable — but of course I was curious as to what he had to say. I'd suggested to Seth that we take every opportunity to speak with anyone involved in Alvaro Vasquez's life, and Karl Westerkoch fell into that category, although I had no idea in what way they'd been connected. His name had not appeared on the possible list of investors Seth had copied from Al's file. I took a seat on the bench, leaving as much space as possible between us, and waited for him to explain his presence. When he did, what he said surprised me.

"You and your companion, the good Dr. Hazlitt, have ended up involved in a rather nasty business."

" 'Nasty business'? I haven't the slightest idea what you're talking about."

"Oh, I think that you do, madam. I can't imagine that someone whose brain is fertile enough to craft murder mysteries — and I understand that you've done quite well with your novels — would miss what's been going on over the past few days."

"If you mean Dr. Vasquez's death, I'm well

aware of it. Dr. Hazlitt and I were there when he died, as were you. Remember?"

"How could I ever forget? But you see, dear lady, there is more to it than his unfortunate passing. There is, as you also know, the fruits of his efforts in the laboratory."

*Where is he going with this?* I wondered, and decided to offer nothing. Let him set the agenda for this unexpected conversation in an equally unexpected setting.

"You've spoken with Oona," he said flatly, a statement, not a question. He pulled a long, slender cigar from his sport jacket — or is it called a cigarillo? — lit it, blew the smoke up in the air so it curled over my head, and looked at me while waiting for an answer.

I didn't provide him with one.

"And I'm sure that Oona made it clear to you that Dr. Vasquez's untimely demise has potential ramifications far beyond the death of one individual," he continued.

I thought back to my meeting with Oona and tried to recollect what she'd said. "She did indicate that there was interest on the part of the government in seeing that Dr. Vasquez's research not fall into Cuban hands."

"And she was absolutely correct," he said.

"Oona has a way of being direct, much to her credit."

"That's always an admirable trait," I said, "and I would appreciate it if you would exhibit the same directness."

Westerkoch gave me a crooked smile before taking puffs and exhaling the smoke into the air, a satisfied expression on his gaunt face. I checked my watch.

"Oh, yes," he said as though his wandering mind had suddenly been brought back to the present moment, "being direct. Frankly, I thought I was."

"How about this for a starting point, Mr. Westerkoch? Seth Hazlitt and I have nothing to do with Dr. Vasquez's research. He and Seth had struck up a friendship, nothing more than that. I was in Florida on a book tour and decided to extend my stay and spend time with Seth here in Tampa. I certainly understand why the government would be interested in where Dr. Vasquez's research ends up, in whose hands it falls, but that has nothing to do with Dr. Hazlitt or me."

"Under ordinary circumstances I would agree with you, Mrs. Fletcher, just a small-town physician and his mystery-writing friend enjoying the good weather here in

Tampa. But you see, there is a complication."

"I'd like to know what that is," I said, even though I suspected he was about to bring up Vasquez's missing laptop.

He was.

"Dr. Vasquez's approach to medical research was unusual at best. He built an outsized reputation in his native Cuba, which traveled with him to Florida, where he established himself as an important citizen, albeit a controversial one."

"Controversial? Why?"

"Because of the way he conducted his research. Don't you find it strange that he worked in almost total secrecy, only one assistant, with progress reports on his work nonexistent? That's hardly the protocol one expects from a medical researcher whose work promises — and I stress 'promises' — such great results."

He was right. I had found Vasquez's methods to be strange, and I had voiced that to Seth. Then again, I could claim to know nothing of how medical research worked, which would have been the truth. Because Seth had become such an unabashed champion of Vasquez's work, it served to temper any doubts I had, at least initially.

236

"Of course," he said, "his methods aren't the most important thing." He cackled. "The mad scientist at work, hey? No, his methods aren't at issue here, Mrs. Fletcher. What *does* interest the government is what he managed to achieve in his laboratory."

"Again, what does that have to do with Dr. Hazlitt and me?"

"You asked me to be direct, and I will be. But I expect the same from you. No one seems to know what he achieved through his research, and we would like to know."

"Just who is 'we'?" I responded. "You? The government? I'd like to know what role you play in all of this, Mr. Westerkoch."

"Let me just say that I have a vested interest."

"A financial interest?" *Perhaps that list we'd found was incomplete,* I thought.

"Do I strike you as being that crass, Mrs. Fletcher?"

"Protecting one's financial interests isn't crass unless the gains are ill-gotten."

It was obvious that Vasquez's missing laptop was at the core of Westerkoch's interest, and that posed a dilemma for me. As far as I was aware, only Seth and I knew what had been on that laptop. Unless, of course, Detective Machado had attempted to decipher the material on the thumb drives we'd turned over to him. My hunch was that the detective had probably dumped them in an evidence locker until he could find an expert to interpret them. Nevertheless, before releasing them, Seth and I had transferred every word from the drives to my computer and then put them on the three new thumb drives, which now rested in the safe in Seth's hotel room. Then again, the information was still on the hard drive of the computer in my room, which was sitting out in plain sight. I had a sudden urge to return to the hotel.

"I have to be getting back," I said.

"To meet with your Dr. Hazlitt?"

"I don't see how that's your concern."

"Plans for the day?"

"I would say it was nice to see you again, Mr. Westerkoch, but I'm not sure that's true," I said, standing.

"A word of advice, Mrs. Fletcher?"

"I'm listening."

"Don't discount the seriousness of the matter in which you and Dr. Hazlitt find yourselves enmeshed. The stakes are high, *very* high."

"I'll keep that in mind."

He looked down at my leg and cooed, "You have a boo-boo."

"A plant with sharp leaves."

238

"Saw grass," he said. "Nasty things, those leaves."

"It's nothing," I said.

"We'll have to continue this little chat another time," he said, rising and stamping out his cigar on the dirt.

I started up the path out of the wooded wetland but was struck with a thought. "Are you responsible for having people follow me and Dr. Hazlitt?" I asked.

"Me? I'm simply concerned with your well-being, Mrs. Fletcher."

"That may be true, but I assure you we don't need someone watching after us."

He ignored my comment and said, "Careful walking back to the hotel, Mrs. Fletcher. Avoid the saw grass, and keep a sharp eye out for alligators. They have a voracious appetite."

I had to work at steadying my nerves on my way back to the hotel, not because he'd been threatening, but because I disliked him so.

Why had he bothered seeking me out that morning? All he'd done was to corroborate what Oona Mendez had said, that the United States government was interested in Vasquez's research and in seeing that it not fall into Cuban hands.

I'd had the feeling after my conversation

with Oona that she was aware that the laptop on which Vazquez's notes were stored was missing, although she hadn't said as much.

But if Seth was correct, it was all moot anyway. His reading of Al's notes said that Vasquez had failed to come up with any conclusions that might lead to a cure for Alzheimer's. The problem seemed to be that neither Oona nor Westerkoch, nor other governmental types, knew what Seth and I knew, and that begged the question: Was it incumbent upon us to let it be known?

I wasn't the one to answer that. It would be Seth's call, and his alone.

I filled Seth in about my conversation with Westerkoch during the ride to the Vasquez home on Davis Island.

"I don't like it," Seth said. "I don't like this fellow trailing behind you."

"I'd prefer that he didn't, too. Do you have an agenda this morning?"

"Nothing specific, but I think it's time I asked some direct questions."

I was pleased to hear him say that because I had a few direct questions of my own.

# CHAPTER SEVENTEEN

The Vasquez daughter, Maritza, answered the door.

"How is your mother today?" Seth asked once we were inside the house.

Maritza twisted her hand from side to side, a nonverbal "so-so" reply. "She's asked for you, Mrs. Fletcher."

"Oh? I'd enjoy very much seeing her."

Maritza's brother, Xavier, joined us.

"How was your trip to Key West?" I asked.

He lifted his brows, apparently surprised that I knew he'd been away. "Great," he replied pleasantly. "Smooth flight both ways. How have you been?"

"We've been fine," I said, aware of the change in his demeanor. During our first meeting, he'd been sullen, perhaps even rude, but on this day he seemed more relaxed and there was warmth in his voice.

Maritza, who'd left us, returned with her mother as Xavier disappeared into another part of the house. Mrs. Vasquez looked stronger than she had the last time we'd seen her. She'd abandoned the blanket and was now stylishly dressed in a taupe skirt, teal blouse, and sandals.

"Hello, Mrs. Vasquez," I said. "I'm Jessica Fletcher."

"Of course," she said. "How good of you to come. Please won't you join me in some coffee or tea?" She sank gracefully into a chair in front of a small table and waved at her daughter.

"Coffee would be fine," I said, and Seth opted for the same.

We settled on a love seat across from Ivelisse as Maritza gave instructions to the housekeeper to fetch *"café con leche"* as Mrs. Vasquez had requested.

"Have you seen the newspapers?" I ventured, deciding to be direct.

"I never read those scandal sheets," Ivelisse said placidly, smoothing her hair with one hand.

"Mami," her daughter said. "I read the story about Papi to you this morning, don't you remember?"

Ivelisse looked momentarily confused. Then she closed her eyes and slowly shook her head. "How could they even think such a thing? To say that Alvaro was murdered

242

upsets me. The American press, with all its liberties, abuses them at times, don't they?"

"I'm afraid they do sometimes," I said. "It's a price we pay for freedom of the press."

"There is no free press in Cuba. The government controls what is written and broadcast. But when I hear what your reporters say about Alvaro, I wonder whether it isn't better in Cuba."

Seth joined the discussion. "Government-controlled press is never better," he said with gravity.

Ivelisse cocked her head at him. "I suppose you're right, Mr. . . . ?"

"Seth Hazlitt," Seth answered.

"Yes, yes, you were a friend of Alvaro's."

"That's right. I'm a physician."

"Like Alvaro."

"He was a great physician and a fine gentleman," Seth said.

Her smile was part agreement but somewhat cynical. "Alvaro was a handsome man, yes?" she said to me.

"Yes, he was very handsome," I replied, a vision of him flashing in front of me.

"So many women," she said, as though casually commenting on the weather or a pretty flower.

Seth and I looked at each other as Maritza said, "I don't think we need to talk about that, Mami."

Ivelisse's face was blank, serene, and a tiny smile came to her lips and stayed there while the housekeeper set out a tray with cups of strong Cuban coffee, sugar cubes, and a pitcher of hot milk.

"Have the police been in contact with you again?" Seth asked, adding sugar to his cup and stirring.

"The police?" she said in a startled voice. "Oh, them," she said. "The police. Why would they be here?"

"I thought they might want to talk with you about their suspicion" — Seth hesitated before continuing — "that Al was murdered."

Her serenity morphed into a hard mask. "Murdered? I will not stand to hear that. No, there will be no talk of murder in my house."

"Maybe you'd better rest again, Mami," Maritza said.

"I do not want to rest," she said. "I want to talk with Mrs. Fletcher. She is a writer."

"That's right," I said, "although I have to admit that I do write *about* murder."

"Murder in books is all right," she said.

Xavier returned and gave his sister a piercing look.

"I'm afraid Mami is getting tired," said Maritza, rising.

"I am not," Ivelisse said sternly.

Maritza motioned to Xavier and they walked from the room.

"My daughter is studying medicine," Ivelisse said.

"You must be very proud of her," Seth said.

"What sort of doctor are you?" she asked Seth.

"General practice."

"Alvaro was a respected research scientist," she said.

"Yes, I know," said Seth. "Did he talk to you about his research?"

"Oh, no, and I didn't want to know about it. He was trying to find a cure for . . ." She trailed off.

"For Alzheimer's disease," I filled in.

"That's right, for Alzheimer's disease," she said. "For the brain."

"For the brain," Seth concurred.

"Alvaro liked women," she said.

"Did he?" Seth said. "I do, too."

"Do you cheat on your wife?"

Seth sat back on the couch as though having been shoved. "I'm not married," he said.

"You, Mrs. Fletcher? Do you see other men?" she asked.

"Other than my husband? I'm widowed, Mrs. Vasquez, have been for a number of years. But my husband, Frank, and I had a wonderful marriage. We were devoted to each other."

"That's nice," she said dreamily. She looked at Seth and her brow furrowed. "I'm sorry, but you are?"

"Dr. Seth Hazlitt, Alvaro's friend."

"Of course, yes, yes, yes, I know that. What do they call it when you get older and forget?"

"A senior moment?" I suggested, although the term "Alzheimer's" was at the front of my thoughts. "I have those senior moments myself now and then."

She didn't respond to my comment. She adopted a dreamy expression as she said, "Alvaro was such a handsome man, a Cuban Casanova. He was proud of that. Cuban men are hot-blooded, Mrs. Fletcher. We accept that when we marry them."

Seth cleared his throat and asked, "Did Al ever share with you what was on the laptop he brought home with him every night from his laboratory?"

"No," she said sharply. "I already told you that. He never told me about his work."

"I'm sorry," Seth said. "You did tell me that."

"Where is Maritza?" she asked, swiveling her head left and right.

Her daughter immediately reappeared as though she'd been poised to be summoned.

"Is it time?" Ivelisse asked.

"Yes."

Maritza explained to us, "My mother likes the Spanish soap opera *La Casa de al Lado*. She never misses it."

"What does that mean in English?" I asked.

" 'The house next door,' " she replied. "Please excuse us."

After they'd left the room, Xavier returned and took his mother's place, sitting in the chair she had vacated and pouring himself a cup of coffee.

"Your mother is a lovely woman," I said, adding, "and a proud woman."

"Yes, she is both of those things, and more."

"How is she handling your father's death?"

"She is very strong and doesn't show her emotions easily," Xavier said, sipping his coffee, "certainly not to strangers."

I had thought she was open in expressing her opinions, but I wondered how much her memory problems affected her understanding of the current circumstances.

"Behind her closed door," Xavier continued, "my mother is able to express herself to her family. She is very sad, of course. She and my father were married a long time."

"There are a lot of questions about your father's research," Seth said.

"Yes. I've heard about the missing laptop," Xavier said flatly.

"That's right, his missing laptop. I was told that he brought it home with him every night from the lab, but it doesn't seem to have shown up. Got to be an answer for that."

"You aren't suggesting that I might have done something with it, are you, Dr. Hazlitt?"

"I'm not suggesting anything, Xavier, but it is strange that something as visible — and important — as that laptop would go missing."

"Well," Xavier said mildly, "maybe it'll show up one of these days." He turned to me. "So," he said, "you told me at the party that you have a pilot's license. I love to fly. I received my license in Cuba." He laughed. "My father was supportive, but my mother was certain it would mean an early death for me. So far, she's been wrong."

"Thank goodness for that," I said. "What sort of certificate do you have?"

"I am instrument rated and have started

working on my instructor's license."

"What's your goal?" Seth injected. "To fly big commercial planes?"

"Yes, I would like that someday. When I am in my plane, I feel free, more free than at any other time in my life. Do you feel that way, too, Mrs. Fletcher?"

"Yes, I do," I said. "There's something liberating about being up there all alone, looking down at the earth, seeing your town from the air. I'm sorry that I'll never get much further beyond my basic private pilot's license, although I never intended to."

"Mrs. Fletcher doesn't have a driver's license," Seth said, chuckling, "but she can fly a plane."

Xavier smiled broadly. "That is funny," he said.

I returned his smile. "That's what all my friends say."

"Which includes me," said Seth. "Frankly, I thought she was crazy when she said she was going to take flying lessons."

"I thought for a while that maybe I *was* crazy," I said, "but once I started I knew I'd made the right decision."

"How about going up for a spin with me, Mrs. Fletcher?"

"That sounds appealing," I said.

Seth fixed me with a hard look.

"Maybe we can do that one of these days," I said, keeping it vague for Seth's sake.

"You know," Xavier said, changing the subject, "my father surprised everyone when he welcomed your friendship, Dr. Hazlitt. He didn't have many close friends."

"Then I'm proud to have been among the few."

"I'm thinking maybe we don't even need my father's laptop to know how his research was going."

"Meaning?"

"Well, it's just that since you became one of his close confidants, I figure you'd know a lot about his progress."

" 'Fraid I can't help you there," Seth said, slapping his knees and standing. He put a hand out to help me up.

I thought of the three thumb drives back at the hotel, and that not only did Seth know everything that Vasquez had noted about the research, but I did, too, although without the medical background to truly understand it.

Xavier said that it was good of us to have stopped by and repeated his invitation for me to go flying with him.

"Before we go," Seth said, "I wanted to ask you about Ms. Mendez and Mr.

Westerkoch."

"What about them?"

"They were friends of your father's, too, and I wonder what the basis of the friendship was."

Xavier shrugged. "The 'basis'? He liked them."

"I know that Ms. Mendez works for the Cuban American Freedom Foundation here in Tampa," Seth said. "Was she helpful in your father's defection and application for asylum?"

He thought before answering. "Oona Mendez has her own agenda where my father was concerned. Sure, she was involved in those things, but she also had a more personal interest in dear old Dad."

Maybe I'd been right when I speculated that Oona might have lost a lover in Dr. Alvaro Vasquez. I also found Xavier's expression "dear old Dad" to be disparaging. I'd sensed tension between father and son during the party and now wondered about the extent of their animosity toward each other. Ivelisse may have been in mourning, but was Xavier also sad to have lost his father?

"And Mr. Westerkoch?" Seth asked. "He seems to be — well, he's demonstrated a keen interest in your father's research."

"Did he?"

"What does he do?" I asked. "He says he's a consultant, but he never said what company he consults for."

"He's — look, I really don't care about Westerkoch." He flashed me an engaging smile. "Last chance, Mrs. Fletcher. I'm flying to Key West first thing in the morning."

"So soon again?"

"The weather report is good," he said, ignoring my question, "so it should be nice flying weather. You can get in some flying time, and you can meet my girlfriend. You, too, Dr. Hazlitt. It's a four-seater, a really nice plane. Game?"

"Can I think about it and call you later?" I said.

"Sure. I'll be around all day. Now, if you'll excuse me, I'll go see how my mother is."

When Seth and I were alone, he said, "You aren't really considering going flying with him, are you?"

"It's tempting," I said. "He invited you, too."

"It's not tempting to me."

"It would give us a chance to spend some uninterrupted time with him, Seth. I know that you're not a fan of flying, especially in small planes, but he sounds like a responsible pilot. He has his instrument rating and is going for his instructor's license. That means he's a serious pilot. Besides, you've flown with Jed back home in his small planes and you made it out alive."

He curtailed that topic of conversation by making a show of looking at his watch. "Let's go get lunch," he said, "and meet up with Dr. San Martín. I'm eager to find out why he wants to see us."

I found Maritza and told her that we were leaving.

"My mother was happy to see you, Mrs. Fletcher. She admires authors. Are you sure you wouldn't like to stay awhile?"

"I'd love to," I said, "but we're meeting someone for lunch at a place called the . . . What is it, Seth?"

"The West Tampa Sandwich Shop."

"Oh," Maritza said, "everybody raves about the Cuban food there — very authentic, I'm told. Xavier goes there a lot."

"I'm ready for some good Cuban food," Seth said.

"Enjoy it," Maritza said, "and please come back anytime."

We managed to find the West Tampa Sandwich Shop, a small, nondescript former house on the busy North Armenia Avenue, across from a large church. Seth had to circle the block a few times to find a parking space, and we ended up two blocks from the restaurant.

"Doesn't look much like a place to find good food," Seth said as we approached the building.

"Looks can be deceiving," I said. "Time to enjoy real Cuban food."

"We'll see," he said as we came upon what looked like a run-down carport attached to the house. There were a few tables beneath the canopy, and a group of six older men, dressed in colorfully patterned shirts and Cuban guayaberas, sat at a long table smoking cigars and talking.

"Is this the entrance?" I asked Seth.

"Beats me," he replied.

Conversation stopped when we entered the sheltered space, but we saw that there was a door leading inside.

*"Hola!"* one of the men said to us.

I returned his greeting as we passed their table and pulled open the door to a small room where a dozen tables were covered in lacy white cloths with clear plastic sheets over them. A short counter with a few backless stools occupied one side of the restaurant opposite a TV set. The white walls were covered with hundreds of photographs collected inside large black frames, presumably pictures of regular customers. Above the collages were individually framed portraits of those I gathered were especially honored guests. Two waitresses, well familiar with the routine, scurried among the tables, all of which were occupied, pushing carts that were transporting diners' meals. All in all, there was a sense of controlled frenzy.

Dr. San Martín sat at a table only slightly removed from the next, where five gray-haired men engaged in a loud, friendly argument on the benefits of vitamin supplements. I was surprised to see that the ME was with another man, considerably younger, who looked out of place with his dark suit, shirt, and tie. San Martín saw us and waved.

"Good to see you," he said as we joined the table. "Welcome to Tampa's best-kept secret."

I looked around and laughed. "Judging from the business they're doing," I said, "I think the secret's gotten out."

"Much to the chagrin of the owners," said San Martín. "There are so many regular customers, they don't want to fill the place up with tourists." He turned to his companion. "Mrs. Jessica Fletcher and Dr. Seth Hazlitt, this is Harry Guterez."

"A pleasure to meet you," Guterez said as we shook hands. "I'm certainly aware of your books, Mrs. Fletcher, although I admit I haven't read one."

"We'll have to rectify that," I said pleasantly.

"Are you with the medical examiner's office?" Seth asked.

"No," Guterez replied. "I'm FBI."

His simple statement had the effect of silencing Seth and me.

"Agent Guterez has something he'd like to discuss with you," San Martín said, "but I suggest that we enjoy a good Cuban lunch before we get into that conversation."

I knew that Seth shared my thought at the moment — we'd rather not have to wait to hear what was on his mind. But that wasn't the way Dr. San Martín had choreographed the meeting. So lunch it was.

"Everything is good here," San Martín said, "and the portions are large. They make a superb Cuban sandwich, although it's misnamed. What people today call a Cuban sandwich was actually born in Tampa in the late eighteen hundreds. Cigar workers, who settled here from many parts of the world, brought their own favorite foods, some of which went into the sandwich. The Spaniards contributed the ham, the Italians the Genoa salami, the Cubans *mojo*-marinated pork, and the Germans and Jews added the pickles, mustard, and Swiss cheese. Of course, without good Cuban bread, it falls flat. Cuban bread is the best. When you make the sandwich, you butter the outsides of the bread and brown it up in a pan or press, like a grilled-cheese sandwich, only better." He pressed his fingertips to his lips and blew a kiss into the air.

"What is *mojo*?" Seth asked.

"It's a Cuban concoction used to marinate pork. When I make it, I use garlic cloves, salt, black peppercorns, oregano, and sour orange juice."

"Never heard of sour orange juice," Seth said.

"Not easy to find," said San Martín. "There are lots of sour orange trees in Cuba. You can substitute regular orange juice and add a little lemon or lime."

"Sounds like you know your way around the kitchen," Seth said.

"I love to cook," San Martín said. "I did all the cooking when my wife was alive and still find fixing myself supper to be relaxing after spending the day probing dead bodies."

I made a face and he apologized for inappropriate table talk.

I was willing to order anything as long as it came quickly and allowed us to get to what Agent Guterez wanted to say. After

San Martín's enthusiastic description, we all settled on Cuban sandwiches, and he insisted that I try a mango milk shake.

Guterez didn't have much to add to the conversation, and once the food came, we all stopped talking anyway. I found myself, as I often do, studying the scene in the restaurant, a habit I imagine a lot of writers — and probably just as many nonwriters — have. Frequent people watching gives me insights into human behavior — at least I hope it does — and many an unwary diner has ended up as a character in one of my novels.

At the neighboring table, the group's loud discussion about a variety of health topics paused for a moment while one man told a joke: "So I went to the VA hospital and this nurse at the desk says to me, 'I already called your name. Didn't you hear me?' And I said to her, 'If I could hear you, I wouldn't be here.' " The story brought forth hearty laughs from his companions and made me smile, too.

The men were momentarily distracted by a shapely young blonde who sashayed through the room wearing tight jeans and a low-cut sweater. At the table next to ours, a middle-aged woman with jet-black hair also eyed the new arrival and registered her opinion of the blonde to her tablemate by pushing out her lower lip and rolling her eyes.

While all of this captured my interest, my main thought was that I wanted lunch to end so we could get to the reason for having gotten together. The sandwich was good, the mango shake sweet, but the combination was too filling. The waitress asked if I'd like to take home the other half of my sandwich and I declined.

Finally, Dr. San Martín paid the bill and suggested that we leave. Seth and I looked at each other in surprise.

San Martín caught our exchange and said, "It's a little too crowded in here for privacy."

Once outside, Agent Guterez led us to the church's parking lot, where a black limousine stood, engine running. A man dressed like Guterez got out on the driver's side when he saw us approach, and opened the rear door.

"I thought a little ride after lunch might be in order," said Guterez.

"A ride?" Seth said. "We have our car parked a coupla blocks from here."

"We'll bring you back, Dr. Hazlitt; just a short drive for us to talk."

Seth and I climbed into the rear of the car and sat on the bench seat. There were two fold-down seats, which Guterez and San Martín took, allowing them to face us. The driver exited the parking lot and drove slowly down North Armenia Avenue, destination unknown to us. Dr. San Martín provided nonstop conversation during the trip, commenting on his love of cooking, the political situation in Tampa and its relationship to what was happening in his native Cuba, and his love of the city's own national league football team, the Buccaneers. It was almost as though he was attempting to head off any questions we might have about where we were going and why.

Twenty frustrating minutes later we arrived in a suburban area, its sign announcing that we were in Citrus Park. The driver parked beneath a tree, turned off the ignition, and got out of the car. San Martín and Guterez made no move to leave their seats.

"Are we getting out and taking a stroll?" Seth asked, not bothering to mask the annoyance in his voice.

Guterez smiled as he said, "No, Dr. Hazlitt, but we will have a chat if it's okay with you and Mrs. Fletcher."

"Do we have a choice?" I asked.

"Probably not," Guterez said. "Why don't you begin, Dr. San Martín."

San Martín came forward on his seat and

260

said, "I know this is confusing, and I must admit that I was against hijacking you this way. But Agent Guterez and his colleagues decided that making it a bit of a social event would be more conducive to accomplishing what it is they *wish* to accomplish. The truth is that the two of you have placed yourselves in an awkward situation."

"Really?" I said. "How so?"

San Martín crossed his legs and thought before continuing. "Let me start by congratulating you, Mrs. Fletcher, for being astute. I suppose that writing murder mysteries has sharpened your powers of observation."

I cocked my head. "I appreciate your kind words, Dr. San Martín, but I have no idea what you're referring to."

"Cigars," he said.

"Cigars?"

"Yes, the one you picked up on the day that Dr. Vasquez died."

"I'd forgotten about that," I said, not entirely truthfully. It simply hadn't been on my mind that day.

"I almost did, too," he said. "You left it in my office and I ignored it for a few days. I'm surprised I didn't toss it away. At any rate, I was looking at it one day and got to thinking about whether it might shed any light on Vasquez's death. It had been beaten up a bit and was still a little soggy since I'd left it in the plastic bag, but I ran it through some preliminary tests to see what it contained. The usual chemicals were present; a cigar contains thousands of poisons, like nitrosamines, ammonia, cadmium, hydrogen cyanide, carbon monoxide, and, of course, nicotine. I expected to find those elements along with others, and I did."

"Are you suggesting that one of those poisons found in cigars killed Dr. Vasquez?" I asked.

He smiled like a kindly uncle correcting an honest misunderstanding. "No," he said, "none of those are capable of killing someone, at least not from one cigar. A lifetime of smoking them might do you in, but none of those poisons are found in sufficient quantity in one cigar to be lethal."

"So did you find something else that might have contributed to Dr. Vasquez's death?" Seth asked.

"I certainly did," said San Martín. "The neurotoxin botulin."

"Is that related to botulism caused by spoiled foods?" I asked.

"That's correct, Mrs. Fletcher. One and the same."

"Is it possible he'd eaten something that contained that toxin?"

"No. As I said, I found it in that cigar you left at my office."

"I take it that botulin isn't usually found in cigars," Seth said.

"Not in my experience," San Martín replied.

"Now, wait a minute," Seth said, holding up his hand. "I've treated my share of patients who ended up with botulism either through something they ate or an infected wound. I've had a few babies who came down with botulism poisoning because of honey their mothers gave them during their first year. I tell every new mother to not give their babies honey until they're older."

"I'm aware of the problem with honey and newborns, too," San Martín said.

"But I've never lost a patient who had botulism poisoning," Seth said. "I had one young fella who waited too long to come in to see me and ended up in the hospital on a breathing machine for a few weeks, and I had another patient who worked for a dermatologist who breathed in too much Botox. But as you know, it takes a few days for the symptoms to show up. With Al Vasquez, his death was pretty darn fast, almost instantaneous. Doesn't figure that inhaling smoke from a single cigar would do him in like that."

"You're right," San Martín said, "provided that what he'd inhaled was common, run-of-the-mill botulin. It wasn't!"

He had our full attention. Alvaro Vasquez had offered his cigar to Seth that night, and when Seth declined, he'd offered it to me. If either of us had accepted — I shuddered at the mental picture it brought up — Seth or I could be dead right now.

Next to me, I felt Seth stiffen, and I was certain his line of thinking followed mine. Both of us had suspected Vasquez had been murdered, but we never realized how close to our own deaths we might have come.

San Martín spoke, breaking into my horrified thoughts. "I took it upon myself to personally deliver that soggy cigar to the lab at the Institutes of Health in Washington, D.C., and waited until they'd analyzed it. Didn't take long. That cigar was full of *C. botulinum,* one of the most powerful known bacteria to secrete toxins. A single microgram is lethal to humans. It acts by blocking nerve function and leads to respiratory and musculoskeletal paralysis. Still, it would not have killed Dr. Vasquez that quickly."

"Then what?" I asked.

"It had been chemically enhanced, no easy trick. It would take a highly sophisticated lab to accomplish that."

Agent Guterez, who'd said nothing during San Martín's explanation, now entered the conversation.

"You might wonder why the FBI is now involved," he said. "Initially this was considered a local matter, something for the Tampa police to handle. But what Dr. San Martín has uncovered changes the landscape. Obviously, someone injected the botulin into the cigar that Alvaro Vasquez was smoking when he died. We believe Dr. Vasquez was the intended victim. Because the toxic substance is, as Dr. San Martín has explained, highly sophisticated, we're going on the assumption that a government could be involved."

Seth and I said in unison, "The Cuban government?"

"Or someone in our own," Guterez said grimly.

Dr. San Martín spent the next ten minutes further explaining what he'd found in the cigar, and the nature of the chemical enhancement that had turned Dr. Vasquez's favorite pastime into a lethal weapon. When the ME was finished, Seth asked, "This is all fascinating, but what does it have to do with us?"

"A good question, Doctor," said Guterez. "The fact is that we feel it would be better if you and Mrs. Fletcher returned to your home in Maine."

"Do you mind if I ask why?" I said.

"I can't be too specific," said Guterez. "National security. Just let me say that Dr. Vasquez's murder and the missing results of his research have spawned a budding problem between our government and that of Cuba. It has the makings of an international incident."

"National security," Seth muttered. "Always a good excuse to not be straightforward."

Guterez said, "I'll be direct. The agency insists that you leave Tampa by tomorrow."

"Is that an order?" Seth asked.

"If you'd prefer to view it that way," Guterez said, his heretofore pleasant, non-combative demeanor replaced by a steely tone and expression.

We were driven back to where Seth had parked the car a few blocks from the West Tampa Sandwich Shop.

"I assure you that the bureau appreciates your cooperation," Guterez said.

Seth was furious that he'd been ordered to leave Tampa by the agent, but I didn't necessarily share his anger. While my mind was swirling with questions — and I knew they would bedevil me for some time to come — I was actually relieved that we'd be leaving.

Cabot Cove had never been so appealing.

come — I was actually relieved that we'd be
leaving.

Cabot Cove had never been so appealing

# CHAPTER EIGHTEEN

"National security my foot," Seth grumbled
as he started the rental car and pulled away
from the parking spot.

"It doesn't matter," I said. "Agent Guterez
and his people know a lot more than we do.
Maybe it's best that we give up trying to
make sense out of your friend's death and
go back to what we know and where we're
comfortable."

"I never thought I'd hear you give up on
something, Jessica."

"I'm not giving up on anything, Seth.
We've been spinning our wheels trying to
find answers. Maybe if it were just a local
matter, a homicide without international
repercussions, we'd be successful. But that's
not reality."

His mood was glum and tinged with ir-
ritation as we headed back toward the hotel.

"Let's pack, have a nice dinner at that
restaurant I've been dying to visit, the

Columbia, and get on a plane tomorrow," I said. "Frankly, I can't wait."

"I suppose you're right, Jessica, but it's gravel in my craw. I want to go back to Al's house first. The least I can do is say goodbye. I also think I should level with Xavier about his father's research notes."

"Tell him what?"

"Tell him that I've read Al's notes and intend to bring those thumb drives back with me to Maine. Al asked me in his letter to show them to the researchers I know up north, and that's what I intend to do."

"But you said that from what you read, he hadn't made much progress in finding a cure for Alzheimer's."

"Ayuh, that's right. But it's not up to me to make that decision. As I told you, they may find his mistakes useful, save them from following an unproductive path. Or even suggest a different way to go."

"But don't you think that if Xavier knows you have those thumb drives, he'll want to read his father's notes?"

"He can get them from the police and Detective Machado. Besides, Al's laptop has to be someplace. It's bound to show up one of these days. I'd just feel better being straightforward with Xavier."

"I hope he'll appreciate it," I said.

Xavier answered the door. "Didn't expect to see you again so soon," he said.

"We've had a change of plans," Seth said, "and wanted to see your mother one last time before we leave Tampa."

"She's resting right now, but she should be up soon. When are you leaving?"

"Tomorrow. I'd like to have a few minutes with you, too, Xavier."

"A problem?"

"No, but there's something you should know. Can we go to your dad's study?"

"Sure. Let me get Maritza."

He returned with his sister, who invited us to have coffee with her on the deck until Ivelisse awakened.

"Dr. Hazlitt and I have something to discuss," Xavier said.

"Only be a few minutes," Seth assured her.

I wasn't certain that what Seth intended to tell Xavier was the right thing to do, but it was, after all, his decision. I walked outside to the deck — not far from where Alvaro had collapsed after smoking the poisoned cigar — and waited for Maritza to bring us small cups of strong Cuban coffee and sugar cookies.

"My mother has become very upset that people are saying that my father was murdered," she said after we'd settled in comfortable cushioned white chairs at a white round table, a red umbrella providing a bit of shade.

"I can understand that," I said. "Hopefully the police will do their job and identify who might have done it."

"That Detective Machado came by earlier," she said.

"What did he have to say?" I asked.

"He just wanted us to know that he and his department are working on the case."

"Did he mention anything else?" I asked, thinking of what we'd just been told by Agent Guterez, and that Detective Machado had a set of the thumb drives from Dr. Vasquez's laptop.

"No," Maritza said. "Is there something else we *should* know?"

"I don't know," I said. "We'll give him a call before we leave."

We passed the next fifteen minutes with small talk until Xavier and Seth reappeared. Both men seemed in good spirits. If what Seth had confided in Xavier had upset the young man, it didn't show.

"Bad news on flights," Seth said. "All the flights out of Tampa tomorrow are booked

271

solid. There's a big convention that ends tonight. That's probably the problem."

"What about flights from other cities?" I asked.

"That's the good news," Xavier said. "I can get you on a flight from Fort Lauderdale to Boston tomorrow afternoon if I book it right away. This is high season in Florida, but the airlines have cut back on the number of flights. If you want, I'll book the last two remaining seats for you."

"That's good of you, Xavier," I said, "but what about getting from here to Fort Lauderdale? It's a long drive."

"Easy," he said, grinning. "I told you I'm heading for the Keys tomorrow morning. No problem dropping you off in Lauderdale on my way."

I looked to Seth for his reaction. His initial expression was one of dismay, but it soon morphed into reluctant acceptance.

"Here," Seth said, handing his credit card to Xavier, "use this to pay for the tickets."

Xavier returned ten minutes later with our printed boarding passes. "How about we leave at eight in the morning?" he said.

"Sounds fine," I said.

Ivelisse Vasquez joined us just as we were about to leave. We again expressed our condolences, wished her well, and thanked her for her hospitality.

"You are welcome in my home anytime," she said.

"I hope we see you again soon," I told her, though I suspected that we never would.

Once in the car, Seth used his cell phone to call Detective Machado.

"Thought we might have a chance to see you again before we leave Tampa," Seth said, and went on to tell him of our plans for the next morning.

"I'd enjoy that," he told Seth. "Free for dinner?"

We were, and Seth arranged to meet him at seven at the famed Columbia Restaurant in Ybor City.

We spent time at the hotel packing and — in my case — napping before heading out for dinner. Knowing that we'd be leaving had siphoned away some of my adrenaline, and I'd felt a wave of fatigue roll over me. I awoke groggy and in need of a shower to wake me up. Refreshed, I met Seth in the lobby.

"How do you feel?" I asked. He'd looked drained, too, when we'd parted a few hours earlier.

"Fair to middlin'," he said. "I suppose what I'm really feeling is disappointment at having to leave without the answers I wanted about Al's murder."

"I understand, Seth, but it's beyond us. You meant well and tried, but sometimes we have to accept what we can't change."

He agreed, and we left the hotel and went to the car.

"Look," I said, pointing across the small parking lot to where a young man wearing a hoodie stood next to a car, smoking a cigarette. "That's the same person who was following us before."

Without saying a word, Seth walked in the young man's direction.

"Seth," I called after him.

He ignored me and picked up his pace, actually breaking into a labored trot. The young man saw him, dropped his cigarette, and started to walk away.

"Hey, young fella," Seth called. "I want to talk to you."

The man paused before darting out of the lot and up the street.

I came to Seth's side.

"Just wanted to know who he was and why he's been following us," Seth said, out of breath.

"It doesn't matter now," I said.

"Matters to me," he said. "People've got no right to be following other people."

"Forget it," I said. "Let's go to dinner.

I've been looking forward to an evening at the Columbia. Everyone we've met has raved about it. Besides, I'm eager to hear what Detective Machado has to say. As far as we know, he didn't have much to offer when he visited the Vasquez house today. Maybe he'll open up more to us."

The Columbia restaurant on East Seventh Avenue takes up a city block in Ybor City, between Sixth and Seventh avenues and Twenty-first and Twenty-second streets. We parked in a lot across the street and stood admiring the elaborate facade, hundreds of Moorish-style tiles in a wild variety of colors. A larger tile sign spelled out the restaurant's name and included the date it had been established, 1905. An ornate white overhang spanned the entire length of the building, reaching from one corner to the next, where the Columbia Gift Shop was situated.

"Some fancy building," Seth commented.

"Makes me feel like I'm in Spain," I said.

"Or Havana," Seth said. "I read that Cubans founded the restaurant."

"Does it look like buildings in Havana?" I asked as we crossed the street.

"Like they used to be, I suppose. Everything seems to be falling down there these days."

■ ■ ■ ■

A young woman greeted us in the opulent entranceway, also a colorful mosaic of tiles punctuated with heavy chairs and myriad works of art covering the walls.

"We're meeting Detective Machado," Seth told her.

"Oh, yes, he's already here waiting for you in the Café Room. Follow me."

Machado, dressed in suit and tie, sat at a table in a corner of the handsomely furnished and appointed room, which was both dining room and bar. He kissed me on the cheek like an old friend, shook Seth's hand, and waved for the waitress. A pitcher of sangria sat in the middle of the table; the glass in front of him was half-consumed.

"Welcome to the Columbia," he said. "Been here before?"

"No, we haven't," said Seth, "but Mrs. Fletcher has been dying to come."

"Oldest restaurant in Florida, oldest Spanish restaurant in the U.S. This is the original room built in 1905. There're fifteen rooms now, seats almost eighteen hundred people. What are you drinking?"

"That sangria looks appealing," I told the

waitress, who delivered glasses for Seth and me.

"Best sangria in all of Tampa," Machado said. With our glasses full, he raised his in a toast. "Here's to you two," he said. "Not often I get to have dinner with a famous author."

"I wish we'd met for a different reason," I said, touching my glass to his.

"That's the problem with being a homicide detective," he said, grinning. "I get to meet interesting people, but only because somebody's been murdered."

"I don't always get to meet people under the best of circumstances either," Seth said. "Too many times it's because someone got sick and is dying."

Their comparison of the grimmer aspects of their professions was interrupted when a man came to the table. "I see that our favorite detective is taking good care of you," he said. "I'm Richard Gonzmart, manager of the Columbia."

"It's a spectacularly beautiful restaurant," I said.

"Thank you. It's been in my family for more than a hundred years. Casimiro Hernandez, my great-grandfather, opened the room you're sitting in back in 1905 to serve fellow immigrants who worked in the cigar factories. He kept adding rooms, including the first air-conditioned dining room in Tampa."

"Do you have time to join us?" I asked.

"Only for a moment. Carlos told me that he was having dinner with the famous writer Jessica Fletcher. I am honored to have you here this evening. We've had many celebrities dine with us," he said. "We had a strip steak on the menu, 'The Bambino,' named after Mr. Babe Ruth, who came here often back in the twenties and thirties." He laughed. "I've been told that he would eat two fourteen-ounce steaks at a single sitting."

Seth made a face. "Hate to see *his* arteries," he said.

"So many of your baseball greats made the Columbia their home when in Tampa," he continued proudly. "Baseball is the Cuban national pastime, just as it is in America. Joe DiMaggio and his wife, the beautiful Marilyn Monroe, also used to come here." He became conspiratorial. "One night they had quite a row at the table, and Ms. Monroe went to the restroom and confided in the attendant there. Word has it that she returned to the table a much happier woman."

He went on like that for another ten

minutes, telling tales of the famous who'd dined at the Columbia. I was impressed with the obvious pride he demonstrated, not only in the restaurant, but in his family as well.

"Enjoy your dinners," he said as he stood to leave. "I have reserved a special place for you in the Patio Room. Nalda will show you to your table when you are ready to eat. ¡Buen provecho!"

"What a charming man," I said.

"I've been coming here for years," Machado said. "I feel like a member of the family." He sat back, a satisfied smile on his face. "So, how has your stay in Tampa been so far?"

"Frustrating, to say the least," Seth answered. "We're leaving in the morning."

"So you said when you called. A last-minute decision?"

"Yes," I replied without elaborating. "We understand that you stopped by the Vasquez house today."

"That's right."

"Maritza Vasquez said that you told her that you were continuing the investigation into her father's death."

Machado sighed and slowly shook his head. "It seemed to be the right thing to say," he said.

Seth and I looked at him quizzically.

"I'll be honest with you," Machado said, "because you've been honest with me. You didn't have to bring me that letter the doctor wrote to you, or the flash drives you gave me. To put it simply, the investigation is out of my hands now."

"I'm sorry to hear that," I said. "Someone new has taken over?" I asked, knowing the answer.

"You might say that," Machado said.

"The FBI," Seth said flatly.

"That's right," Machado said, shaking his head. "We're not completely out of the picture," he added. "Whether they admit it or not, they still need our local expertise. But I'd say there are several more layers of authority above us, and I don't see wasting the department's money and the time of our detectives investigating a murder if the federal government is pushing us out of the way."

Seth looked at me questioningly, and I knew what he was thinking. "We'd like to give you something else we found," I said in a low voice. "We're not sure if it's relevant to Dr. Vasquez's death, but since we can't follow up on it, perhaps you'll find it useful."

Machado looked from me to Seth.

"What's that?"

Seth pulled out the lined paper on which he'd written the list of names we'd found on Al's thumb drive. "We think these people may have invested money with Dr. Vasquez without the knowledge of K-Dex and Bernard Peters," he told Machado as he handed over the page.

"It's just a hunch we have," I added. "We know that Carlos Cespedes gave Al a considerable sum in expectation of sharing in the profits when Al's research bore fruit. His name is on this list. The others here may have done the same thing."

Machado scanned the list and smiled. "I recognize some of these names. They're among our wealthier citizens." He pocketed the paper Seth had given him. "Thanks. I'll check it out. It'll be nice to have another piece of information the feds don't know about. By the way, how did you know they'd taken over the case?"

"That's why we're leaving Tampa," Seth said.

That got Machado's attention. "Can you talk about it?" he asked.

"I suppose I'm not telling tales out of school," Seth said, "when I say that we've been asked to leave Tampa by the FBI."

"Is that so? They tell you why?"

"National security," Seth said disgustedly.

"I'm not sure that's what it is, although that's the excuse they used," I said. "I think they feel that our being here gets in their way. Since Dr. Vasquez's death seems to have moved from a local homicide to something with international consequences involving our government and the Castro regime, I also wonder whether they feel that we might be in some sort of danger."

Machado's smile was small but telling. "Who've you been talking to?" he asked. "An agent named Guterez?"

I looked at Seth before saying, "Yes."

Machado lowered his voice. "You know," he said, "the CIA's involved, too."

I didn't hesitate to say what immediately came to mind. "Would that be Karl Westerkoch?"

Another smile from the detective. "Our resident spook," he said. "He's hardly the sort of invisible spy who stays undercover."

Seth chimed in. "I still don't understand why they're sending us home to Cabot Cove," he said. "Seems to me that . . . Wait a minute, you say that Westerkoch works for the CIA?"

Machado leaned close to Seth and said sarcastically, and with mirth in his voice, "Not so loud. You'll blow his cover. More sangria?"

For dinner we were seated in the absolutely spectacular Patio Dining Room. A huge glass ceiling that could be opened covered the large space patterned after classic outdoor patios found in Andalucía in the south of Spain. Machado ordered a wide variety of tapas for us, including scallops, lobster, crab cakes, shrimp, stuffed peppers, and chicken. It was a veritable feast, accompanied by another pitcher of sangria.

"This was wonderful," I said as one of two attentive waiters cleared our table. "I don't think I'll eat another thing for a week."

"You must have dessert," Machado insisted, and so he ordered a flan and key lime pie and three spoons. It was over cups of powerful Cuban coffee that the conversation came back to the death of Dr. Alvaro Vasquez and the multiple law enforcement agencies that were now involved.

"What have you done with those thumb drives we gave you?" Seth asked.

"They're under lock and key," said Machado.

Seth's doubtful expression prompted the detective to add, "And I mean locked away. You're aware of the leaks from our department, but I assure you that those devices will be handed over to the appropriate people."

"Other medical researchers?" Seth asked. "Al — Dr. Vasquez specifically asked me to show his notes to other physicians engaged in Alzheimer's research, and that's what I intend to do."

"Then I assume you made a copy before giving the originals to me."

"As a matter of fact, I did."

"And you will carry out the doctor's wishes?"

"You bet I will. I know some top-flight researchers in Boston who'll make good use of what's on those thumb drives."

I understood why Seth didn't want to share with Machado his belief that what was on the thumb drives wasn't especially promising. As he'd told me, he didn't feel it was his place to make such judgments. But I wasn't sure it was wise not only to have copied evidence, but also to admit his actions to a police detective.

It was an evening for candid exchanges, however. I was surprised at how forthcoming Detective Machado had been. Perhaps our pending departure made him feel free to discuss the Vasquez case and others. Up until that point, he had dominated the conversation, telling amusing, interesting stories about fighting crime in Tampa. "Actually," he'd said, "we have a pretty solid record in lowering the crime rate. Of course there's always drugs and gangs, but Tampa is a relatively safe city." Then he shifted the conversation to what I assumed he had been planning to talk about all evening. "We do have occasional problems with some of the zealots in the Cuban American Freedom Foundation."

"I've heard they're a group of Cuban exiles who are against the Cuban government," I said.

"Right you are. On the other side are Castro loyalists in Tampa and Miami who get their marching orders from the Cuban Comités de Defensa de la Revolución. That's the organization inside Cuba that recruits and runs the CDRs, neighborhood spies. It's a very active and wide-sweeping organization that reports to the Cuban national police, who work for the Ministry of the Interior. We know that they have agents in Florida who report back on what the members of the Freedom Foundation are up to. That's why we work with the CIA and FBI on occasion. Real cloak-and-dagger stuff. I was wondering if you two have learned anything about the Cuban exile group while you've been in Tampa."

"I don't think we can help you there," I said. "The closest we've come to cloak-and-dagger stuff, as you put it, is that we've been followed almost every day. I thought it might be one of your men."

Machado laughed. "No, not us. We'd have no reason to follow you. Maybe it was Westerkoch. I get the feeling that he enjoys following people. Makes him feel like James Bond."

We parted on the sidewalk outside the Columbia.

"Travel safe tomorrow," Machado said.

"We intend to," I said, checking Seth's reaction, which was noncommittal.

In the car on the way back to the hotel, I caught Seth smiling.

"What are you thinking?" I asked.

"I'm thinking that what we need in Cabot Cove is good Cuban food. What do you think if I ask Ed Kim whether he'd be interested in opening a Cuban restaurant?"

Kim was a Chinese American entrepreneur who'd recently opened two small eating places in Cabot Cove, one specializing in Thai food, and the other a Spanish tapas place.

I hesitated before saying, "The problem, Seth, is that we don't have a Cuban population in Cabot Cove to support it."

"Doesn't matter," he said. "Not sure we have many Thai or Spanish people there either."

"True."

"Once people taste authentic Cuban cuisine, they'll flock to it. Besides, it's healthier than lobster chowder and whoopie pies."

I laughed. "What made you think of whoopie pies?"

"You can't be a true Down-Easter and not think of whoopie pies every now and then," he said. "I could go for one right now."

"Even after flan and key lime pie?" I asked.

Seth didn't answer, but he looked a little embarrassed.

I've never developed a taste for whoopie pies, but they are a quintessential Maine dessert staple. I teased Seth. "Then I guess you won't be unhappy about going home tomorrow if you have a package of whoopie pies in your cupboard."

"I do."

As Seth and I parted in the hotel lobby, I said, "What a wonderful evening, a perfect final farewell to Tampa. The restaurant is superb."

"Ayuh, it certainly is. Well, I'd better get some shut-eye before we take off tomorrow, provided I can get visions of a small plane out of my head."

"The flight will be fine," I assured him. "I have plenty of faith in Xavier Vasquez."

# CHAPTER NINETEEN

We were up early the following morning and in the dining room by six thirty, our packed bags checked with the concierge. Seth had turned in the rental car at the hotel when we returned from dinner; we would take a taxi to the Vasquez house on Davis Island to meet up with Xavier.

I'd picked up a copy of the *Tampa Tribune* on my way in to breakfast and showed it to Seth. Peggy Lohman had written an article about the Vasquez case in which she reported that "people inside the investigation who wish to remain anonymous" told her that the FBI was now an active participant in the investigation and that Dr. Vasquez's death had been officially labeled a homicide, the method of death poison, the cause of death acute respiratory failure.

"Keeping a secret seems to be out of the question with the police here in Tampa," Seth commented after reading the piece.

"I wouldn't blame the police," I said. "With so many agencies involved — the medical examiner's office, the FBI, and even the CIA — the sources of the leaks could be anyone, even the family."

Seth pondered that for a few moments before saying, "Dr. San Martín said that the sort of *C. botulinum* they found in Al's cigar had to have come from a very sophisticated laboratory. Remember that story about how the CIA developed a virulent strain and used it in cigars to try to assassinate Fidel Castro?"

"I do remember. But it didn't work. He'd stopped smoking by the time the CIA tried it."

"The point is, Jessica, Dr. San Martín and Agent Guterez suggested that the laboratory might have been a government-run one."

"But why would our government want to kill him?" I asked.

"Maybe it wasn't *our* government," Seth replied. "Al told me Mr. Castro and his government were pretty upset when Al defected with all the research he'd conducted there."

"Do you think those Castro agents here in Florida might be responsible?"

"Could be. Let's get over to the house

before I lose my nerve about flying in that stupid little plane."

I smiled but didn't say anything. All I hoped was that Xavier would take into account that he had a white-knuckle flier onboard and would make all his maneuvers slow and easy.

Xavier appeared to be angry when he greeted us at the door.

"Bernard Peters is suing us," he said without prompting. "I never liked him, never trusted him."

"He's basing his suit on having financially supported your father's research?" I asked.

"That among other things. I don't know what he's complaining about. He had some sort of insurance policy that paid him off in the event my father died."

*Which could have provided a strong motive to kill Alvaro Vasquez,* I thought, *especially if Peters somehow learned that the research had hit a dead end.*

"You ready to fly?" Xavier asked.

"Ready as I'll ever be," Seth said.

Maritza appeared carrying a small suitcase.

"You're leaving, too?" I asked.

"I'm going with you," she said. "I'm spending a few days with Xavier in the Keys before going back to Havana."

"Will you have trouble going back?" Seth asked as we gathered up our luggage and headed for the taxi that would take us to the Peter O. Knight Airport at the tip of Davis Island, where Xavier housed his plane.

"No," she answered without elaborating.

Xavier's Cessna 172 aircraft was a more recent model of the popular aircraft. It sat shiny and bright in the morning sun, its red and white paint glistening.

"It's a beauty," I told Xavier.

"My baby, Mrs. Fletcher. It's the R model, with a Lycoming fuel-injected engine and a Garmin avionics package, top-of-the-line, ADF, GPS, transponder. It's even got added fuel capacity in the wingtips and extra baggage compartments."

"There are four of us with our luggage," I said, aware from my days as a student pilot how critical weight was with a smaller aircraft.

Xavier grinned and asked how much Seth and I weighed.

I told him but Seth hesitated, finally admitting his heft. Xavier did a fast mental calculation, taking into account the luggage. "We should be fine," he said. "It's got a gross takeoff weight of over twenty-five hundred pounds. We'll be below limits, though it may slow us down a little."

Seth, who'd been listening, said, "If there's a weight problem, I'll be happy to volunteer to stay back and find another way to Fort Lauderdale."

"Seth," I chided.

"Just bein' generous," he said.

Xavier carefully loaded our luggage into the baggage holds and wedged a few small pieces behind the two rear seats. "Let's see," he said. "Mrs. Fletcher will want to do some of the flying, so she'll sit up front with me. Maritza, you and Dr. Hazlitt sit in back."

"I'd love to fly," I said, "but I think Seth would be more comfortable up front. There's more leg room."

Seth eyed the cramped rear seats and said, "If you wouldn't mind, Jessica."

"I wouldn't mind at all," I said, slightly disappointed that I wouldn't be taking the yoke and flying the plane, but more concerned with my friend's comfort.

The plane's leather seats were comfortable, and there was more room in the rear than I'd anticipated. I settled in next to Maritza, fastened my seat belt, and observed as Xavier walked around the plane to visually check its condition. He climbed into the pilot's left-hand front seat, ran down a printed checklist, cracked open his window and yelled "Clear!" to inform anyone nearby that he was about to start the engine, and set the propeller into motion.

Seth sat ramrod straight, as though to move would in some way cause the plane to blow up. He watched everything Xavier did, including using another checklist to run down various engine settings. He handed Seth a set of earphones attached to a tiny microphone. "Thought you might like to listen in," he said. Seth reluctantly put it on, and I was glad that Xavier had thought to offer the set to Seth. It would occupy his thoughts and take his mind off his anxiety. Or so I hoped.

Xavier donned his own microphone and earphones and informed the tower that he was starting his taxi to the runway. He was cleared and slowly moved down a taxiway until reaching the runway in use that morning, its designation based upon the wind's direction. Planes take off and land into the wind whenever possible. After some more chatter with the tower, he turned onto the runway, advanced the throttle to the firewall, released the toe brakes, and started his takeoff roll. I was concerned at how long it took us to become airborne, but I chalked up the extended takeoff to the weight of the plane. Eventually we lifted off. Xavier banked, affording us a view of downtown Tampa. From my seat behind the pilot, I could see Seth squeeze his eyes shut when the plane tilted in the air.

Xavier continued his climb until he'd reached his desired cruising altitude. He adjusted the controls and looked back over his shoulder. "There's a sectional chart in the pocket behind my seat," he said. "We'll be heading down the west coast until we reach the Naples beacon, then fly due east to Lauderdale."

Xavier's plane was considerably quieter than the older model in which I'd taken my flying lessons in Cabot Cove from Jed Richardson, and its smooth flight through the air at five thousand feet was almost hypnotic. I noticed that Seth nodded off a few times, snapping his head up when he realized that he had. I, too, had to fight dozing off despite Xavier's occasional commentary pointing out sights along the coast and on the ground. Maritza didn't contribute to the conversation during the trip. She'd barely said a word from the time we'd taken off until we reached Naples, where Xavier was to alter his course.

In order to head for Fort Lauderdale, we would have to fly due east, which was what Xavier had said he intended to do. But as I followed our course on the aeronautical chart, I was aware that we were now flying southeast, which would take us south of Miami. I debated asking Xavier about it but held back. This was, after all, his plane, and he was the pilot in command. He'd probably changed course because of the weather forecast for east of Naples, or perhaps he'd been instructed to alter his flight plan by air traffic control.

But as we continued in the southeasterly direction, I decided to ask why we'd changed course.

"Weren't we supposed to fly east when we reached Naples?"

He didn't answer my question.

"Xavier, I'm just curious why the change in our course," I said louder.

When there was still no reply, I leaned forward and tapped him on the shoulder. That was when Maritza tapped *me* on the shoulder.

I turned to see her pretty face set in a scowl. Then I noticed the small handgun she held. It was pointed directly at me.

# CHAPTER TWENTY

"What are you doing?" I said in a loud voice to Maritza.

Seth heard me, turned, and peered through the space between the front seats. "What's the matter, Jessica?"

I started to explain, but the words wouldn't come. I pointed to the gun in Maritza's hand.

It was difficult for Seth to see, but when he did, he said, "Gorry, what in the world is going on?"

"I'd suggest you both be quiet," Maritza said coolly and calmly. "I won't hesitate killing you."

"This is . . . this is outrageous," I said, realizing as I did that it was a pathetically weak response to a powder-keg situation.

"Just what do you think you're doing?" Seth asked Xavier.

The young pilot turned to Seth, and I could see the traces of a smile on his face.

"Righting a wrong," he said.

"*What* wrong?" Seth demanded.

"You'll know soon enough," he said.

I looked out my window in search of a landmark below but saw nothing distinguishable that would indicate where we were.

"Xavier," I said loud enough to be heard over the engine's noise, "where are you taking us?"

When he didn't answer, I asked the same question of Maritza.

"Just shut up, Mrs. Fletcher, and enjoy the ride."

"No, that's not good enough," I said. "This is kidnapping. Don't you realize that you're engaging in a criminal act?"

She laughed and said, "The only criminal act was taking my father's research away from the people it belongs to, the Cuban people."

"What does that have to do with us?" I asked.

She guffawed. "You and your doctor friend got a lot closer to my father than anyone else. If anybody knows what stage my father was at in his research, it's you two."

"That's absurd," I said. "I'd just met your father before he died."

298

She waved her gun at Seth. "But your friend was in Tampa a lot and spent plenty of time with my father, reading his notes, hearing about the research."

Seth, who'd undone his seat belt and was trying to twist around, was admonished by Xavier in no uncertain terms. "Sit still! I don't need you wiggling around while I'm flying this plane. Do you want to end up in the water? My sister means business. She'll blow your brains out before we land if you don't cooperate. Now, put your seat belt back on."

Seth and I fell silent until I said, "You're taking us to Cuba, aren't you?"

"How did you ever guess?" Xavier said as he banked sharply, pushing Seth against his door and exerting g-forces on everyone. Seth quickly buckled his seat belt. We looked at each other and simultaneously came to the same unstated conclusion. There was nothing to be gained by fighting the situation, not at five thousand feet above the ground, and with someone holding a lethal weapon in her hand. We settled back in silence as Xavier continued en route to Cuba.

When he'd reached a point over the ocean off the Florida coast, he communicated with Cuban air traffic control. I knew from having read about others who'd violated Cuban air space without permission that the result could be deadly. Xavier obviously knew the system, complete with language that would allow him to approach the island without fear of being intercepted by Cuban military aircraft.

We could do nothing but sit stoically and watch the island of Cuba come into view. I was aware that the José Martí International Airport, Cuba's main aviation hub, was located close to Havana, and that many international airlines scheduled flights into it on a daily basis. Xavier continued radio contact with Cuban air traffic controllers in Spanish, although English is the universally accepted aeronautical language throughout the world. Maybe the sort of special dispensation he had with Cuban authorities called for all communication to be in his native tongue.

I saw a Virgin Atlantic jet approaching for a landing, and a KLM jet turning to enter the pattern. Xavier was in a holding pattern, waiting for his landing instructions, which would come once the jets had landed and had cleared the runway. Getting too close to the jet blast from a large aircraft could flip a smaller plane like the Cessna.

Finally, Xavier banked sharply, and the

runway appeared ahead. It looked to be very long; our plane would take up only a fraction of its length. We touched down, bounced up, then settled on the asphalt runway and stayed there. Xavier turned off the runway at the first available exit and taxied for what seemed an eternity to a large building on which a sign said TERMINAL 2. He headed for an isolated area at the far end, where a half dozen armed men wearing drab green pants and shirts, and with wide belts over their shirts, stood. Five of them wore black berets; the sixth's hat had a visor, and I assumed that he was an officer. All had automatic weapons slung over their shoulders.

"So," Seth growled at Xavier, "what are we supposed to do now?"

"Just do as you are told," he replied. "Do not try anything foolish."

"You realize that kidnapping us will kick back on you," I said. "I called people back home to tell them of our travel plans. Once they haven't heard from us, they'll want to know why."

Neither Xavier nor his sister said anything in response as he brought the Cessna to a stop, shut off the engine, opened his door, and hopped down. He was approached by the soldier who I assumed was in charge and they had a spirited conversation, with Xavier frequently pointing back at the plane. The officer issued a command to the five soldiers, and they surrounded us. One opened Seth's door and ordered him to get out. Maritza said to me, "You're next," and waved the handgun for emphasis. I pushed Seth's seat forward and managed to slide past it, then step down on a foothold and to the ground.

The officer barked an order in Spanish, but because we didn't understand what he said, we didn't move.

"That way," Xavier said, pointing to a doorway at the end of the terminal.

With the soldiers flanking us, Seth and I followed Xavier and Maritza to the door. Another uniformed man opened it, and we entered a large gray cinder-block room with a table and four chairs, the only furniture in the room. We were told to sit, which we did, of course, and waited for what would come next. Maritza had disappeared, but Xavier took one of the chairs. After five minutes of silence, I decided to take a stand.

"I insist on seeing someone from the United States embassy," I said in as strong a voice as I could muster.

"You have no embassy here in Cuba," Xavier said.

"But we have offices under the auspices of the Swiss embassy, as I understand it."

"You have nothing to talk to them about," he said.

"We're American citizens who've been taken hostage. We've been hijacked to Cuba."

"Look, Mrs. Fletcher," Xavier said, "we've brought you here for a good reason. All you have to do is cooperate and everything will be all right. No one wants to see you or Dr. Hazlitt hurt. Just do what you're told, tell the officials here what they want to know, and everything will be fine."

Seth, who'd said nothing during my exchange with Xavier, now spoke up. "Your father must be turning in his grave the way his only son has turned out."

"My father knew how I turned out, Dr. Hazlitt, and frankly I couldn't care less what he thought. He was a traitor and deserved what he got."

"A traitor?" Seth barked. "A traitor to what?"

"His people."

"And just who are they?"

"The people of Cuba, *my* people. My esteemed father abandoned them while he went looking for money. The Cuban government financed his research here, and what did Papi do? He ran away with the results, defected, sold out to Yankee imperialism, to that fraud Peters."

"But you defected, too," I said.

"In preparation."

Seth and I were about to ask him to explain, but it was obvious that none was needed. He'd been sent to Tampa by the Cuban government to spy on anti-Castro people and to sabotage any actions they might contemplate against the Castro regime.

Seth and I asked a few more questions without receiving answers. The next ten minutes seemed like hours as we sat in our uncomfortable chairs, eyes nervously going to the armed soldiers standing in each corner of the room. What could they possibly have in mind for us?

While Seth knew something about Alvaro Vasquez's research, it hardly represented the sort of knowledge that would be of help to someone in Cuba. Besides, Seth had come to the conclusion that Vasquez's research efforts hadn't led to anything tangible or useful in the fight against Alzheimer's disease. Would they believe Seth if he told them that? And if they didn't believe him, to what extent would they pursue it? Torture?

Would I ever have thought it possible that Seth and I would end up being kidnapped and brought to Cuba? The more I pondered the situation, the colder the room seemed to get, and I wrapped my arms about myself.

I was deep in those thoughts when the door opened and a middle-aged man entered. He wore a gray suit, white shirt, and tie and carried a briefcase. Xavier jumped up, greeted him, and pointed to his chair, which the man took. He opened his briefcase, laid a stack of papers on the table, smiled at me, and said in good English tinged with his Cuban accent, "Welcome to the People's Republic of Cuba, Mrs. Fletcher." He said the same thing to Seth, referring to him as Dr. Hazlitt.

"Who are you?" Seth asked.

"I represent the Ministry of the Interior," he said matter-of-factly.

"No, what's your name?" Seth insisted.

"That is of little interest," he countered. "I am authorized to welcome you and to see that your stay is a pleasant one."

"Our *stay*?" Seth erupted. "Mrs. Fletcher and I are not staying here one more minute. I demand that you either see to it that we are delivered back to the United States or put us in touch with U.S. officials here in Cuba."

The man listened impassively. When Seth was finished with his demands, the Cuban official said in the same flat, low-key voice, "All in due time, all in due time — provided that we have your cooperation."

"Cooperation in what?" I asked, buoyed by Seth's spark. "Why have we been brought here? Who ordered that we be brought here?"

The man sighed, slipped the papers back into his briefcase, and stood. "You obviously do not intend to make this easy for yourselves. Surrender your cell phones to the officers. Enjoy your stay in Havana. I will see you again this evening." He barked an order at Xavier, who jumped up and followed him.

"You have no right to take our phones," I called out as they walked from the room, to be replaced by two other men wearing army uniforms. They were older than the soldiers we'd first encountered. One of them repeated the order that we turn over our phones. Seth balked, but the menacing look on the soldiers' faces, fortified by the weapons in the hands of the others, won out, and we handed them over.

"Get up!" one of them ordered us.

"Now, look," Seth said, "I won't tolerate this. I demand to know under what authority we've been brought here."

The men came to either side of Seth and placed their hands on his arms. Seth tried to shake them off, but they tightened their grip.

"Seth," I said, "don't argue. I'm sure that they'll see that this has been a big mistake and let us go."

My advice calmed him, and he allowed the men to escort him from the room. I followed, flanked by two of the younger soldiers, though they refrained from touching me. We went down a dank concrete corridor and out a different door from the one we'd previously entered. Waiting was a gray Mercedes limousine with blackened windows.

We were instructed to get inside the car. We followed their orders and settled in the backseat, with two of the soldiers facing us on jump seats, another in the front passenger seat. The doors were closed and locked, and the driver sped off, destination unknown.

It was a bumpy ride that lasted twenty minutes. I was certain we were in the city because of the sounds heard outside the limo: music, laughter, cars with loud, damaged mufflers, and even a few loud reports that might have been gunshots. Eventually we came to a stop, the doors were opened, and we got out, blinking in the glaring sunshine. One of the men popped open the trunk and I was relieved to see that our luggage had been transported with us. However, I noted that a piece of red tape had been affixed to each of our bags, suggesting to me that they had probably been searched before being put in the trunk.

We stood in front of a small cabin with a porch situated on a lovely stretch of pebbly beach. I looked around in search of others, but Seth and I and our handlers — if that's the right word — were the only people there.

The door to the cabin opened, and a tall man wearing a bright yellow guayabera, tan slacks, and sandals sans socks emerged. He walked down the two steps leading to the porch, came directly to me, and extended his hand. "Mrs. Fletcher, welcome to Havana."

Seth came to us and said, "I want you to know, whoever you are, that I'll report this outrageous situation to our State Department."

"I hope there won't be any need for that," the man said in perfect English. "My name is Dr. Eduardo Rodriguez of the Health Ministry. I will be your host while you are here."

"Host?" Seth said. "You make it sound

like we're on a vacation."

Rodriguez's laugh was gentle. "I would hope that you would consider it a vacation of sorts," he said. "You have been here before, Dr. Hazlitt, and can appreciate our country. And once you have satisfied the authorities with answers to their questions, I am authorized to extend to both of you some days of sun and relaxation, at the expense of our government, of course."

At that moment, I was dazzled with confusion. We'd been forcibly kidnapped and brought to Cuba by Xavier and Maritza Vasquez, who said that their father's research on Alzheimer's disease had to be returned to the Cuban people. We'd been greeted by armed soldiers, made to sit in a stark concrete room, threatened if we didn't cooperate, piled into a car and driven to a cabin on a beach, and now offered an all-expenses-paid vacation in sunny Cuba.

I could do nothing but laugh, which seemed to please Rodriguez, who laughed along with me.

"You say you're a doctor?" Seth said. "So am I. This is one heck of a way for a physician to act."

"I am sure that after a good night's sleep, you and your charming companion will be in better spirits, Dr. Hazlitt. You'll find that your accommodations are satisfactory, and that the meals provided for you will be likewise. A car will pick you up here this evening at six. In the meantime, I suggest that you relax, freshen up, perhaps nap. *Buenas tardes.*"

He walked to where the officer stood with his men and issued instructions in Spanish. Two of the soldiers came to us and indicated through hand gestures and a smattering of English that we were to go into the cabin.

"Let's do as they say, Seth," I said.

The cabin consisted of one room, a pair of twin beds with colorful bedspreads, a small desk and chair, a brown vinyl easy chair that had seen better days, and a tiny bathroom lighted by a single wall lamp, in which towels and a bar of soap were provided. There was a window at the front. I looked out and saw that the soldiers had taken up positions thirty feet away. They sat on a low stone wall, their weapons resting on the ground, their faces mirroring their boredom.

Seth plopped himself in the easy chair and I sat at the desk.

"Are you all right?" I asked.

"Tired, that's all. You?"

"I'm fine. I'm just wondering whether there's a way for us to escape."

"Don't be foolish, Jessica," he said. "We're on an island. Where would we escape to? More of Cuba?"

"Can we get word out to someone that we're being held prisoner?"

"I don't see how. They took our phones."

"It's stuffy in here," I said. "I'm going out on the porch."

As Seth followed me through the door, two soldiers picked up their weapons and approached. They shouted orders, presumably for us to return to the cabin, but I said to Seth, "Ignore them."

We defiantly sat in two green-and-white webbed chairs and stared at the soldiers, who continued to issue orders and used their weapons to indicate that we were to go inside.

"Ignore them," I repeated. "We're here because someone up the chain of command wants something from us. They won't dare hurt us."

I wasn't sure that my bravado was appropriate, but it seemed to work. The soldiers muttered under their breath and returned to their colleagues. They must have found the situation humorous, because they all laughed at something one of them said.

"The way I figure it, Jessica, there's one good thing to come out of this," Seth said.

"What is that?"

"The question of Al's death and why he died. I keep thinking about what Dr. San Martín said about the strain of *C. botulinum* that was in that cigar Al smoked just before he died, that it had to have been concocted in a sophisticated lab, most likely a government lab. Seems to me that it had to have come from here in Cuba."

"That makes sense to me," I said.

"And if that's true, the next question is, who gave Al that cigar?"

"He said it was a gift from someone special."

"He also said it was Cuban," Seth said.

I knew where he was taking this and didn't want to acknowledge it.

"Had to be one of his children," he said grimly.

I started to respond, but he cut me off.

"That's not a pleasant thought to have," he said, "that a son would kill his own father."

"Or a daughter," I said. "But we don't know whether that's true."

"How else could it have happened? They obviously resented their father defecting with the research he'd started in Cuba, research that was paid for by the Castro government. Xavier said it himself, that the research belongs to the Cuban people. The cigar was Cuban. He must have had his sister get it for him. Who else would have easy access to Cuban cigars? The way I see it, Xavier and Maritza murdered their father to get hold of his research and return it to his native land. I'll bet you all the moose in Maine that Xavier also ended up with Al's laptop."

"Then why bring us here?" I asked. "If they have the laptop, they have the research."

"I don't have an answer for that, Jessica, but I suppose we'll find out tonight when they come get us. Meantime, I'm going back inside and catch me a nap. I suggest you do the same. We'll need our wits about us this evening."

I was too keyed up to nap. I stayed on the porch for an hour before returning inside, where Seth was snoring loudly on one of the beds. I curled up on the other and tried to fall asleep but to no avail. The afternoon passed slowly, but eventually it was time to get ready for whatever the next step would be. We walked out on the porch just as the gray limousine arrived. Dr. Rodriguez came to us, a wide smile on his face. "Did you rest?" he asked.

"No," I said.

"I am sorry that you didn't. I imagine that you and the doctor are hungry."

"As a matter of fact, I am," said Seth. "Hungry and damn mad."

"I assure you that we will feed you well," Rodriguez said. "Come. Time to go."

After another bumpy half-hour drive, we came to a stop in a sprawling square, where an intimidating eight-story building stood, its drab front dominated by a soaring black metal mural bearing the unmistakable face of Fidel Castro's revolutionary colleague Che Guevara. The mural dwarfed the building, and the setting sun gave the picture an eerie three-dimensional look.

"Where are we?" I asked Rodriguez.

"The Plaza de la Revolución," he answered pleasantly as two of the soldiers took up positions on either side of Seth and me. Back home in the United States, the sight of two obvious foreigners being escorted across a wide plaza by armed soldiers would have aroused the interest of onlookers, but no one — and I judged there to be a couple of hundred Cubans milling about — paid us any notice.

"Why are we being taken here?" Seth asked, indicating the building.

"We are not going there," Rodriguez replied. "That is the Ministerio del Interior building. We are going to *that* building." He pointed to an equally uninspired gray building across the square. "It is where our Maximum Leader, Prime Minister Castro, and his brother have their offices. It is also headquarters for the Central Committee of the Communist Party." He laughed. "See?" he said. "We treat you especially well. Few tourists see the inside of our most important buildings."

"I don't care whether we see the inside of your buildings or not," Seth said. "I already saw this square when I visited Havana as part of a tour. Now I wish I'd never come."

"Follow me," Rodriguez said. "Just a few questions to be asked."

"And what if we don't want to answer them?" Seth asked, sticking out his jaw for emphasis.

Rodriguez's reply was a noncommittal shrug.

We were propelled toward the gray building, where armed guards checked papers that Rodriguez carried before allowing us entry. He led us down a long hallway past dozens of closed doors until reaching an elevator manned by an armed guard. It carried us up two floors, where we exited and walked across the hall into an office's anteroom. An open door led to a conference room with a small black-and-white marble table and six comfortable yellow swivel armchairs. A large color photograph of Fidel Castro dominated one wall. The other walls contained smaller framed photographs of Havana scenes intermingled with pictures of Castro posing with foreign dignitaries. There was only one person in the room when we entered, the gentleman from the Ministry of the Interior who'd spoken with us at the airport. While Rodriguez was congenial and treated us politely, this man had the look of someone who'd been absent when genes of sympathy and sensitivity had been handed out. He sat ramrod straight at the head of the table, small, rimless eyeglasses catching the light from overhead fixtures.

"Please take a seat," Rodriguez said. "I will be back in a moment."

As Seth and I took chairs, our eyes went back and forth between the interior ministry representative and the only item on the table, a laptop computer. I was about to whisper something to Seth when the door opened and Rodriguez entered, followed by Dr. Pedro Sardina.

"I believe that you know Dr. Sardina," said Rodriguez.

"Yes, we do," I said, "and I must admit,

I'm shocked to see him."

Sardina said nothing as Rodriguez took a chair, opened the laptop, and sat back as though expecting us to comment on what it revealed.

"Is that Alvaro Vasquez's laptop?" Seth asked.

"It is," replied Rodriguez.

I looked at Sardina, who sat stoically, his eyes fixed on the tabletop.

"How did you get it?" I asked.

"That really doesn't matter, Mrs. Fletcher," Rodriguez said. "What's important is what is not on it."

"*Not* on it?" I said.

"Yes. Unfortunately someone — and I must assume it was Dr. Vasquez — saw fit to remove the hard drive from the computer. There is nothing on the computer. Absolutely nothing."

I addressed my next question to Sardina. "Did you bring the laptop here to Cuba? Or was it Xavier?"

Sardina looked at Rodriguez, who told him, "You can answer her, Pedro."

"We brought it here together," Sardina said.

"Even though there was nothing on it?"

"We didn't know that. The night Dr. Vasquez died, Xavier hid it away until we left for Havana. We assumed it had all the notes."

"You can imagine how disappointed we were when we turned it on and nothing worked," Rodriguez said. "We had been awaiting its arrival with great expectations. Needless to say, it was a very unpleasant discovery."

"So why have you brought us here?" Seth asked. "You've broken the law, United States law and international law, too. Believe me, sir, someone will pay for this."

Rodriguez told Sardina to bring Xavier to the room. They returned a moment later and joined us at the table.

"Dr. Hazlitt wishes to know why we have invited him and his lovely friend to join us," the Cuban doctor said to Xavier. "Perhaps you are the one to explain."

"I'd like to wipe that smirk off your face," Seth said to Xavier.

"Big, tough talk from you," Xavier said. "You're just like my father, lots of talk, always talk."

"And your father is dead," Seth said, "murdered by the son he loved."

I felt a wave of disgust come over me, and I knew that Seth was feeling the same thing.

"I don't care what you think of me," Xavier said, the nasty smile still on his lips.

"What is important is that my father's research benefits the Cuban people."

"And if it doesn't do that," I said, not trying to hide my anger, "was it still worth murdering your father?"

"It wasn't murder," Xavier said. "It was a political assassination. If he hadn't defected, he'd still be alive. But no, he was greedy, and he and my mother abandoned everything good that was given to them and done for them here in Cuba, to sell out for a fancy home and a boat. You know what we call Cubans who defect to the United States? We call them *escoria*. Scum. Worms."

"You are a despicable young man," Seth said. He turned to Dr. Rodriguez. "And you would sanction a murder in order to get your hands on Alvaro Vasquez's research?"

"It was not my decision," Rodriguez said, casting a glance at the Ministry of the Interior's representative. "While I agreed that Alvaro Vasquez's work belonged here, I would never have suggested that he be killed in order to accomplish that. There are always other ways of recovering what has been lost. Alvaro was a friend and colleague. Whenever he mentioned defecting, I always tried to talk him out of it, but I never took him seriously. When he actually did leave for the United States, I was shocked and dismayed."

He looked again at the other official at the table, and I had the feeling that Rodriguez was concerned that he might be viewed as having cooperated with Vasquez's defection. I believed him when he said that he would not have wanted Vasquez assassinated and had tried to talk him out of defecting. Whether the Ministry of the Interior official also believed him was something Rodriguez would have to deal with.

Rodriguez picked up where he'd left off. "But this has all resulted in a blind alley, I am afraid. That Alvaro's laptop has been dismantled leaves us in a quandary, Dr. Hazlitt, one that we hope that you will resolve."

I hadn't seen Seth smile since we left Tampa, but now he shook his head in response to the situation, an ironic smile playing on his lips.

"I can't believe how wrong you people could be," he said. "You assumed that because Al Vasquez and I became friends, and because we spent time together over the past few years discussing his research, that I would know anything about his results. Want the truth?" Seth stood and leaned on the table. "Here's the truth. Al never told me anything that could be used by anyone else because every avenue he explored in his research resulted in a dead end. There were no advances toward finding a cure for Alzheimer's. Every test failed. Every hypothesis proved a waste of time."

"You lie!" Xavier snapped. "He was paid millions by Bernie Peters. You think Peters paid that money for nothing?"

"That's exactly what I think," said Seth. "No one liked and admired your father more than I did. I treasured the friendship that developed between us, was thrilled that such a great man would call me — an insignificant country doctor — his friend. But since his death, I've also come to realize that he was not the man I thought he was. Maybe I even suspected it before then — or at least I should have."

"Meaning what, Dr. Hazlitt?"

Seth drew a deep breath before continuing. "From what Mrs. Fletcher and I have learned, Alvaro might not have been completely honest when it came to his business dealings. I believe that even though his research wasn't getting anywhere, he continued to claim that he was on the verge of a breakthrough, encouraging his backer, Bernard Peters at the pharmaceutical company K-Dex, to keep funding his laboratory. Despite his failures, he manipulated Peters

— and maybe others, too — into giving him more and more money without any hope of a return."

"That's quite an accusation," Rodriguez said.

"And I hate making it."

"But why should we believe you?" Rodriguez asked. "Everything that Alvaro ever told me, and based upon information we've received from friends in Tampa, he was making progress."

"Friends in Tampa?" I said. "You mean those spies your government has in Florida who report back on Cuban American citizens living there?"

The man from the Ministry of the Interior cleared his throat but said nothing.

"Please sit down, Dr. Hazlitt," Rodriguez said.

"No, I won't sit down," Seth said. "You've brought Mrs. Fletcher and me here under false pretenses in order to satisfy your government's need to benefit from Alvaro's work. Let me tell you something, Dr. Rodriguez. I've always had mixed emotions about where Alvaro's research should end up. On the one hand, I was pleased that a cure for Alzheimer's would come out of the work of someone in America. At the same time, I felt sympathy for Cuba, where Alvaro's work started and was supported for years. But the truth is that neither of our countries will benefit from his research because it didn't lead anywhere, not to a deeper understanding of the disease, nor to a cure. In other words, Dr. Alvaro Vasquez was murdered by his son for no good reason."

"That's not true!" Xavier shouted, pounding the table with his fist. "My father told me he'd already made significant progress." He looked at the official from the Ministry of the Interior, who was scowling at him. "He did. He even said that he had developed an outline of the next steps in his research that would lead to discovering a cure. Everything was in place. He just had to follow his outline."

"He lied to you," Seth said. "There was no outline."

"Perhaps he simply wanted you to have faith in him," I added.

"No! It's true. It must be true."

Rodriguez turned his mild gaze on Vasquez's son. "That's enough, Xavier. We will talk later," he said. To Seth, he said, "I would like to believe that you have told us the truth about Alvaro's research results, Dr. Hazlitt, but I have only your word for it. That is not enough for our government."

"Then maybe these will convince you."

Seth reached into his sport jacket pocket, withdrew the three thumb drives, and slapped them down on the table with such force that everyone jumped.

# CHAPTER TWENTY-ONE

"What are those?" Rodriguez asked once the initial surprise had dissipated.

*"These,"* Seth said, "are what was on Alvaro's laptop, all his notes on his research."

I felt myself release a sigh of relief. When I'd seen the red tape on our luggage, I'd feared that our captors had found the thumb drives and confiscated them. I hadn't said anything to Seth. Our situation was upsetting enough. But he had been wise enough to keep such valuable items on his person, and we'd been fortunate not to be searched personally.

"Where did you get those?" Dr. Rodriguez asked.

"Dr. Vasquez left these thumb drives to me in a letter he wrote."

Rodriguez asked Xavier, "Did you know about these?"

"Yes," Xavier replied. "*He* told me about

them yesterday. My father must have transferred his notes to them before he destroyed the laptop."

"You have the letter?" Rodriguez asked.

"No," Seth said. "I left it with the Tampa Police Department, along with a set of the thumb drives."

"The police?" Xavier said. "You never mentioned them."

"That's right," said Seth, "the same police who'll see that you pay for your father's murder."

Xavier's smile grew wide. "Don't count on it, Dr. Hazlitt. Your police don't have jurisdiction here in Havana."

Rodriguez had picked up the thumb drives and held them as though weighing their contents. "You say that Alvaro's research led to nothing?" he said.

"That's right, Doctor. As much as I hate to admit it, Dr. Vasquez's work didn't cast any new light on Alzheimer's disease. I wish it weren't so."

"Don't believe him," Xavier said. "I bet that when you see what's on those thumb drives, you'll know that he's lying."

Rodriguez said to the Ministry of the Interior's representative, "I'll look at what these devices contain overnight, sir, and report to you in the morning."

"What about us?" I asked.

The ministry official answered, "You must remain overnight until Dr. Rodriguez completes his examination."

"In that case, I need to make some calls back home," I said.

"That will not be possible," the ministry official said in a voice that warned arguing with him would be a waste of time.

Nevertheless, I tried. "There are people who are expecting us to arrive home today," I said. "They'll be worried when we don't show up."

"Depending upon what Dr. Rodriguez reports, you may be allowed to use your phone tomorrow. For now, this meeting is ended." He stood and picked up his brief-case from the floor. "I will expect to hear from you, Dr. Rodriguez, no later than eight o'clock in the morning."

Rodriguez was visibly relieved to see the ministry official leave. There was little doubt who was in charge, and I had the sense that the doctor was feeling pressure to come up with something that would please the taciturn government bigwig. "Well," he said, "there is nothing more to do here except to see that you have a satisfactory dinner and a good night's sleep."

When the limousine pulled up in front of

the cabin on the beach, a small white panel truck was also there. A young man wearing a wide-brimmed straw hat and smoking a big cigar hopped out and came to where Seth and I stood with Dr. Rodriguez and our military escorts.

"Your dinner is here," Rodriguez said. "I took the liberty of ordering for you."

The young man removed an insulated box from the truck.

Rodriguez spoke to him in Spanish, and the driver carried our dinner into the cabin.

"A hot meal and a bed," I said with a laugh. "I suppose we should be grateful we're not in a jail cell."

"You have a water view, too," Rodriguez said, smiling. "I trust the accommodations are satisfactory."

"And if they weren't?" I said half seriously. "Would we be given suites in your best hotel?"

"I am afraid not," he replied. "Please, go inside and enjoy your dinner before it gets cold."

Seth, who by this time had calmed down, trudged to the cabin.

"I assume, Mrs. Fletcher, that I needn't underscore the importance of you and Dr. Hazlitt remaining here for the night," Rodriguez said. "We will not take it well if you decide to go for a stroll on the beach."

He needn't have bothered to remind me. I was hardly going to plan an escape, not with the armed soldiers in attendance. Even if we had been able to evade their notice, where would we go? We had no access to a vehicle by land, sea, or air. And while we knew we were close to Havana, the nearest American refuge was all the way on the other side of Cuba at the Guantanamo Bay Naval Base.

"Good night, Dr. Rodriguez," I said.

He leaned close to my ear and said, "Believe me, Mrs. Fletcher, this is not how I would have arranged for your visit. Have a good night's sleep. Tomorrow the sun will shine. It always does in Cuba."

Seth had his suitcase open on the bed when I came into the cabin. He grunted. "Nothing seems to be missing." He turned to me. "Pretty clever of me to keep the thumb drives in my pocket, don't you think?"

"I do," I said. "Thank goodness, too, because they're our only proof of what took place in Dr. Vasquez's laboratory. Without them, the Cubans wouldn't have believed you. I'm sorry to say, though, I doubt if the man from the Ministry of the Interior will let Dr. Rodriguez return them to you after he's examined them."

Seth heaved a great sigh. "You're probably right. Shame I can't follow up with my plan to give them to the researchers in Boston. At least Al's work might have saved someone else from following a wrong turn in the future."

I set our dinner out on the desk — slices of pork in a marinade, served over black beans and white rice, lettuce drenched in a tart dressing, and two loaves of bread with olive butter. Also in the box were two slices of key lime pie, and napkins and utensils. The big surprise was a bottle of red wine and two plastic glasses.

"I suppose we should count our blessings for little things," I said. "It's a nice meal."

"The least they could do for us."

I uncorked the wine, filled the glasses, and held mine up in a toast. "Here's to our visit to the People's Republic of Cuba," I said.

"Not funny, Jessica."

"Better than wallowing in the situation we're in," I said. "Come eat. The pork looks good."

After dinner I settled in the easy chair and got back to a novel I'd started in Tampa. I'd resolved to try to maintain a sense of humor and to keep things in perspective. There was nothing we could do about our situation, at least not until morning, and that wonderful Serenity Prayer popular with members of sobriety groups came to mind: "Grant me the strength to accept things I cannot change, the courage to change the things I can, and the wisdom to know the difference."

Seth had moved outside to sit on the porch. When my eyes started to close, I dog-eared the page I was on in the book and joined him. "I'm going to bed," I said.

"Seems like a sensible decision," he said. Then he started laughing.

"What's so funny, Seth?"

He waved his arm at the door. "Here we are about to share a cozy little cabin in Cuba. Can you imagine what those nosey parkers back in Cabot Cove would have to say about that? Mara's Luncheonette would be buzzing."

"It would certainly give them something to gossip about."

"Don't want to give 'em any ammunition. I'll sleep out here on the porch if you'd like."

"Don't be ridiculous," I said. "We both need a good night's sleep. Let's just not tell them about this part of our adventure. I'll go inside and wash up, change, climb into bed."

"And I'll be along shortly."

I was almost asleep when I heard him come in, use the bathroom, and get into his bed, and I realized how fortunate I was to be in this predicament with someone like Seth Hazlitt. That was my final thought as I drifted off.

To my surprise, I awoke refreshed, having slept solidly. I'd heard Seth get up and shower, and he was on the porch when I emerged dressed and ready for what the day would bring.

"Good sleep?" he asked.

"As a matter of fact, yes. You?"

"Not bad, not bad. Wonder what kind of sleep *they* had." He tipped his head toward the guards who were sitting on the ground under a tree. "When do you think they'll be by to get us?"

"I have no idea." My watch said eight thirty.

"Could use some breakfast," he said, patting his stomach.

"Blueberry pancakes at Mara's?"

"Don't be cruel, Jessica."

The limousine came up the road and pulled to a stop.

"You both look rested," Dr. Rodriguez said.

"Dinner was good," I said.

"Speaking of that, how about breakfast?" Seth said. "Expect breakfast at a five-star resort like this."

"I don't blame you for being irritable," Rodriguez said. "Please pack your suitcases and come with me. We have a stop to make, and then we will make sure you don't go hungry."

Seth and I looked at each other. *Did packing our bags mean we would be leaving Cuba?* I asked.

"Yes, but we have things to do before you go."

Since we'd never really unpacked, we were back on the porch within ten minutes. Two soldiers took our luggage and put it in the limo's trunk as Seth and I got in the back along with Rodriguez. A solitary armed soldier shared the front with the driver.

"Where are we going?" I asked.

"To an interview."

*Interview?* I hoped it wasn't an interrogation.

"Could we roll down these darkened windows?" I asked.

"Yes, but why?"

"So I can at least see something of Havana."

"Of course," he said, and the rear windows

were lowered halfway.

As we drove slowly through the streets, Havana came alive to me. I saw the island's fabled vintage American automobiles, kept running by ingenious Cuban mechanics; stall after stall of vegetables, fruits, and cigars; uniformed police directing traffic with a flourish; horns blowing in a cacophony of sounds; men, women, and children walking with purpose; street musicians performing for donations; and fascinating buildings, most in desperate need of repair but painted in gaudy island colors, augmented by the strings of laundry drying on the balconies. I was filled with a sense of what a wonderful island this used to be before it fell to Fidel Castro and his Communist leanings. But even though its present-day plight was evident, the spirit of its people was palpable. I wanted to return of my own volition and soak it in.

Seth, too, responded to the vivid scenes outside our windows. "The Cubans are lovely, warm, and gracious people," he said. "They couldn't have been nicer to me and the other physicians that last time I was here. Mind you, I'm not talking about the officials we spent most of our time with, our so-called 'handlers.' " He glanced at Rodriguez. "No offense," he said.

"None taken," Rodriguez replied.

"I had a chance to mingle with some of the common folk on the island," Seth continued. "Despite the lousy situation they're in thanks to Mr. Castro, they're filled with pride and optimism that it'll change one day."

Rodriguez cleared his throat.

"No sense in contradictin' me," Seth told him. "That's what I believe, and you're not going to convince me otherwise."

"I wouldn't dream of it," Rodriguez said, "but I would suggest perhaps that you refrain from expressing such thoughts at our next meeting."

We pulled up in front of the building in the Plaza de la Revolución where we'd been taken the day before. The limousine doors were opened by soldiers, and we stepped out into the bright sunshine. Rodriguez led the way into the building, where we were met in the lobby by a half dozen people, including armed guards in different uniforms from those the soldiers wore, and the same somber representative from the interior ministry we'd met the day before. We were greeted in Spanish and led to a section of the building where security was especially tight; we had to pass through a gauntlet of armed men as we entered a spacious ante-room.

"Please wait," the interior ministry official said as he disappeared through a door. A minute later he reappeared and motioned for us to enter.

The office was huge. Seated behind a massive desk was a man wearing a white suit, white shirt, and white tie with narrow blue stripes. He'd been looking down at papers on his desk through half-glasses. We came to a halt in front of the desk. He looked up, removed his glasses, stood and said, *"Buenos días."*

Rodriguez said, "Dr. Hazlitt and Mrs. Fletcher, may I present to you Señor Raúl Castro, president of the Council of the State of Cuba, and first secretary of the Communist Party."

Seth and I were momentarily speechless. President Castro smiled and shook our hands.

"I have asked that I be allowed to meet you," he said through an interpreter, "and to address a mistake that has been made. My brother would also be here, but he is busy with other matters."

*Or too ill to be here,* I thought. News of Fidel Castro's failing health was widely reported, which had led to his brother Raúl taking over the reins of government.

"Please accept my apology for the inconvenience you have suffered," the interpreter said after waiting for the president to finish his speech in Spanish. "Sometimes such mistakes are made even when the motivation is pure. I trust that when you return to your United States, you will not look back with too much resentment at what has occurred." He sat down at his desk again. Our interview was over.

Seth and I were hurried out of the office and taken downstairs. Once we were back in the limo, Dr. Rodriguez told the driver, "The airport."

We stopped in front of Terminal 2, and Rodriguez escorted us inside the cavernous building, where he was greeted by two men in suits. After conferring for a few minutes, Rodriguez and the men led the way to a restaurant with a table far removed from others.

"Time for your breakfast, Dr. Hazlitt," Rodriguez said. "My apologies that you had to wait."

"You folks sure do things different here in Cuba," Seth told Rodriguez as he finished up a platter of eggs, bacon, and Cuban bread, which I'd grown to love. "You kidnap us, then you bring us to meet your leader and buy us a big breakfast."

"It is the least we can do," he said. "But as President Castro said, sometimes mistakes are made even with the best of intentions."

"That doesn't excuse the fact that —"

I nudged Seth under the table. This was no time to get into an argument.

Rodriguez gave us back our cell phones. "You are scheduled to leave in one hour," he said. "We have a plane reserved only for you, and all clearances have been obtained from your government and air traffic control to fly to Tampa."

"I hope it's bigger than that puddle jumper we flew in to get here," Seth grumbled.

"It is." Rodriguez's smile faded. "Before you go, I must take a moment to explain something to you. While I was charged with the assignment of learning what progress Dr. Alvaro Vasquez had made in his research, I never once considered that he would be killed. Xavier Vasquez is employed by the Ministry of the Interior, and I was told to use him as a means of determining the status of his father's work. Unfortunately, the young man had his own agenda. He had access to scientists in the interior ministry and was able to obtain the lethal poison he used to murder his father. He and

Dr. Sardina conspired to steal his father's laptop, which, as you know, contained nothing."

"Al must have suspected something was up when he stripped the hard drive from his computer," Seth said.

"That's probably when he wrote you the letter saying where the thumb drives could be found," I put in.

"When you informed Xavier about the drives, he decided to bring the two of you to Havana at gunpoint," Rodriguez said. "He is a foolish, brash young man who does not represent me or the Cuban government. He took for granted that bringing you here would make him heroic in our eyes, that he was doing something good for the Cuban people. Instead, he has embarrassed me and all my people. Believe me, President Castro does not often apologize — to anyone!"

"Xavier is a murderer," Seth said. "He has to pay for his crime."

"I'm sure the Ministry of the Interior will have something to say to him."

"That's not enough!"

"What about his sister, Maritza?" I asked.

"I don't know who will undertake to discipline Maritza Vasquez." He looked at Seth. "You know she is a very talented medical student."

"So I've heard," Seth replied.

Rodriguez shook his head regretfully. "Her place in the university was in jeopardy when her parents defected. I often wondered if they realized the pressure put upon their children by their decision to leave."

"Is she still in medical school?" I asked.

"Oh, yes. She managed to convince the authorities that her loyalty lay with Cuba, not with her mother and father. I believe it is she who introduced her brother to the idea of using a powerful poison to take their father's life."

"What a calamity for the Vasquez family," I said. "The children conspire to kill their father and leave a mother struggling with the disease he was trying to cure. Poor woman."

"Perhaps Ivelisse will be fortunate enough not to comprehend all that has taken place," Rodriguez said. "But come. Your plane home is waiting."

The plane that would take us back to Tampa and the United States was, as Rodriguez had promised, larger than Xavier Vasquez's Cessna 172. It was a vintage Russian-built twin-engine piston-prop Ilyushin 11-14 that had probably been built in the 1950s to replace America's workhorse DC-3. A pilot and copilot greeted us from the cockpit as we boarded and took two of the eighteen vacant seats.

"You sure this thing will fly?" Seth asked Rodriguez.

He laughed. "It hasn't crashed yet," he replied. As he shook our hands, he said, "One last favor."

"Yes?"

"You will undoubtedly be asked by many about your unexpected stay here in Cuba. All I ask is that you report that you were treated with dignity while here as our accidental guests."

"Accidental?" Seth said. "I'd hardly call being kidnapped and threatened an 'accident.' "

I quickly said, "We will report that considering the circumstances, you did all you could to provide for us."

*"Buen viaje,"* he said. "Spanish for have a good trip."

*"Gracias,"* I said.

"Oh," he said, "please remember me to Ivelisse Vasquez. A lovely woman. I am afraid that her deteriorating condition was very much behind Alvaro's interest in Alzheimer's disease, and perhaps why he felt he needed to leave Cuba to continue his research in the United States. *Adiós,* my new *amigos.* "

# CHAPTER TWENTY-TWO

The flight to Tampa in the vintage Russian-built aircraft was smooth and without incident. It was a strange feeling being in a plane with nothing but empty seats around us — except for those on which we'd piled our luggage — but we used the time to discuss everything that had happened to us over the past twenty-four hours.

"I only hope that Xavier gets what's coming to him, and not just some bureaucratic reprimand," Seth said. "His father may not have been the most honest man, but he certainly didn't deserve to be killed."

"When he introduced me to Xavier at the party, Al sounded so proud of him," I said.

"Al told me that their relationship suffered when he and Ivelisse defected, but he was still fond of his son and hoped they could patch things up. Too bad it's too late."

"It's too late for Alvaro Vasquez," I said, "but it's not too late for you to give those thumb drives to your research colleagues in Boston."

"What do you mean? I never got them back from Rodriguez."

"I know, but I've got this." I reached into my shoulder bag and pulled out my laptop. "We copied Al's files onto my computer before transferring them to the second set of thumb drives," I said. "All his notes are still in here."

"Jessica, you're a marvel!" Seth said, delight on his face. "But I wonder why the Cubans didn't confiscate your computer when they searched our luggage."

"I was wondering that myself," I said, "but I think I've figured it out. You're the one who was so close to Vasquez. They were probably more interested to find out which bag was yours, and when they opened mine and saw women's clothes, they just closed it up again. Besides, I had this buried in the bottom of the suitcase wrapped inside my robe."

"Talk about a cover-up," Seth said, chuckling at his own joke.

The pilot landed smoothly at Tampa airport and taxied to a remote corner of the field, far removed from the terminal and other commercial aircraft. We had naturally speculated about the behind-the-scenes machinations that had transpired before this flight from Cuba to Tampa could be arranged and had hoped that it wouldn't involve us in any official role. But we were to be disappointed.

I looked out my window and saw a dozen or so people awaiting our arrival. Included in the group were Oona Mendez and Karl Westerkoch. I pointed them out to Seth, who muttered, "I could do without having to see them again."

The copilot came back and opened the door, and two members of the ground crew pushed a wheeled set of stairs to it.

"Thank you," I told the copilot.

"My pleasure," he said, smiling as he lowered our luggage down the stairs to waiting hands. *"Buenas tardes, señora."*

Seth preceded me down the staircase. Before going to the door, I poked my head into the cockpit and gave a final good-bye to the pilots, who wished me well. I stepped outside and looked to where Seth now stood speaking with Oona and Westerkoch. I carefully negotiated the narrow metal steps, holding on to the railings, and stepped onto the tarmac with a sense of overall relief. It was good to be back on U.S. soil.

"Come with us," Oona said and led us through the knot of people into a small hangar reserved for private aircraft. We were ushered into an office with a round wooden table and folding chairs. We'd no sooner been seated when two others arrived who were introduced as representatives of the U.S. State Department.

"We need to debrief you on your recent trip to Cuba," one said.

"Nothing to tell you," Seth said grumpily.

"Just a few questions," the fellow from State said. "We promise we won't take too much of your time."

"We're happy to cooperate," I said. "We're just glad to be home."

Our official debriefing took almost an hour, and I was concerned that Seth's patience would wear thin and that he would rebel. But he maintained his composure and even made a few witty comments about Cuba and our meeting with Raúl Castro. The State Department officials were especially interested in our evaluation of President Castro's physical and mental health.

"He looked in good physical shape to me," Seth said.

"What did he say about his brother Fidel Castro?"

"Just that he was busy," I said.

"Nothing about his health?"

"No," I said. "That never came up."

We were asked about the other people with whom we met, including Dr. Rodriguez and the unnamed representative of the Cuban Ministry of the Interior. We told them everything we could think of, including that Dr. Rodriguez was especially solicitous and that we were treated with respect considering the circumstances.

Naturally, the questioners were also interested in how we were kidnapped by Xavier and his sister and what we knew of their life and connections in Cuba.

"One thing we do know," Seth said, "is that Xavier Vasquez murdered his father using a poison he'd gotten from someone in the Cuban government. You can find out more about it by talking to Dr. San Martín, the medical examiner here in Tampa."

"We'll be sure to do that. You say that Xavier Vasquez murdered his father. Is it your belief that he plans to remain in Cuba?"

"No doubt about that. Practically bragged to me that he was out of the range of our criminal jurisdiction."

"We'll see about that."

"Dr. Rodriguez led us to believe that the Ministry of the Interior would deal with him," I added. "I don't know whether they're concerned that he's a murderer, but he's obviously upset those he works for."

"We have ways of keeping track of what happens to him in Cuba," Westerkoch said.

*I'm sure you do,* I thought.

"What about the daughter?" we were asked.

"After her father's death, she was allowed out of Cuba to see her mother," I said. "She's still a student, so I imagine that she'll remain in Cuba, too. It's so sad."

"What is?"

"Dr. Vasquez's widow, Ivelisse. She's all alone. Aside from not being well, she has a son who murdered her husband and a daughter who may have abetted him. This entire affair defines the word 'tragedy.' "

The State representatives thanked us for our information and cautioned us to be careful about what we said about our experience.

"Staying in Tampa for a while?" Oona Mendez asked.

Seth was quick to answer. "No," he said, "we'll be leaving on the first available plane home."

Westerkoch chuckled, actually chuckled. "Intrigue not your cup of tea, Doctor?"

"No! It certainly is not," Seth said.

"There's a plane leaving for Hartford,

Connecticut, in an hour and a half," Oona said.

"Do they have seats?"

"They have seats," she replied.

"Then we'll be on it," said Seth. "Right, Jessica?"

"Right," I said.

I called Jed Richardson at the Cabot Cove airport and arranged for him to meet us in Hartford and ferry us home, and Seth called his office and told his distraught nurse — "Where have you been? I've been frantic." — to tell her our plans.

We were waiting in the boarding area when the reporter from the *Tampa Tribune,* Peggy Lohman, breathless and talking as fast as ever, entered the lounge and came to where we sat. "I missed you before," she said, "but they told me you were leaving for home and —"

"Why don't you sit down, Ms. Lohman," Seth suggested, "and *slow* down."

"Thank you. I was afraid I'd be too late. Wow, what a twist to the Vasquez story. Is it true that you were hijacked to Cuba by Dr. Vasquez's son and daughter?"

I checked Seth before responding. "Actually, Ms. Lohman, I think that Dr. Hazlitt and I would prefer not to comment on what happened."

She looked horrified. "You can't say that," she said. "This is big news. My editor told me to tell you that if you stay in Tampa for a few days, we'll pick up all your expenses, every one of them, for an exclusive."

"I understand that you've a job to do, Ms. Lohman," Seth said, "but Mrs. Fletcher and I are anxious to get home and put this behind us."

"But what about your experiences in Cuba? We have a big Cuban American readership in Tampa. Would you at least comment on that?"

Seth gave me a look that said that I should answer the reporter.

"All we have to say about our experiences in Cuba — and bear in mind they only lasted a day and a night — is that we were treated decently, were not abused in any way, and we found the Cuban people to be warm and friendly. As for the government, Dr. Hazlitt and I fervently wish that the Cuban people will be free one day, and we expect that will eventually happen. Other than that, Ms. Lohman, it was truly a pleasure meeting you. You'll have to excuse us. They're boarding our flight."

# CHAPTER TWENTY-THREE

We arrived in Cabot Cove that evening in
Jed Richardson's twin-engine Cessna. Seth
was so grateful to be going home, he refrained from commenting about the size of
the plane. Jed kindly dropped us off at
Seth's house before driving himself to town.
Seth's office is in his home, and his nurse
and receptionist had left him a sheaf of pink
papers detailing calls to be returned. While
I rustled up food from Seth's freezer —
we'd never had any lunch — Seth excused
himself to retrieve something from his office, returning a few minutes later with a
big fat cigar.

I couldn't help but laugh. "Where did you
get that?"

"Al gave it to me the first time I was in
Cuba," he said pulling out a chair at his
kitchen table.

"You're going to start smoking cigars?" I
asked.

"I might," he said, sitting back and admiring the cigar he held in his fingers. "That's between you and me, Jessica. Cuban cigars are illegal here in the States. This is a Hoyo de Monterrey Double Corona, one of the best, robust and full-bodied."

"You sound like an advertisement," I said. "If it's illegal, how did you get it into the country?"

"Slid it in one of my shoes that I packed and crossed my fingers that Customs wouldn't check my bag. Didn't want to end up a prisoner in Guantanamo."

"Or stateside," I added. "You know that won't go over big with those patients you're always urging to quit smoking."

He came forward in his chair and used the cigar the way Groucho Marx would, twirling it in front of me. "I'll just have to smoke it sub rosa," he said, "sneak a puff now and then the way I did as a teenager with cigarettes behind the barn."

"I never knew that about you, Seth Hazlitt," I said.

"Lots you don't know about me, Jessica," he said with a smirk.

He was joking, of course, about intending to smoke that cigar.

Or was he?

"We toured a cigar factory there," he said.

351

"Amazing how many steps are needed to create a truly good cigar, a painstaking process. In fact, Castro once said — at least that's the story — that it's easier to produce a fine cognac than to produce a good cigar." He grinned.

"Sounds like a bit of Castro braggadocio to me," I said. "I'm surprised that a bunch of American physicians would end up visiting a cigar factory."

"All part of the experience," he said, his smile fading. "We spent most of the time meeting with Cuban doctors. Amazing, Jessica, how much good research was being conducted in that poor country, first-class medical research. That's when I met Al."

I put my hand on Seth's arm. "He was a good friend, even if he wasn't all he professed to be."

"That he was."

We ate at Seth's kitchen table. He insisted upon cleaning up since I'd prepared the meal. Afterward, he sat down again to go through the pink notices of all the calls he'd received while he'd been away. Some of them were from media, and one, of course, from Evelyn Phillips, editor of the *Cabot Cove Gazette*. "The last thing I want," he said, "is to have to deal with another jojeezly reporter."

"I share your feelings," I said, "but I'm afraid that we won't have much of a choice. Our little adventure is bound to be big news. After all, it wasn't a PTA bake sale or a cat rescued from a tree. Maybe it won't be as bad as we're anticipating. How about giving me a ride home? I'm ready to fall on my nose."

As it turned out, it *was* as bad as we'd anticipated. Once word got out, we were bombarded with calls from media, local, regional, and national. Even the British Broadcasting Corporation got in touch and requested an interview. We turned to a friend, Sanford Teller, who had a public relations agency in town, and he urged us to hold a press conference and get it over with in one fell swoop. It was a standing-room-only event held in our city hall's meeting room. Seth and I answered the questions as best we could, careful not to stray into editorializing or venturing into the political arena. Everyone seemed satisfied as the room cleared, and Teller congratulated us on putting on a good performance. I didn't consider that we'd *performed;* I was just glad that it was over and we could get back to our normal lives. The only positive thing that came out of all the attention was that the sales of my latest book increased dramatically, which pleased my publisher, Vaughan Buckley, and my agent, Matt Miller.

Thankfully, media interest in our Cuban experience soon faded, replaced in newspapers and on television by more pressing issues of the day, which didn't mean that either Seth or I would ever forget it. What stayed with us most was the exasperation that Xavier Vasquez had murdered his father and was getting away with it by living in Cuba.

I copied Dr. Alvaro Vasquez's research notes from my laptop to a set of thumb drives, which Seth delivered to a colleague in Boston, who confirmed that Alvaro Vasquez's research on the mysteries of Alzheimer's disease hadn't led to anything medically useful. We both maintained occasional contact with Dr. San Martín, and with Detective Machado, who informed us that Tampa PD had charged Xavier Vasquez with his father's murder and had listed him as a fugitive. Seth telephoned Ivelisse Vasquez a few times and reported that a full-time caregiver had been hired. Our hearts went out to her. She'd not only seen her husband murdered; she had to live with the knowledge that her son had been his killer. If her advancing disease allowed her

354

to forget, it was a blessing.

I had a few things to remind me of the adventure Seth and I had been on. I used the cigar lighter I'd purchased from the peddler in Ybor City to light kindling in my fireplace, and I'd framed a photo of the Columbia Restaurant and hung it over my desk, along with other photos from past trips. But in time our forced visit to Cuba receded in our memories, replaced by the activities of our day-to-day lives.

Then, six months after we'd returned to Cabot Cove, I received a phone call from Tampa.

"Oona Mendez," the caller said. "Remember me?"

"Of course. How are you?"

"Doing splendidly. I'm calling with a bit of news. As you know, Xavier Vasquez has been charged with his father's murder."

"Dr. Hazlitt and I were pleased to see that those charges had been filed, not that anything will come of it."

"That's where you might be wrong."

"Oh? Tell me more."

"A remarkable thing has happened. I don't know if you're aware that we have an extradition treaty with Cuba."

"We do? I had assumed the opposite."

"And for good reason. The original treaty

355

was signed back in 1904 but rescinded in 1926. Then, in 1959, a new extradition treaty was signed. Of course, because of the frayed state of relations between Washington and Havana, that more recent treaty hasn't been used since it was signed into law. However —"

"Are you saying that it's about to be put to use where Xavier Vasquez is concerned?"

"That's exactly what I'm saying. It's been a long, tough process to bring it to this point, a lot of strings being pulled, and a real diplomatic push by the State Department among others, but it looks like it's worked. The Cuban government has agreed to extradite Xavier to the United States."

"That's wonderful! Justice will be served."

"Believe me, I share your enthusiasm. Yes, justice will be served, but there's a much larger meaning to all this. Raúl Castro, who now runs the government, is less hard-line than his brother Fidel. We've learned through diplomatic channels and intelligence agencies that he's leaning toward opening up new lines of communication with us. It's the opinion of those in State and the other agencies that his acceptance of extradition might be a signal that he's serious about making those changes."

"Well," I said, "I'm pleased to hear that. I

know that Dr. Hazlitt will be delighted to know that Xavier might soon be facing trial for murder."

"I knew that you'd be interested," Oona said. "I get the feeling that your unexpected and unwelcome visit to Cuba might have helped pave the way for this to happen. Maybe you charmed Raúl Castro."

"I don't see how," I said. "I don't think I said two words to the man, but thank you for sharing this news. I can't wait to tell Seth."

Seth and I celebrated that night with friends. Toward the end of the evening, our travel agent, Susan Shevlin, said that she was putting together a State Department–approved person-to-person trip to Havana. "How about you two signing up?" she said.

Seth looked at Susan quizzically and said, "You can't be serious."

"Oh, but I am," she said. "How about you, Jessica?"

"I don't think so," I said, but I felt doubts creeping in.

What little I'd seen of Cuba had been as a reluctant visitor. Experiencing it as a willing tourist was appealing. I tucked that thought away, but in my heart I knew: One day I would return.

# ABOUT THE AUTHORS

**Jessica Fletcher** is a bestselling mystery writer who has a knack for stumbling upon real-life mysteries in her various travels. **Donald Bain,** her longtime collaborator, is the writer of more than one hundred other books, many of them bestsellers.